FREE-MARKET ENERGY

FREE MARKET ENERGY

The Way to Benefit Consumers

Edited by S. Fred Singer

Foreword by William E. Simon

A Heritage Foundation Book

UNIVERSE BOOKS
New York

8-7-84

Published in the United States of America in 1984
by Universe Books
381 Park Avenue South, New York, N.Y. 10016

84 85 86 87 88 / 10 9 8 7 6 5 4 3 2 1

Printed in the United States of America

Library of Congress Cataloging in Publication Data
Main entry under title:

Free market energy.

    Bibliography: p.
    Includes index.
    1. Energy industries – United States – Addresses, es-
says, lectures. 2. Energy policy – United States – Ad-
dresses, essays, lectures. I. Singer, S. Fred (Siegfried
Fred), 1924–
HD9502.U52F743 1984   338.2'0973   83-040563
ISBN 0-87663-443-9

CONTENTS

FOREWORD

This Heritage Foundation book assembles responses of a dozen energy experts to *Energy Future, Global Insecurity*, and other works of doom and gloom, which forecast a continuing energy crisis. It argues—and demonstrates—that an energy crisis is usually the result of government regulation usurping the free market.

Deregulation of oil prices in January 1981 was the first important step in the Reagan administration's battle to restore free markets to the nation's energy business. Contrary to the fears of vociferous Chicken Littles, the sky did not fall. Instead, an oil glut has lowered the world price of oil, and OPEC is on the verge of collapse.

But further steps are needed if we want to restore a more competitive market to benefit the energy consumer in the United States and elsewhere. This volume explains the policies required to achieve this goal and shows how to minimize government involvement in energy production. It also points out how national security improves and how standards of living rise as imports decline and oil becomes cheaper.

Written by specialists, this book is intended for the nonspecialist. It rises above the angry rhetoric of the zealots and "know-nothings" who have been so antagonistic toward the energy-producing sectors of our economy. Oil, gas, coal development, and especially nuclear energy have been hamstrung by complicated, costly, and often conflicting government regulations that, contrary to their stated intent, have done little to improve environmental quality or safety.

This book provides a solid basis for a better understanding of the energy problem. It is the authors' belief that Joe and Jane Citizen are smart enough to figure out what is going on once they have the facts. Consumers will then act to protect their own best interests—the fundamental basis of Adam Smith's "invisible hand" that directs the free market.

William E. Simon

PREFACE

This book had its origin in a series of informal discussions arranged by The Heritage Foundation in the fall and winter of 1982–83 at the suggestion of Burton Yale Pines, vice-president and director of research at Heritage. The general topic was natural resources management and the appropriate governmental role. We talked about nonrenewable resources like energy and minerals, about renewable resources like fisheries and agriculture, and about owned-in-common resources like air and water. Each has its particular problems. For example, there clearly is a governmental role in the management of air and water. For energy, however, the governmental role can be, and should be, greatly reduced and reliance placed on market forces if we want to achieve three major objectives:

1. Lower energy costs for the consumer now and in the future
2. Improvement in national security through the reduction of oil imports
3. Less governmental overhead, to benefit the taxpayer

Not everyone agrees that a free-market approach is the best way of achieving these objectives. In particular, some energy industries prefer to further their individual objectives. Many so-called consumer advocates call for more, not less, governmental regulation. Bureaucrats and politicians have a built-in bias for political rather than market approaches. Lawyers thrive on government regulation.

This volume was conceived as an effort to present to the open-minded person the advantages of free-market solutions for resources management. By March 1983, a number of experts had been asked to contribute to a volume that would express our philosophy. Each author would be responsible only for his own chapter, in which he could express his own views and frustrations. As editor, I would preserve the individuality of the authors' expressions and not homogenize their opinions, particularly on controversial topics where different views still exist.

Much to my surprise, almost everyone immediately agreed to write a chapter in spite of the tight deadlines—one month for the first draft and another four weeks for the final manuscript. By mid-May, all papers had been received and edited.

The plan of the book has not changed. We try to identify the major issues where, realistically, some improvement can be made. These are discussed by specialists in the individual chapters. I have added an overview and a prologue that sets the stage in terms of the price of oil—the parameter that determines what else happens in the field of energy. The epilogue was originally written by my colleague Richard Holwill for The Heritage Foundation volume *Agenda '83*.

To make the book generally useful, I have included appendixes on units and energy statistics, as well as a glossary and an index.

It is my hope that this book will be not only useful and informative but also readable and exciting. The subject, after all, is one of the most important that we have been dealing with, especially during the past decade. It has absorbed the time and attention of many in the White House and in Congress. It affects the welfare of us all. If we can deal with our energy problems in a sensible and effective way, we will all be better off. On that point there can be no disagreement.

I am grateful to The Heritage Foundation for its hospitality and support during the academic year 1982–83, particularly to Ed Feulner, Phil Truluck, and Burt Pines, and to the University of Virginia, which granted me a research leave. I am especially grateful to the contributors who have been willing to write their chapters under tight deadlines and editorial duress; I hope they will be proud of their handiwork. I also thank Stephen Eule and Judith Hydes for their capable and effective assistance.

The usual disclaimer applies: The authors express their personal views and not necessarily those of their respective institutions or of The Heritage Foundation. Any errors should be blamed on the editor; he has broad shoulders.

S. Fred Singer

ABBREVIATIONS

AAPS: All-Alaska Pipeline System
ACRS: Accelerated Cost Recovery System
AFUDC: Allowance for Funds Used during Construction
AGA: American Gas Association
ANGTS: Alaska Natural Gas Transportation System
ANS: Alaskan North Slope
ASTM: American Society for Testing and Materials
BACT: best available control technologies
bbd: billion barrels per day
BCF: billion cubic feet
b/d: barrels per day
BEIR: National Academy of Sciences Committee on Biological
    Effects of Ionizing Radiation
BPA: Bonneville Power Administration
Btu: British thermal unit
CANDU: Canadian power reactor, nuclear
CES: cost escalation factor
CF: cubic foot
CLC: Cost of Living Council
CPI: Consumer Price Index
CRBR: Clinch River Breeder Reactor
CWIP: construction-work-in-progress
CWM: coal–water mixtures
CWP: coal worker pneumoconiosis
CWS: coal-water slurry
DOE: U.S. Department of Energy
EAA: Export Administration Act
ECCS: emergency core cooling system
EEDB: energy economics data base
EIA: Energy Information Agency (of DOE)
EIS: environmental impact statement
EPA: U.S. Environmental Protection Agency
EPAA: Emergency Petroleum Allocation Act of 1973
EPCA: Energy Policy and Conservation Act of 1975
EPRI: Electric Power Research Institute
ERDA: Energy Research and Development Administration
    (absorbed by DOE)
ESC: cost escalation factor
ESP: Economic Stabilization Program
FBC: fluidized bed combustion

FERC: Federal Energy Regulatory Commission
FOGCO: Federal Oil and Gas Corporation (proposed)
FPC: Federal Power Commission (now FERC)
FUA: Power Plant and Industrial Fuel Use Act of 1978
GAO: U.S. Government Accounting Office
GNP: Gross National Product
GRI: Gas Research Institute
Gw: gigawatt (10^9 watts)
hp: horsepower
HTGR: high temperature gas-cooled reactor, nuclear
HVAC: heating, ventilating, and air-conditioning market
IAEA: International Atomic Energy Agency
ICC: Interstate Commerce Commission
ICRP: International Commission on Radiological Protection
IEA: International Energy Agency
INFCE: International Nuclear Fuel Cycle Evaluation
ITC: investment tax credit
kcal: kilocalorie
kg: kilogram
kw: kilowatt
kwh: kilowatt-hour
LMFBR: liquid-metal-cooled fast breeder reactor
LNG: liquified natural gas
LOCA: loss of coolant accident
LOFT: Loss-of-fluid test, nuclear reactor
LWR: light-water reactor
M/B: market-to-book ratio
mbd: million barrels per day
mbtu: million British thermal units
MCF: thousand cubic feet, natural gas
MHD: magnetohydrodynamic
mrem: millirem
MSHA: Mine Safety and Health Administration
MSR: molten salt reactor, nuclear
mty: million tons per year
NAAQS: National Ambient Air Quality Standards
NEA: National Energy Act of 1978
NGPA: Natural Gas Policy Act of 1978
NPC: National Petroleum Council
NRC: Nuclear Regulatory Commission
NSPS: new source performance standards
OCS: Outer Continental Shelf
OECD: Organization for Economic Cooperation and
    Development
OETO: OPEC exports to OECD

OPEC: Organization of Petroleum Exporting Countries
PSD: prevention of significant deterioration
PURPA: Public Utility Regulatory Policy Act of 1978
R&D: research and development
RD&D: research, development, and demonstration
REA: Rural Electrification Act of 1936
SFC: Synthetic Fuels Corporation
SIP: State Implementation Plan
SPR: Strategic Petroleum Reserve
TAGS: Trans-Alaska Gas System
TAPS: Trans-Alaska Pipeline System
TCF: trillion cubic feet
TFTR: Tokamak Fusion Test Reactor
TVA: Tennessee Valley Authority
UMW: United Mine Workers
VIF: variable import fee
w-hr: watt-hour
WPT: windfall profits tax

PROLOGUE: THE WORLD OIL OUTLOOK

S. Fred Singer

Summary

The world oil market changed dramatically as a result of the 1979–80 price rise. Enhanced conservation is likely to decrease the demand for oil, especially in the industrialized nations. Price decreases are then inevitable, with widespread economic and political consequences. The path of oil prices is certain to have a strong influence on the use of other energy resources and on energy policy generally.

Price of Oil

The single most important parameter of world oil is the price,* which determines to a large extent, but not completely, the price of substitute fuels. Because of internal competition, coal and gas can often be sold at a much lower price on a heat-equivalency basis. Nuclear energy is considerably less expensive and produces the lowest-cost electricity in many parts of the world. Table P-1 shows the increasing prices paid by U.S. electric utilities for oil, gas, and coal, as well as the consumer price of electricity during the last ten years.

Future Price

The future price of oil is a very important quantity.[1] People base their investment decisions, whether for oil or competing energy sources, in conservation, in industry, or for personal use, on their expectation about the future price. The general expectation nowadays is that the price of oil will continue to increase, perhaps after a hiatus of a few years, but certainly

---

* World price usually refers to the price of Light Arabian oil, f.o.b. Ras Tanura, even though there are various contract prices and free prices (spot prices) in Rotterdam or New York depending on the quality of the oil (sulfur content and gravity).

17

Table P-1
Cost of Fossil Fuels Delivered to Steam Electric Utility Plants[a,b]

Year	Coal	Residual oil[c] (cents per million BTU)[c]	Natural gas	All fossil fuels[d]	Retail electricity prices (cents per k Wh)[e]
1973	40.5	78.8	33.8	47.5	1.96
1974	71.0	191.0	48.1	90.9	2.49
1975	81.4	201.4	75.4	103.0	2.92
1976	84.8	195.9	103.4	110.4	3.09
1977	94.7	220.4	130.0	127.7	3.42
1978	111.6	212.3	143.8	139.3	3.69
1979	122.4	299.7	175.4	162.1	3.99
1980	135.2	427.9	212.9	189.3	4.73
1981	153.3	529.0	282.8	223.0	5.46
1982[f]	164.7	475.8	335.9	223.5	6.13

[a] Source: *Monthly Energy Review*, U.S. DOE.
[b] Geographic coverage: fossil fuels—the lower 48 states and the District of Columbia; electricity—the 50 United States and the District of Columbia.
[c] Includes the price of all liquid fuels used at utility plants.
[d] Includes small quantities of coke oven, refinery, and blast furnace gas.
[e] Average price for total sales to ultimate consumers.
[f] Average price for months January through October.

after 1985 (Figure P-1). This widely held view, which I do not share, is based on the general statement that oil is a finite resource that is being depleted. Yet the same statement could have been made at any time during the last century. The pattern of oil prices shows quite violent swings (see Figure P-2). Discoveries of very low-cost oil resources in the Caucasus, then in east Texas, and finally in the Persian Gulf have depressed the price at various times. Further discoveries can be expected in view of the high price—although not on the scale of those in the Middle East.

The simple picture of constantly increasing prices may not hold for a number of other reasons:

1. While the Organization of Petroleum Exporting Countries (OPEC) can affect the price by restricting production, it cannot control the consumption of oil, which is entirely under the control of consumers, who are sensitive to the price. In the final analysis, the price will be determined by supply and demand. And demand is likely to decline over the next decade or so as a consequence of various forms of conservation.

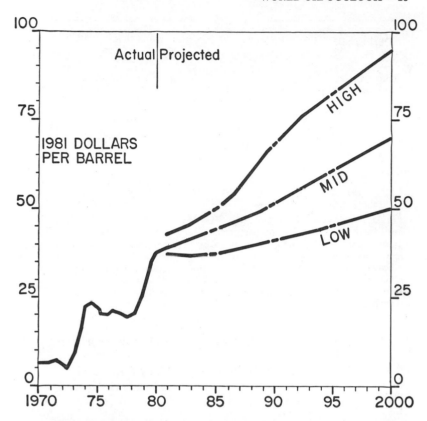

Figure P-1. World oil price projections published by DOE in July 1981, assuming high and low GNP growth rates, high and low OPEC production, and high and low non-OPEC production. (The individual scenario forecasts are not shown.) The price is expected to increase under all combinations of assumptions and to reach between $50 and $100 in the year 2000.

DOE's 1982 *Annual Report to Congress* (Vol. 3, p. 118) projected the world oil price for 2000 as $75 (in 1980 dollars per barrel), following a reasonably flat price out to 1986.

A compilation dated December 1982 was produced by the Stanford (University) Energy Forum's International Energy Workshop. Of the 17 respondents listed, 16 gave the real 2000 price as greater than the 1980 price. The median was 56 percent higher, although one expert gave 240 percent. The Institute of Energy Analysis, Oak Ridge, gave the 2000 price equal to the 1980 price.

2. Even though the present (mid-1983) price, about $29 per barrel, is more than ten times what it was ten years ago (in nominal dollars), and although oil is being depleted, there exists today a large reservoir of low-cost oil. Much Arabian oil can still be lifted for less than $1 per barrel. If

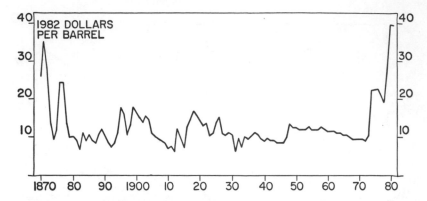

Figure P-2. U.S. crude oil prices, 1870–1981 (in constant 1982 dollars). Note both the high prices and intense price swings before 1930 when large discoveries created "gluts" and growing demand restored prices. Prorationing (regulation of production) in Texas and other states stabilized prices after 1932. The discovery of low-cost Middle Eastern oil put downward pressure on U.S. prices after World War II, in spite of a system of oil import quotas. The price events after 1970 are described more fully in the text. Source: A. R. Tussing, *Univ. Alaska Inst. Soc. Econ. Res.*, 1982, p. 5.

there were a competitive price for oil in the world today, it would likely be on the order of $10 per barrel; in other words, the last incremental barrel of oil that would just satisfy demand would cost about $10 to produce.*

3. But the main reason why oil prices cannot increase indefinitely is the possibility of substitution by other fuels. Even before the use of costly synthetic oil becomes feasible, much cheaper gas, coal and nuclear energy can replace oil in heat and steam applications—as is indeed happening now.

Rise of OPEC

Price of Oil from 1970 to 1978

From the end of World War II onward, the price of oil (in constant dollars) had been falling steadily (Figure P-2). Consumption had been increasing in every part of the world, but

---

* This is a difficult subject; marginal cost of oil in a given well is much less than marginal cost in a given field, or marginal cost if a new field has to be discovered and developed.

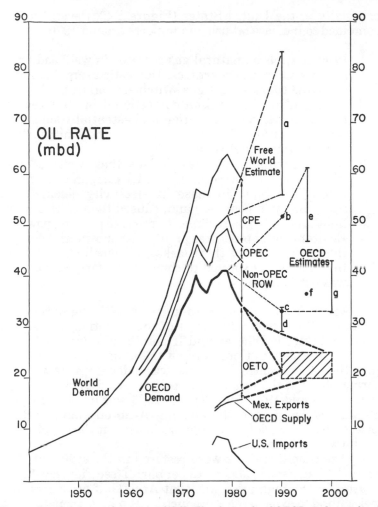

Figure P-3. Oil demand—rise and fall. The demands of OECD (industrialized nations), non-OPEC rest-of-the-world, OPEC members, and the CPE (centrally planned economies) are shown cumulatively and add up to the world demand. Also plotted are OECD supply, OECD supply plus Mexican exports, and U.S. imports (for comparison). Line *a* encompasses seven different estimates (made between 1977 and 1980) for Free World demand in 1990; *b* through *g* show the wide range of OECD demand estimates. Microsectorial analysis (see text) yields a much lower range of demand estimates. The rectangle (cross-hatched) indicates the approximate uncertainty in the eventual OECD demand and in timing. Note that the estimated end point is *not* an extrapolation of the 1979–1982 downward trend of OECD demand. The long-term estimate is therefore independent of cyclical recessions and recoveries, and of stockpiling and destocking. The quantity OETO is "OPEC exports to OECD" and is expected to decline drastically as OECD demand drops and more reliable sources become available. Source: Reference 1.

especially in the United States (Figure P-3). Several factors combined to increase oil use and imports around 1970:

1. Price controls on natural gas resulted in wellhead prices so low as to cause shortages. Domestic users and industries that could not get gas switched to oil instead.
2. Price controls imposed on domestic oil in 1971 reduced the incentives for exploration and essentially subsidized the importation of oil. Production in the United States peaked around that time.
3. Severe environmental restrictions that were suddenly imposed on coal forced many utility companies to switch to oil. The rapid increase in electricity demand, and particularly in peak demand, caused the utilities to use light fuel oil in gas turbines to produce peaking power.
4. Environmental pressures produced a substantial delay in the construction of the Alaskan oil pipeline, so that oil that had been expected to come to the forty-eight states was not available.

It was difficult for supply to keep up with the burgeoning demand. OPEC took advantage of the situation; in negotiations conducted in Teheran and Tripoli in 1971, Iran and Libya (and later others) put pressure on the major oil companies to give them more control over production decisions, as well as a larger share of the profits. The price started to rise and had doubled by mid-1973. The October 1973 war involving Egypt, Syria, and Israel, and the ensuing Arab oil embargo drove prices up more rapidly, as Arab producers cut the supply of oil to the world market.

By the time these cuts were restored in the spring of 1974, the price had risen to about $12 per barrel (see Figure P-4). The reason for the high price was that Arab Gulf producers were not willing to expand their production at a high enough rate to meet the increasing demand for oil but preferred to act as a monopoly.*

* The Arab Gulf producers, with the largest reserves and the ability to expand or reduce production as they wish, exercise the greatest control over the price of oil. These producers, also members of the Gulf Cooperation Council, include Saudi Arabia and the Gulf sheikhdoms of Kuwait, United Arab Emirates, Qatar, Bahrain, and Oman, which generally go along with Saudi direction. They can be referred to as the "core" of OPEC, since they are the ones who really exercise monopolistic control over prices. The other members of OPEC can be termed the "price takers," since they generally sell as much as possible at the world price in order to get the largest amount of

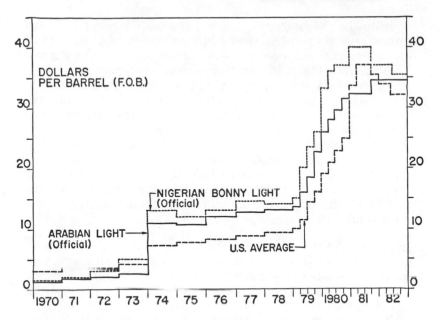

Figure P-4. Crude Oil Prices, 1970–1982. OPEC prices are set for the so-called benchmark or "marker" light oil sold by Saudi Arabia, a high-quality oil that is more easily refined. Actual market prices, however, have varied by as much as several dollars per barrel, depending on quality and production location as well as general market conditions. The benchmark price, nevertheless, is the general guide against which other prices are determined.

Nearly all the oil in the noncommunist world, including non-OPEC production from countries such as Mexico and Norway, is sold at the OPEC-set price. Most of this oil is sold under long-term contracts, but some is sold on a one-time basis on a "spot" market. As a result of the "oil glut" and the buyer's market, a larger percentage of the world's exports is now sold on a spot basis.

Bonny Light oil sells at a premium with respect to Saudi Light because it is of lower gravity and higher quality, yielding more product per barrel. The average U.S. price is lower than the world price because of the price regulations. U.S. oil prices were deregulated in January 1981.

The 1980 two-tier pricing system was as follows: Saudi Arabia, $32.00 per barrel; the other 12 OPEC countries, $36.00 base price with special surcharges in some cases taking prices to $41.00 per barrel.

The pricing system was unified on October 29, 1981, as follows: $34.00 per barrel for all member countries with up to $4.00 surcharge allowed; price freeze through 1982.

Source: DOE/EIA, *Monthly Energy Rev.* (various issues).

revenue. Their reserves are smaller, and they are, therefore, not as deeply concerned about the more distant future. Libya and Iraq, with large reserves, should be found among the ranks of the core, but more often than not are among the pricetakers. Libya and Iraq, for example, did not participate in the Arab oil embargo, although they expended a lot of rhetoric on it.

It can be shown that a monopolist should aim for an "optimum" price that will maximize his profits over time.[2] The 1974 price was at about the right level for the Arab Gulf producers; their immediate need for money was not great, but their reserves of oil were large enough to make them concerned about maintaining a strong future market. They could have decreased the price of oil by increasing their production, but this would have depleted their oil more rapidly and over the long term would have given them lower profits. Or they could have increased the price further by cutting their production, but this higher price would have encouraged oil consumers around the world to develop alternatives and would have destroyed the long-term oil market—again, lowering profits over time.

The Oil Price after 1979

The Shah of Iran was overthrown in 1978, and Iranian oil production slowed to a halt. These events triggered strong panic buying among oil consumers who visualized another oil crisis and another spurt in oil prices. Their actions created a self-fulfilling prophecy. Even though the decline in Iranian exports was made up by increases from other countries, especially from Arab producers in the Gulf, the excess purchases of oil for stockpiles put upward pressure on the price. The rise in spot price convinced consumers that oil prices were going to rise further and intensified buying, which in turn drove up the price even more. The contract prices set by OPEC members lagged behind the increases in the spot price, but not by very much.

The linchpin of the price increase was provided by the Saudis, who acted in a peculiar manner. On the one hand, they increased their production to make up for the shortfall in Iranian exports; on the other hand, they slashed production suddenly at crucial times, which doubled the spot price in both cases: first on January 21, 1979, just before an important oil auction in Abu Dhabi, and then again on March 31, 1979, after the conclusion of the Egyptian-Israeli peace treaty. These short-lived cutbacks by Saudi Arabia seemed to affect the nervous spot market sufficiently to drive up prices and convince consumer nations that further price increases were inevitable. The increase also persuaded producer nations to withhold oil from the market in the hope of getting higher prices in the future—another self-fulfilling prophecy. By 1980 the spot price had reached the vicinity of $40 per barrel, and the official OPEC price was $36. Saudi Arabia held out for a price of $32, but eventually raised it to $34 (see Figure P-4). The 1979–80 price increase was remarkable and unexpected. Unlike the situation in 1973, there had been no fundamental change in the world oil market. In 1973 the producer nations took over production decisions from the major concession holders and by restraining production were able to push up the price. In 1979 there were no further changes in the control of production decisions, nor was there any appreciable shortfall in world oil production, in spite of the Iranian cutback. Furthermore, actual oil consumption did not change, although oil demand increased because of hoarding and stockpiling.

Accordingly, as soon as it was recognized that the price had peaked, it should have come down again to the 1978 value.[3] As consumers emptied their stockpiles, producers would have resumed full production before the price fell. Consumers, in dumping their stockpiles quickly before the price fell, would of course cause a downward spiral in the price, roughly parallel to the upward spiral that had been experienced earlier during the hoarding phase.[4]

But this anticipated turn of events did not happen—at least, not at that time. Instead, general concern about the availability of oil continued to frighten consumers. A number of events in the Middle East made them extremely nervous. There was the takeover of the Great Mosque in Mecca (November 1979), the Soviet invasion of Afghanistan (December 1979), disturbances in the eastern oil province of Saudi Arabia (November 1979), and the invasion of Iran by Iraq (September 1980), which led to a sharp decline in Iraqi oil exports. Throughout this period, there was agitation and concern over the occupation of the U.S. embassy in Teheran.

Whatever the causes, the price of oil held firmly at around $34 per barrel, and it became accepted wisdom that the price of oil could only go up. The general view in 1980–81 was that the real price of oil would rise between 3 percent and 10 percent per year. The official projections of the Department of Energy (DOE) placed the price in the year 2000 between $50 and $100 per barrel (in 1981 dollars). As a result of these expectations, the consuming nations set about firmly to conserve oil. Demand for oil started to fall, particularly in the industrialized nations (OCED) where the peak of oil consumption was reached in 1978–79 (see Figure P-3).

Future Oil Demand

Misjudgments about the Price Rise of 1979

Saudi Arabia and the other "core" producers made two principal mistakes: (1) in 1979–80 they permitted the price of oil to rise; and (2) they did not bring it down right away. There appear to be a number of misjudgments involved relating to economics, although the political factors may have been predominant. In retrospect, it is quite clear that Arabian producers should have done everything possible to bring down the price of oil just as soon as it shot up in 1979. They could have done this by reassuring the world that they were willing to produce more oil and by actually investing in increased capacity and producing to the maximum extent possible.

The Saudis apparently misjudged the degree of conservation the price rise would precipitate among consumer countries. They at first ascribed the decline in demand to the dumping of stockpiles; later they ascribed it to the worldwide recession, but it is evident that they did not fully understand the effect of price elasticities of demand or the possibility of fuel substitutions. In this respect, they may have been misled by Western publications and assertions.

For example, the economic literature is replete with articles showing that the price elasticity of demand is extremely small, on the order of 0.1 or less. (This means that a 10 percent increase in the price would reduce the demand by only 1 percent.) But this literature is based on the early 1970s experience with much lower prices and many of the studies refer to gasoline for which there is no feasible substitute. Furthermore, the methodology itself is suspect. It has been shown that a different mathematical approach to the same data can give much larger values of elasticity.[1]

Be this as it may, three forms of conservation followed from the higher prices. (1) Reduced demand from changes in habits and life-styles is only one aspect. Another is the long-range change in demand that comes about from the introduction of (2) fuel-efficient devices and (3) fuel substitutions. Here, the expectation of future prices plays an important role. When consumers expect prices to rise, they will install insulation and other heat-saving devices or buy more fuel-efficient cars. Once the investments have been made, they are not likely to be reversed, even if oil prices go down. Industry has been particularly effective in making such changes, introducing new and efficient processes and replacing old machinery with energy-saving equipment, with particular emphasis on substituting other fuels for oil. Figure P-5 shows how industrialized countries have succeeded in reducing energy consumption coefficients with time. This process is still going on in the expectation of higher prices in the late 1980s and 1990s.

Substitution of Other Fuels for Oil

Even more important has been the replacement of oil by fuels such as coal, gas, and nuclear energy in heat and steam applications. This replacement makes economic sense, since coal and gas have cost less than half the price of oil for a number of years, at least in the United States. Once the capital investments have been made in coal and gas transportation systems, these fuels become more competitive worldwide. For example, gas pipelines, liquified natural gas (LNG) terminals, and LNG tankers are spreading the use of gas throughout the world. New coal technology has developed a combustible coal-

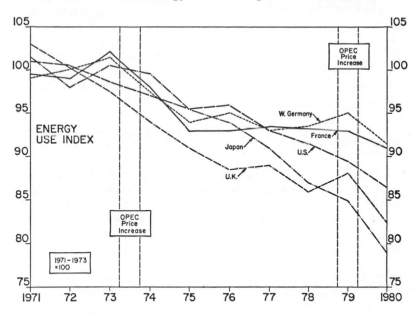

Figure P-5. Energy efficiency in selected countries. This measure of energy efficiency is referred to as the energy/GNP ratio. Put simply, it is the ratio of an economy's total energy consumption to its real economic output (correcting for inflation). Even more striking is the decline in the ratio of oil use/GNP. Japan reached an index of 67 in 1980, and 61 in 1981. Source: Japan Institute of Energy Economics, 1982.

water slurry (CWS) that can replace heavy fuel oil directly in existing oil-fired installations. Its commercial availability and low transportation and handling cost should turn CWS fuel into a world fuel able to "back out" oil, i.e., replace it.

Nuclear power clearly is a potential world energy source. France and countries in the Far East (Japan, South Korea, and Taiwan) have only recently committed themselves to nuclear energy, but they have shown the most active growth in this area. In France, 60 percent of electricity will soon be supplied by nuclear energy.

Another misjudgment, both by Western nations and by OPEC, was that petroleum would be replaced by synthetic oil, manufactured from coal or extracted from oil shale. It is generally accepted that synthetic oil is not competitive with crude oil, even at the present high price. OPEC, therefore, felt safe in letting the price of oil go up toward what it regarded as the much higher price of "backstop" technology. OPEC, however, neglected to note that more than one-half of the oil throughout the world is used simply to produce heat and steam. It is natural to replace fuel oil by much cheaper coal, gas, and nuclear energy, where available, as has indeed been happening. This trend greatly accelerated after 1979 as more people in more nations found it profitable to make the large up-front capital investments required to replace oil. At the same time, some forward-thinking refiners made the necessary capital improvements to turn heavy fuel oil, which would soon become surplus, into transportation fuel. In fact, the trend now is for more of a barrel of oil to be refined to transportation fuel—in essence, doubling the amount of available oil.

The important thing to note is that once these major capital investments have been made, they are essentially irreversible, even if the price of oil should fall. For example, a nuclear power plant in Japan would not close down unless the price of oil fell below $4 per barrel. Analogous calculations can be made for gas pipelines, LNG terminals, coal-handling facilities, or refinery modifications. The point is that once the price of oil has stayed at a high level for a long enough time and once people expect prices to go even higher in the future, these capital investments will be made, permanently depriving oil of a substantial part of the longer-term market.

Microsectorial Analysis of Future Demand

How large can the savings in oil consumption become? It is possible to derive an *upper limit* by analyzing in detail just how

Table P-2
World Energy Consumption, 1978[a,b]

	1	2 Market economies	3	4 Centrally planned economies[d]	5	6	7	8	9	10	11	12 OECD	13	14	15	16
	World[c]	Dev.[e]	LDC	Total	E. Europe	USSR	China	Total	Canada	U.S.	Japan	W. Europe	France	FRG	Italy	U.K.
I. Population (mil)	4258	—	—	—	372	262	933	754	24	218	115	374	53	61	57	56
II. 1977 GNP ($B)	—	4935	1186	—	—	—	—	4793	200	1879	564	2768	381	516	196	244
III. Total energy requirements	6030	3448	671	1912	1405	991	469	3900	209	1940	392	1239	193	273	146	214
IV. Solid fuels	1829	729	92	1008	603	347	370	712	18	357	52	239	33	74	10	70
a. Electr. gen.	—	—	—	—	—	—	—	452	—	270	13	136	17	49	2	46
b. Industry	—	—	—	—	—	—	—	180	—	77	25	60	9	15	5	10
c. Residential	—	—	—	—	—	—	—	44	—	5	3	26	4	5	1	9
V. Liquid fuels[f]	2791	1811	468	511	429	341	80	2105	88	972	290	699	117	145	101	97
a. Electr. gen.	—	—	—	—	—	—	—	234	5	94	61	76	10	7	21	13
VI. Gas	1217	769	84	364	351	285	13	717	43	471	20	175	20	44	23	38
a. Electr. gen.	—	—	—	—	—	—	—	120	—	75	10	31	1	14	2	1
b. Industry	—	—	—	—	—	—	—	226	—	140	3	66	9	14	12	14
c. Residential	—	—	—	—	—	—	—	158	—	131	5	53	5	9	8	17
VII. Electr. gen. (non-fossil)[g]	194	138	27	29	21	17	6	374	61	139	33	129	24	12	12	9
a. Nuclear[g]	51	46	1	5	5	3	0	132	7	70	15	40	7	8	1	1
VIII. Electr. gen. (fossil)[g]	467	297	34	135	117	85	16	—	—	—	—	—	—	—	—	—

[a] Unless otherwise specified, units are in million tons oil equivalent (mtoe); one mtoe = 10^7 kcal or 11,630 k Wh.

[b] Columns 1–7 are taken from United Nations statistics. Columns 8–16 are adapted from OECD statistics.

[c] Columns 2, 3, and 4 add up to Column 1.

[d] Column 6 is part of Column 5.

[e] Developed countries, including OECD (without Turkey), Israel, South Africa, and Yugoslavia.

[f] Includes NGL (natural gas liquids) and LPG (liquid petroleum gas).

[g] Columns 1–7 give electric *output* in mtoe units; Columns 8–16 give equivalent thermal *input* in mtoe units.

Table P-3
World Consumption of Oil Products, 1978[a,b,c]

	1	2 Market economies	3	4 Centrally planned economies	5	6	7	8	9	10	11	12 OECD	13	14	15	16
	World	Dev.	LDC	Total	E. Europe	USSR	China	Total	Canada	U.S.	Japan	W. Europe	France	FRG	Italy	U.K.
I. Total products	2710	1755	452	503	422	336	79	1740.7	75.1	791.6	218.9	601.0	98.3	125.1	90.7	82.6
a. Total nonenergy	305	—	—	—	—	—	—	113.2	—	49.1	25.4	38.4	6.1	9.9	5.4	5.3
II. LPG	106	83	14	9	9	8	0	75.5	0.6	43.9	14.4	15.4	2.9	2.8	1.3	14.4
a. Nonenergy/petrochem.	—	—	—	—	—	—	—	26.3	—	20.9	—	1.9	—	1.3	0.2	0.1
b. Residential[d]	—	—	—	—	—	—	—	32.7	0.6	20.0	5.4	6.6	1.8	0.3	0.9	0.2
III. Gasoline	652	488	76	89	77	—	—	496.7	27.4	330.2	24.4	100.0	17.6	23.5	11.3	18.3
a. Motor	649	486	75	89	77	63	11	494.5	27.7	328.6	24.4	99.7	17.6	23.4	11.3	18.3
IV. Jet fuel	99	79	20	—	—	31	16	75.9	3.2	49.1	3.3	18.3	2.2	2.7	1.7	4.5
V. Kerosene	119	39	31	49	33	31	—	37.6	1.6	8.5	20.8	5.3	0.1	0.1	1.3	2.7
a. Residential[d]	—	—	—	—	—	—	—	29.8	1.1	8.5	14.7	5.2	0.2	—	1.0	2.2
VI. Distillate	700	440	119	141	118	90	22	463.1	24.2	169.2	35.9	222.2	41.8	63.5	24.6	20.3
a. Bunkers	20	15	5	—	—	—	—	10.4	—	2.0	0.8	7.3	0.8	0.6	0.8	0.8
b. Electr. gen.	—	—	—	—	—	—	—	1.8	0.4	9.8	—	1.3	—	0.1	0.2	0.7
c. Transportation	—	—	—	—	—	—	—	142.8	6.6	61.7	14.5	55.1	10.0	10.5	7.8	7.8
i. Road	—	—	—	—	—	—	—	106.7	3.7	48.4	6.8	46.1	8.2	8.9	7.5	5.9
ii. Rail	—	—	—	—	—	—	—	21.1	2.0	13.4	1.1	3.1	0.6	0.7	0.1	0.8
iii. Water	—	—	—	—	—	—	—	13.0	0.9	—	6.6	5.1	0.4	0.9	0.2	1.1
d. Industry	—	—	—	—	—	—	—	70.9	3.7	30.4	9.7	25.3	6.1	6.2	1.8	3.3
i. Nonenergy/petrochem.	—	—	—	—	—	—	—	19.5	—	16.4	—	3.1	—	0.5	1.0	—
e. Agriculture	—	—	—	—	—	—	—	17.4	1.2	—	1.5	11.7	2.4	1.2	1.6	1.0

f. Pub. serv. & comm.	—	—	—	—	—	—	—	34.2	3.5	—	—	30.6	—	15.8	—	4.2
g. Residentiald	—	—	—	—	—	—	—	180.7	9.5	65.2	8.6	96.3	23.2	29.1	12.4	0.8
VII. Residual	956	566	188	201	171	131	30	504.9	14.9	168.2	94.0	215.3	33.4	24.6	43.8	30.1
a. Bunkers	106	66	35	4	4	4	—	57.1	—	19.2	7.8	28.5	4.1	2.2	4.8	1.8
b. Electr. gen.	—	—	—	—	—	—	—	193.9	3.1	80.1	36.2	73.9	10.6	6.2	21.1	12.7
c. Industry	—	—	—	—	—	—	—	180.0	8.1	35.7	42.1	87.0	14.6	14.4	13.6	12.2
i. Nonenergy/petrochem.	—	—	—	—	—	—	—	1.4	—	—	—	1.4	—	1.2	—	—
ii. Pulp & paper	—	—	—	—	—	—	—	28.1	3.2	11.3	5.6	7.9	1.4	1.1	0.7	1.2
d. Residentiald	—	—	—	—	—	—	—	54.6	0.4	33.2	3.2	17.8	2.0	1.2	4.0	0.5
VIII. Naphtha	—	—	—	—	—	—	—	79.7	2.9	16.3	25.6	34.4	6.1	5.7	5.4	4.9
a. Nonenergy/petrochem.	—	—	—	—	—	—	—	61.1	—	10.0	22.0	28.7	6.1	5.4	3.8	4.9

aFootnotes b to e from Table P-2 apply.
bAll values refer to 1978 except for Turkey and Spain, which are 1979.
cUnits are in million metric tons (mt) for each particular product, each with its own specific gravity and heat content.
dResidential (Rows IIb, Va, VIg, and VIId) includes more items and has been adapted.

oil was used in 1978, the year of peak consumption. The data are given in Tables P-2 and P-3, and the steps are summarized in Appendix P-1. The analysis was carried out on the assumption that oil prices do not decline. But the basic result is quite robust. Even if prices decline, the conservation effort will likely continue, based on the expectation of increasing prices.

The analysis looked at seven different oil products in some twenty applications and in nine different geographic areas, but concentrated mainly on the industrialized nations (OECD).[1] Figure P-3 incorporates the results of this analysis: a consumption level of about 20 mbd (million barrels per day) in the 1990s—one-half of the consumption peak of 1978. Figure P-3 also shows the actual decline in demand between 1978 and 1982. Also shown is the production of oil within OECD and within non-OPEC nations. The difference, denoted OETO (i.e., OPEC oil required by OECD nations), declines rapidly over the medium term and could even reach zero during the 1990s.

The analysis is robust in several respects. (1) It does not depend strongly on OECD demand declining all the way to 20 mbd from its 1978 maximum of 42 mbd. Even a partial backout of oil can drop world demand sufficiently to induce strong competition among oil exporters who will try to increase their own market shares in order to raise their dropping oil revenues.

(2) Should prices drop enough to slow conservation efforts, then even an increase in demand by OECD will not result in increased total revenues to OPEC—thus exacerbating the competition for markets among OPEC members. The upshot of either scenario will be one of loss of financial power by OPEC and a corresponding gain by OECD.

OPEC Options

As discussed earlier, the 1979 price increase pushed the price of oil well beyond the level the market could support; it should have come down immediately once consumers realized that fact. However, further panic buying by consumers and the willingness of OPEC members to restrict supply maintained the high price for a longer time. But the resulting structural changes in world oil consumption are steadily reducing demand, making the price of about $30 per barrel unstable.

OPEC's oil production has already declined from a peak of 32 mbd in 1979 to around one-half that, with a consequent drop in revenues. It is becoming increasingly difficult for the OPEC nations to reduce production further. At this point, there are several (nonexclusive) options open to OPEC, and especially to Saudi Arabia. All of them involve risks, and none of them is pleasant.

1. OPEC can begin to act as a real cartel, allocate production quotas, and continue to cut its production to match the declining demand, in order to hold up the price. The problem with this approach is that almost all of the OPEC nations are running budgetary deficits and are badly in need of larger, not smaller, oil revenues. Therefore, the incentive will be to gain market shares at the expense of fellow OPEC members by cutting prices. This, of course, is the classical problem that makes cartels unstable. There is no reason to believe that OPEC will be able to function as a cartel for any length of time. It tried in March 1982 and did not succeed. It remains to be seen whether the March 1983 agreement can hold up.

2. Another option is for OPEC to persuade Saudi Arabia and the associated sheikhdoms to take the necessary production cuts. This has obvious advantages for countries like Nigeria, Algeria, Venezuela, Indonesia, and Iran, as well as for Mexico and other non-OPEC oil producers. It keeps the price at a high level for a while longer and does not disturb their revenues. However, it reduces the revenues of Saudi Arabia and other nations toward zero. This approach, therefore, may not be acceptable to the OPEC core. In any case, Option 2 and Option 1 are only short-term solutions. The continued high price encourages the further backout of oil and produces a steadily declining demand.

   There is a wide range of opinions concerning the minimum level of oil revenues, and hence minimum production level, required by Saudi Arabia. Some analysts have argued that the Saudis would find it difficult to cut their swollen budget and retrench. Saudi spokesmen have indicated that they could cut back production to 4 mbd but are producing more in order to keep oil prices moderate. This self-serving statement has yet to be tested.

3. Saudi Arabia might decide to restore the long-term oil market and cut the price. The optimum price path can be calculated from theory.[2] The details depend on assumptions; but as shown in Figure P-6, it would mean a rapid decline to about $18 per barrel, followed by a gradual

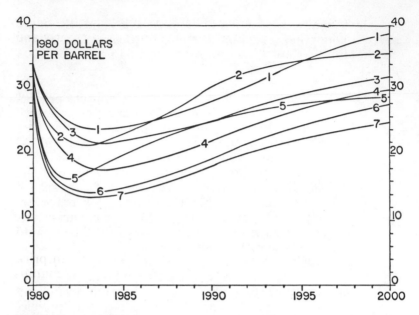

Figure P-6. Optimal price path for OPEC core (calculated under assumptions of appendix cited in Reference 2). The price is expressed in 1980 dollars. To convert to 1982 dollars, multiply by 1.25. The base case (Curve 4) uses the following parameter values (for details, see Reference 2):

Base year: 1980	Intitial price: $P_0 = 34.0$
World demand: $D_{-1} = 22.9$	$D_0 = 21.7$ billion barrels/year
Competitive supply: $S_{-1} = 16.6$	$S_0 = 16.9$
Long-run elasticity of demand:	$\varepsilon = 0.70$
Long-run elasticity of supply:	$\eta = 0.70$
Adjustment parameter:	$\lambda = 0.125$
Discount rate:	$\rho = 0.05$

Demand parameters: $a_0 = 28.56$; $b_0 = 0.45$; $\delta = 0.007$
Supply parameters: $a_1 = -6.33$; $b_1 = 0.35$; $\varsigma = 0.025$
Cost function: $C_0 = 0.50$; $k = 0.09$

Curve 1: Base case, except that demand elasticity is 0.20.
Curve 7: Base case, except that demand elasticity is 2.00.
Curve 2: Base case, except that discount rate is 0.15.
Curve 3: Base case, except that supply elasticity is 0.20.
Curve 6: Base case, except that supply elasticity is 2.00.
Curve 5: Base case, except that adjustment parameter is 0.25.
The results are seen to be quite robust (i.e., independent of the assumed parameter values), a sharp decline in price within two to three years, followed by a slow recovery. The price in 2000 is likely to be less than the initial (1980) price.

recovery in the oil price. To achieve this objective, Saudi Arabia would have to increase its production greatly and maintain it at a high level for several years.

There are several problems with Option 3:

a. It would take away market shares from high-cost oil producers and therefore cause political problems for Saudi Arabia.

b. It could stimulate an oil price war in which other producers with low marginal costs might try to undersell in order to retain their market shares. Under these conditions, prices could dip down to about $10, near the competitive price level. Such a price collapse would probably be brief, followed by a renewed attempt by the cartel to enforce production quotas.

c. The sudden price drop may upset oil producers in consuming countries, especially the United States, who would then clamor for an import fee to maintain price stability in the face of what they would regard as manipulation of the world oil price by Saudi Arabia. (See Chapter 4.) Other industrialized countries that have made investments in alternative technologies might follow suit. The end result would be that even at the lower price there would be no great increase in oil demand, either in the short or in the long term. Consumer countries, afraid that the price might again go up, would continue to make the necessary investments to replace oil with other fuels.

4. It is likely that the Saudis will adopt a combination of the options outlined above: cutting their production further; encouraging other OPEC members to join in production cuts; drastically cutting their budgets by stretching out development projects, reducing foreign aid, and so forth; dipping into their accumulated financial reserves; and letting the price erode in the hope of thereby increasing world demand for oil. Of course, the ongoing inflation serves to erode prices. The $34 price of 1980 is only about $28 in 1983 when expressed in 1980 dollars.

But this "fix" may only be temporary: a termination of the Iraq–Iran conflict could upset this game plan, as Iraq tries to recover its earlier production levels and seeks export markets. And of course, consumers' perception of continuing high prices should serve to continue to cut the world demand for oil and force further production cuts by OPEC. But if demand does not increase appreciably, then a lower price (say $25 to $30) is not stable. It leaves all producers worse off than before. They sell the same amounts of oil but at a lower price. Some of them will therefore attempt price cutting, which may precipitate a price collapse.

Price Increases or Price Decreases

It is always possible, indeed likely, that major interruptions of the world oil supply will occur, either by design or by accident. The conflict between Iran and Iraq in the Persian Gulf may spill over to other Gulf nations and destroy much of their distribution and loading capacity. Sabotage and terrorist activities could interrupt oil shipments. Such events would cause price increases, which would be moderated, however, by the existence of strategic stockpiles in consumer nations.

Nevertheless, it appears to us that the general trend of oil prices during this decade will be downward—providing that ongoing programs of conservation and fuel switching are maintained. This general trend will likely be punctuated by occasional upward price spikes and (temporary) price collapses, as discussed earlier.

A slow recovery of oil prices to about the present level is forecast for the 1990s, as low-cost oil is gradually depleted. But even in the next century, oil prices should not exceed by much the prices of competing and more plentiful energy alternatives.

It must always be remembered, however, that forecasting is difficult because prices are based not only on economics but also on essentially political events affecting the major exporters in the Persian Gulf. In the words of Professor Niels Bohr, the Danish Nobel Laureate in physics: "It is very difficult to make an accurate prediction, especially about the future."

Appendix P-1: A Low Oil-use Scenario

A microsectional analysis has been carried out according to the attached outline (March 1981). This study looks at seven different oil products in some twenty applications and in nine different geographic areas, mainly for the industrialized countries (OECD).

   I. *Assumptions*

      A. The price increase of 1979 has set into motion irreversible economic forces whose consequences will now be felt.

      B. Oil is fungible; in other words, oil displaced (backed

out) at any one location becomes available elsewhere in the world.

C. The world price of oil is set by supply and demand, with supply restricted by the production decisions of Arabian producers. Departures from this market will occur as noted below.

D. Refineries can and will keep up with changing product demands.

E. World coal production can expand to meet whatever is demanded. Alternately, world nuclear expansion rate will reach 25 Gw per year or 1.0 mbd oil equivalent per year.

II. *Base Case* (with world oil prices assumed to be roughly constant)

A. Oil will be replaced in all applications where cheaper alternatives are available, principally for the production of heat and steam. New capacity will not use fuel oil.

B. In the transportation sector, both efficiency and activity will increase with time. Hence total oil consumption must be estimated for each country (region), under the assumption of different technologies. For example, in the medium term consumption may decrease, and in the longer term it may increase.

C. The full supply response throughout the world to the oil price increase of 1979 will be felt in the late eighties. It may involve some production of heavy oil, tar sands, and shale oil.

D. In parts of the world where natural gas is available at low cost, it may pay to produce methanol or gasoline, which would displace oil.

III. *Scenarios of World Oil Price*

A. Likely Price Scenario, short-term, 1981–83

1. Saudi Arabia tries to depress the price (to around $32) by maintaining high production, in order to get OPEC agreement on its price strategy.

2. The realization that prices are dropping (in constant dollars) becomes widespread. This serves as a psychological trigger and produces two consequences:

a. The reduction of oil inventories in the consuming countries.

b. Full-scale production by OPEC members and other producers.

3. The net effect of (2) would be to produce a downward price spiral based on self-reinforcement, abolishing much of the price increase of 1979, which was due to a similar upward spiral.
4. The actual price will depend on Saudi reaction, on their own production decision, and on whether they can enforce discipline, that is, production cutbacks, on OPEC members.
5. Complicating this picture, and reinforcing the price drop, will be the (partial) comeback of production by Iraq and Iran. Both countries will want increased revenues.

B. Likely Price Scenario, medium-term, 1985–95
1. Dropping world demand and rising world production should cause a crunch for OPEC by about 1983–85, if not before. Even if Saudi Arabia persuades other Arab producers to cut production, they may lose control over prices when they cannot cut any further.
2. World price will then fall, unless military actions or sabotage reduces oil exports from the Middle East. For example, it would be to Iran's economic advantage to eliminate Saudi oil exports in order to maintain high prices for itself.
3. Consuming countries may take one of three actions:
   a. Admit the cheaper OPEC oil, delay conversions to coal and nuclear, and give up the investments already made.
   b. Impose tariffs or import quotas to protect some or most domestic investments in higher-cost energy resources such as tertiary oil, shale oil, and nuclear reactors (the most likely choice).
   c. Do both a. and b., thus causing international trade problems.

C. Likely Price Scenario, long-term
1. In the long term, beyond 1995–2000, the medium-cost oil resources outside Arabia may be pretty well exhausted, including those in Alaska, the North Sea, and other OPEC locations such as Indonesia, Iran, Nigeria, Algeria, etc.
2. Arabian oil will gradually reestablish itself, at a price set by the price of coal and nuclear energy, probably close to the present price (in real dollars).

In other words, with alternatives to oil pretty well established, even in the transportation sector, people would not pay a great premium for oil.

IV. *Backout of Heavy Fuel Oil*

A. Residual fuel oil will be replaced in electric utility boilers, industrial boilers, heating installations, and marine transport by cheaper gas, coal and nuclear energy. Replacement of existing oil-fired capacity will proceed at different rates in different countries, depending on economics, availability of capital, environmental and safety considerations, and desire for self-sufficiency. We can extrapolate current coal and nuclear trends toward saturation.

B. Leading technology candidates for speeding up the replacement of oil are:

1. Highly loaded (70%) coal-water mixtures, designed to replace No. 6 fuel oil in existing boilers (including marine boilers).
2. Fluidized-bed combustion units designed for retrofit applications.
3. Mass production techniques being adopted for nuclear reactor manufacture (in USSR and elsewhere).
4. Powdered coal to replace oil in marine diesels.
5. Low-temperature nuclear reactors for district heating (e.g., Canadian "Slowpoke;" Russian systems).

V. *Replacement of Medium Fuel Oil*

A. Medium and light fuel oil will be replaced in residential and commercial heating applications by whatever technology or fuel is cheaper. Candidates are:

1. Natural gas, which should become plentiful once it is deregulated (in the United States) and replaced in utility and industrial use by cheaper coal.
2. Electricity (produced by coal and nuclear reactor); various uses of heat pumps.
3. District heating based on coal, nuclear energy, or geothermal energy (where appropriate).
4. Solar energy (where appropriate).
5. Biomass, municipal waste.

B. It should be noted that conservation in the residential-commercial sector has not yet reached its economic limits and can be relied on to reduce demand further.

VI. *The Transportation Sector*

  A. Transportation, which now consumes about one-third of all world oil—mainly as gasoline, jet fuel, and diesel fuel—should become the major oil user by 1990. Therefore, its worldwide requirements need to be estimated rather carefully.

    1. Private cars: Gasoline consumption will decline greatly in the United States in the short and medium term, with some increase in the use of diesel fuel, as more efficient cars enter the fleet and improve average fleet mileage. Outside North America, where gasoline has always been heavily taxed, the effect will be smaller. If minicars (50 to 60 mpg) become widely accepted, or even electric cars, then consumption can decrease much further, even beyond 1995–2000.

    2. Trucks: Conservation effects will be important but less so than for cars. Vehicle miles will grow.

    3. Aircraft: Conservation effects will be of some importance. Better scheduling and higher load factors could achieve significant savings.

1
ENERGY POLICY IN A FREE-MARKET ENVIRONMENT

S. Fred Singer

Summary

Government regulation of energy resources and involvement in the market is inefficient and counterproductive. Indeed, the "energy crises" of the 1970s were mostly the result of federal regulation of both oil and gas. There are, however, areas in which the government does have a clear responsibility—for example, maintaining competitive market conditions, protecting the environment and public health, managing public lands, and working in the areas of advanced research and national security. Together with a commitment to a free market, these functions constitute a responsible energy policy.

The Reagan administration's approach to energy represents a distinct break with those of the 1970s. Despite the visible success of oil price deregulation, the administration still faces difficult issues in its efforts to reduce government intrusion in the energy market.

Introduction

For twenty-five years, the U.S. government has played an increasingly important role in energy matters, but has pursued an inconsistent policy. On the one hand, it has tried to make energy cheap in order to benefit consumers; on the other hand, it has made energy expensive to benefit producers. For example, natural gas prices have been held at low levels since 1954—most of the time on the order of a few cents per million Btu.[1] During the 1960s, however, oil producers were protected from cheaper imports and were able to sell oil at around $3 a barrel, or 50 cents per mbtu.[2] Had low-cost Middle East oil been allowed to enter without restriction, the price would have been lower by at least a factor of two, and much of U.S. oil production would have closed down.

These contradictory policies of the U.S. government result in political conflict and compromise between oil-consuming states (like Massachusetts) and oil-producing states (like Texas). In trying to please both consumers and producers, the U.S. Congress has produced a quagmire of policies and involved the federal government more deeply in—thus distorting—the energy market.[3]

By 1970 price controls had led to serious shortages of natural gas because of excessive consumption and inadequate incentives for producers. Imported oil was needed to fill the demand. In 1971 the Nixon administration imposed price controls also on crude oil. They remained in effect for ten years, even after world prices rose above the domestic level in 1973. As a result, a domestic market developed that had two basic tiers: cheap, price-controlled domestic oil, and expensive, uncontrolled imported oil.[4] Mind-boggling regulations were required to establish some measure of equity; a large bureaucracy was employed to track oil transactions and prices. A special program had to be established to equalize the price of crude oil to all refineries, regardless of the origin of the oil. The overall results of these policies were (1) greatly reduced economic efficiency; (2) overconsumption of oil because of an effective price subsidy; and (3) a resulting pro-import government policy (see Chapter 3). Oil imports rose from 23 percent of consumption in 1970 to nearly 50 percent by 1978, partly because of price regulation of both oil and gas.

After the 1973 Arab embargo, President Nixon sought independence from oil imports. But it was soon recognized that independence would also mean the excessive costs of substituting alternative fuels—costs well above the prices set by OPEC for world oil. (By 1980, however, oil prices had risen sufficiently to make many substitutions economical.)

President Carter injected the federal government even more deeply into energy matters. He created a Department of Energy, and he plugged hard for conservation and solar energy in his first National Energy Plan of 1977. But because he did not free oil and gas prices, he actually discouraged conservation and the development of solar energy. By 1979, Carter decided to encourage production. He tried to deregulate (or at least raise) the prices of natural gas and oil and pushed for a large, government-backed $100 billion synthetic fuels program.

The Reagan administration drastically reversed the policies

of the previous administrations. One of President Reagan's first acts in January 1981 was to decontrol oil prices. He then began dismantling the vast machinery designed to enforce price regulation. He even proposed abolishing the Department of Energy, which had become a symbol of government intervention in energy markets (see Chapter 3 and Epilogue).

The Free-Market Philosophy on Energy

The Reagan administration's approach to energy is based on the conviction that a free market can allocate scarce energy supplies most economically and efficiently through prices set by market forces. Though the administration has been trying to apply these principles consistently, it has encountered obstacles for both historic and political reasons.

A free market is well suited to supply energy, because energy resources are owned, either by individuals or by corporations. The existence of property rights provides incentives for proper management of resources. Under competitive conditions, management for individual profit also benefits the general population—a principle originally set forth by Adam Smith. In contrast, certain other natural resources such as water and air, which are not owned by individuals or corporations, are not properly managed by them. It is impossible for anyone to own a parcel of air—although an owner of a lake or a pond would probably want to take care of water quality and not discharge wastes into it. Thus there is an argument for government concern about air and water quality, since no real incentives come from market forces to control pollution (see Chapter 7).

While the free-market approach denies a governmental role in setting fuel prices, there are still important functions that the federal government must perform if energy resources are to be used properly. In aggregate, these functions make up a government energy policy. They include:

1. *Guarantee that a free market exists.* The federal government takes action against companies or individuals that inhibit competition. It is also appropriate for government at the state, local, and federal levels to regulate certain natural monopolies, such as electric power transmission and natural gas pipelines and distributors.[5] The government, along with private groups, can provide information (for example, about the energy efficiency of cars) to in-

form consumers so that they can participate more effectively in the market.

2. *Regulation of interstate transportation of fuels and electric power.* The interstate-commerce responsibility of the federal government also includes preventing energy-rich states from taking undue advantage of energy-poor states by exorbitant taxation or other means of price discrimination. This is currently a matter of controversy.

3. *Protection of the environment.* As the guardian of national public health and safety, the federal government, along with the states, sets appropriate quality standards for the ambient atmospheric and water environment. It also licenses energy facilities, such as power plants and nuclear reactors. Much more can be done to streamline the process of achieving the environmental standards and to speed up nuclear licensing (see Chapter 7 and Chapter 9).

4. *Strategic Petroleum Reserve.* Ensuring national security is an important federal function. U.S. dependence on imported oil and the possibility that cutoffs could produce severe economic damage led to legislation for the establishment of a Strategic Petroleum Reserve operated by the federal government (See Chapter 3).

5. *Management of public lands.* When leasing public lands for oil and gas (especially on the outer continental shelf) and for coal and other energy minerals (including geothermal energy), the government acts as a prudent landowner, concerned with maximizing its financial return (see Chapter 8).

6. *Advanced research and development.* In areas where no single industry or group of industries can capture all the benefits of its own research and development investments, the government has a role to carry out basic scientific research for future energy sources, such as nuclear fusion (see Chapter 12).

7. *International energy cooperation.* The federal government has important functions as a party to various international agreements. For example, the International Energy Agency (IEA) was set up in 1974 to operate under the auspices of the Organization for Economic Cooperation and Development (OECD). One of its principal purposes is to provide for oil sharing in case of major interruptions in world oil supply (see Chapter 3). Another cooperative venture is the International Atomic Energy Agency, which is concerned with the exchange of atomic

information, and with safeguards against the spread of nuclear weapons (see Chapter 10).
8. *Owning and operating energy facilities.* Unlike many other nations, the United States does not own refineries or purchase oil on the world market. Even the fuels used by the military are provided by private oil companies under government contract.

The U.S. government is involved in owning and operating certain energy facilities, such as naval petroleum reserves, hydroelectric plants in the Far West, and the well-known Tennessee Valley Authority, which includes hydroelectric, coal, and nuclear sources for the production of electricity. (In a sense, these federal involvements are now an anachronism.) The Congress has never approved, however, the creation of a Federal Oil and Gas Corporation (FOGCO), although some legislators have felt that a federal yardstick should be used to judge the performance of private oil companies.

It is not easy for the federal government to carry out these various energy functions in a consistent manner. The main problem is political: how to satisfy the often conflicting desires and requirements of such different interest groups as energy consumers, environmentalists, owners of oil and gas resources, different kinds of energy companies, and other more specialized interests.

Current U.S. Energy Policy: Oil

The energy policy of the Reagan administration relies on the laissez-faire approach of a free market. Administration decisions, however, are made incrementally. Though the prices of crude oil and oil products have been decontrolled, the administration has left undisturbed the "windfall profits tax" imposed by Congress in 1980. This is really an excise tax based on the difference between the world price (i.e., the market price) and a base price corresponding roughly to the production cost plus a "reasonable" profit. For example, oil discovered before 1978 has a base price of $12.89 per barrel and a tax of 70 percent above that. On the other hand, post-1978 oil and hard-to-produce heavy (high-viscosity) oil is taxed at 30 percent on a base of $16.55. The exact amount of windfall profits tax, which is likely to exceed $200 billion over the next ten years, and how its proceeds are to be allocated are sure to trigger lively political controversy (see Chapter 6).

Past administrations have provided special subsidies to so-called small refiners at the consumer's expense. These benefits are no longer available, however. The changing and shrinking market for oil products is likely to benefit those refiners willing to make capital investments to produce more gasoline and other motor fuels, and less heavy fuel oil. These investments are being made in response to market forces—without any government assistance or direction.

In the leasing of public lands, the Reagan White House has moved more rapidly than any past administration. As a result, the energy industry should be able to make its plans with more certainty—and, therefore, more efficiency. This will ultimately benefit consumers. It is ironic that the windfall profits tax has removed the ready cash of oil companies and so decreased the amount of money they can pay to the U.S. Treasury for oil and gas leases. Some would argue that the best way to tax away a windfall profit is simply to offer more public lands for lease and encourage more oil companies to enter into the bidding.

An important energy policy issue is emergency allocation of oil products in case of a shortage, which usually results from a supply interruption. In a free market this problem disappears. With prices decontrolled, there may be a dislocation but not a long-term shortage. The price will simply rise and dampen the demand to match the available supply of oil. The allocation will also be accomplished automatically, with oil flowing to users who are willing to pay for it. This is the most *efficient* method of allocating during a scarcity and requires no govern-ment action; the allocations are effected by price and not by political influence. Allocation of the available supply by the free market is also fairly *equitable*. Even though the poor are pinched by the higher price, they also suffer under other systems of allocation, such as rationing by coupons (whether per car or per driver, and whether ration coupons are kept or resold), or a political method of distribution without a change in price. The fairest method is to let the price rise and recycle increased tax revenues to provide general aid to the poor without regard to their energy purchases (see Chapter 3).

The legislatively-mandated Strategic Petroleum Reserve (SPR) thus may not really be needed. With decontrolled prices, the allocation of any shortfall could proceed automatically. With an SPR, however, the government, as the owner of the oil, has to develop policies for releasing it: when, how much, and in what manner. This creates uncertainties that discourage oil companies and individual users from maintaining adequate private stockpiles. At present, the SPR exceeds 300 million barrels. Some hard decisions will have to be made before the SPR reaches its announced goal of 750 million barrels, a target that has an annual carrying cost of some $5 billion dollars. National debate can be expected to focus on who should bear this cost and in what manner (see Chapter 3 and Chapter 4).

The Reagan administration has not yet been able to remove the ban on oil exports from Alaska; this restriction was established by Congress in 1973 in the mistaken belief that this would protect U.S. oil security. But with oil prices decontrolled, oil can be bought freely, even though the price of all oil (including Alaskan) would rise in the event of a supply shortfall in the world for whatever reason. Currently, Alaskan oil is creating a glut in California, discouraging production at the margin in both California and Alaska. To avoid a sharp price discount, excess Alaskan oil is shipped through the Panama Canal to the U.S. East Coast at great expense. Permitting the export of Alaskan oil, say, to Japan, would save nearly a billion dollars per year and would encourage greater development of oil and gas resources in the Arctic (see Chapter 5).

Import fees on crude oil or higher federal taxes on transportation fuels are being widely discussed. They are viewed as means of enhancing conservation, decreasing oil imports (with attendant benefits to national security and trade balance), and increasing Treasury revenues. Such fees and taxes might become particularly appealing for stabilization purposes if and when world oil prices should decline drastically—at least for short periods. Short-lived price collapses could raise havoc with the U.S. domestic energy industry and produce disincentives to energy investments as well as to energy conservation (see Chapter 4).

Natural Gas Policy

Natural gas poses a difficult—some would say insoluble—problem. The major conflict is between those who would deregulate the price and those who would simply maintain ceilings. Proponents of ceilings include some consumer advocates (who may only be taking a shortsighted view) and gas pipeline owners (who would like to see the price low and demand high to maximize the shipments of gas). Support for ceilings also comes from importers of costly LNG (liquefied natural gas) and producers of expensive "deep" gas who look on the availability of a large reservoir of price-controlled cheaper gas as an opportunity for "rolling in" (price averaging) their higher-priced gas. "Old" gas under contract still sells for less than 50 cents per 1,000 cubic feet at the wellhead in many cases, while gas from the same region, but from a deeper structure, can sell for as much as $9.

Under the Natural Gas Policy Act of 1978, about half of all gas, and any "new" gas, will be decontrolled in price by 1985. The consequences of this are difficult to predict. For example, intrastate gas (gas produced and sold within the same state) would not be subject to control after 1985; therefore, gas suppliers would prefer to sell to the intrastate market, producing a shortage in the interstate market—the same situation that pertained before 1978.

Another example: The ultimate effects of the existing "fuel pass-through clause" (by which electric utilities can pass on any increases in the price of their fuels) may be to make the electric utilities insensitive to higher gas prices. But if utilities were to stop using higher-priced gas and switch to coal, a large surplus of gas would develop and U.S. gas prices could remain below the equivalent level of oil for many years. Residential and commercial users would then switch to gas more rapidly. The consequences of this sequence of substitutions would be a furthering weakening of demand for oil as a heating and boiler fuel, and a reduction in oil imports.

President Reagan's strategy on natural gas deregulation has recently been announced. The proposed legislation is designed to provide a transition to a free gas market that would protect consumers and provide incentives for producers. As befits this complicated subject, the proposed legislation is complicated, inviting congressional tinkering—with unforeseeable consequences. Some of the bill's main features are: (1) deregulation of "old" gas prices to provide incentives to bring more of this low-cost gas to market;[6] (2) steps to deregulate the interstate pipeline systems to permit different arrangements for moving gas, such as "contract carrying"; (3) various features to eliminate the unanticipated effects of the Natural Gas Policy Act of 1978—the most recent, but misguided, effort to deregulate gas; and (4) elimination of the Fuel Use Act of 1978 and of incremental pricing, which discriminates among end users. Not all of these features are desirable, and in many cases the gas industry is already making the required adjustments (see Chapter 2).

Coal Policy

Of coal, it used to be said that it is a great fuel except that "you cannot mine it and you cannot burn it." The Reagan administration is likely to move farther and faster on coal than previous administrations did. For one thing, it will speed up mineral leasing on federal lands. For another, by simplifying strip-mining regulations and by making the Clean Air Act regulations more flexible, administration actions should make coal much easier to mine and burn.[7] At the same time, land and air resources should not be adversely affected.

Some political battles will have to be fought to achieve these changes. Congress and the public will have to be convinced that environmental regulations can be made more flexible without damaging the land or lowering air quality.

Transportation costs are an important determinant of coal use. Efforts are under way to lower such costs through the use of slurry pipelines, though the railroads oppose this.

Advances in technology undoubtedly will speed the adoption of coal as a boiler fuel. "Fluidized-bed" combustion provides a low-pollution method for the use of coal, without high-cost "scrubbers" for flue gas desulfurization. The development of combustible low-cost coal-water mixtures will make it possible to replace higher-cost fuel oil in existing oil-fired boilers without major capital expenditures.

Nuclear Energy Policy

Regardless of U.S. nuclear policy, other countries are now fully aware of the advantages of nuclear energy. It is cheaper than coal and much cheaper than oil. Nuclear energy, on the whole, is environmentally benign, provided that strict safety precautions are enforced. The Reagan administration changed and reversed drastically the Carter administration's policies. Reprocessing of used fuel elements and disposal of nuclear wastes are being allowed—finally—to commence. The export of nuclear technology will be not only permitted but also encouraged. In addition, work on nuclear breeder reactors may resume, to stretch the uranium resources of the United States and other countries.

The most significant action the U.S. government can take to revive its lagging nuclear program is to streamline the licensing process. Just two steps are necessary: (1) selecting sites for nuclear and other power plants well in advance of need, to build up an inventory of approved sites; and (2) standardizing nuclear plants so that the licensing process can be accelerated. These steps not only will cut the time between planning and the date of operation (and thereby lower the cost greatly), but also make nuclear energy safer (see Chapter 9).

Other Energy Resources and Conservation

With respect to other energy sources, the Reagan administration has taken a laissez-faire approach. Solar energy and synthetic fuels from coal have been left largely to the market, although there exist important tax benefits that provide a kind of subsidy (see Chapter 6 and Chapter 12).

Two shale oil projects and one synthetic gas project have received federal loan guarantees. The Synthetic Fuels Corporation, set up under President Carter, has become less active. It is clear that the government is not going to subsidize the crash program for synthetic fuels that often has been envisioned by high-level policy planners in the past. On the other hand, Reagan is continuing government support for research on fusion energy—a long term program whose impact will not be felt until after the year 2000.

The Reagan administration believes that conservation, whether by fuel switching or by using energy more efficiently, is best promoted by market forces. Higher prices are supposed to achieve the appropriate level of conservation. The often stated idea of encouraging conservation by legislation—for example, by means of a gasoline tax—has not found much favor in Congress, although an economic case can be made for such a tax based on the negative externalities[8] produced by driving (see Chapter 6).

International Oil Policy

On the international scene, the Reagan administration has made some important new departures. As customary, the federal government is staying out of the purchasing of oil and gas, and letting private companies negotiate detailed arrangements with foreign suppliers, both governmental and nongovernmental. An exception to this general policy has been a direct purchase agreement with Mexico for the Strategic Petroleum Reserve.

The administration is likely to deemphasize the role of the International Energy Agency (originally conceived as a counterweight to OPEC). Special sharing arrangements of oil supplies during emergencies will undoubtedly be reviewed. Since these sharing arrangements have never been tested, no one knows whether they will work. With prices deregulated, there may be no need for such arrangements at all. In case of supply shortfall, oil will simply flow to individuals willing to pay a premium.

Many "experts" have worried that an oil "shortage" could break up the Western alliance, as countries compete for oil in a beggar-thy-neighbor manner. Their concerns are unfounded. Wealthy Europeans can always outbid those Americans who cannot afford the higher price. Individuals, not countries, willing to pay the higher price will get the oil. The United States is not likely to provide subsidized oil for its citizens; the policies of other nations have not yet been defined.

A major hope of the West has been to discover oil outside the Middle East, in order to diversify sources and make the supply more secure. From time to time, it has been suggested that oil exploration be subsidized, or even completely financed, by United Nations agencies or by the World Bank.

The Reagan administration prefers exploration by private companies—without subsidies. Since many Third World nations oppose multinational companies, particularly those headquartered in the United States, these nations may well prefer other financing arrangements. One possibility may be an organization for oil development in the Third World that will accept money from OPEC nations—particularly Arab nations with surplus funds. It is likely that much new oil will be found in the next few years with or without U.S. government involvement. As far as the Reagan administration is concerned, it will be without U.S. government involvement.

International Nuclear Energy

With the sharp turnabout in the U.S. government's view on nuclear energy, there will be a freer export of U.S. nuclear technology and of enriched fuel. This is based on the realization that countries wishing to build nuclear weapons are not going to be stopped by the U.S. government. Those countries can build or acquire weapons directly, without first developing nuclear power for electricity production. A number of technically advanced nations are now able to act as suppliers of nuclear technology and fuels, so that the actions of the United

States no longer determine what happens to nuclear power in the world community (see Chapter 10).

Consumers of oil everywhere will benefit from the construction of more nuclear plants anywhere in the world. With world demand for oil thereby reduced, downward pressure will be created on world oil prices.

Conclusion

The Reagan administration is committed to maximum reliance on the forces of a free market and a minimum of government intervention. The price of oil is now so high that oil can be replaced by less expensive gas, coal, and nuclear energy. These cheaper fuels can be substituted in many applications—principally for producing heat and steam—which make up about 60 percent of world oil use. Government policies need not do much more than remove political and institutional obstacles to the use of these alternative energy sources. Economics will do the rest.

2

REGULATION AND DEREGULATION OF NATURAL GAS

Connie C. Barlow
and
Arlon R. Tussing

Summary

The natural gas industry in the United States has suffered a
series of market upheavals in the last decade: acute shortages
in the mid-1970s, a violent price "flyup" between 1978 and
1982, and now an unnatural combination of an acute supply
surplus with prices that continue to rise.

Each of these market shocks was the product of previous
federal and (to a lesser extent) state regulation. Economic
regulation of the industry began with state and municipal
franchising of local gas companies in the nineteenth century.
Federal oversight of long-distance gas transmission is rooted
in the Natural Gas Act of 1938, and in 1954 reached into the
terms of sale between gas producers and interstate pipelines.
Economic regulation culminated in 1978 with extension of
federal authority to gas prices in *intra*state transactions, and
with statutory restrictions on the end uses of gas.

The successive incursions of government into the gas industry were well-meaning and often inevitable responses to
conditions of the time. The current demand, from almost all
segments of the public, that Congress "do something" about the
new crisis is thus understandable as part of a long tradition.
The market problems of 1984 are, however, largely transitory,
reflecting an institutional lag in adjusting to diminished economic control. Though passing, the difficulties are severe, and
congressional leaders are pressed to find a legislative "fix"—
and a fix that, unfortunately, could retard the adjustments now
under way.

We believe that there are no plausible administrative "solutions" to the current crisis, and that the logic of history, rather, points toward a progressive dismantling of much of the existing regulatory apparatus.

An Industry in Turmoil

Only a few years after President Carter and the Congress crafted the Natural Gas Policy Act of 1978 (NGPA), conditions in the gas industry had changed so radically that it was hard to tell up from down. In the early 1980s a critical gas shortage gave way to swelling surpluses. Pipeline companies that had been dragged into court by angry customers whom they had "curtailed" during the 1970s faced another round of litigation in the mid-1980s. This time, gas producers were charging breach of contract when pipeline purchasers failed to "take or pay" for their committed volumes of gas. Finally, large industrial gas users who had been denounced as wasteful consumers of a precious resource and penalized by the incremental pricing provisions of the NGPA, found themselves courted by gas utilities desperate to retain markets.

The problem with this 180-degree shift in the gas business was that almost nobody saw it coming. Pipeline companies, never having experienced demand limitations, signed up for too much gas at too high a price. Gas-production companies, responding to the price incentives Congress built into the NGPA, pointed their drill bits at deeper and more costly reservoirs, expecting to get prices two to four times as high as those allowed from conventional gas sources. By 1983, however, producers were lucky to find buyers for new gas at any price.

Local distributors, too, were caught in the crunch. Faced with a mass exodus of big industrial customers as gas rates approached and exceeded the price of residual (No. 6) fuel oil, they were caught between the proverbial rock and a hard place. The loss in sales meant that the utilities had fewer customers over which to spread the "fixed" costs of their plant. But if they tried to recover lost revenues by raising their rates, they risked losing even more of their customers. Aggravating the companies' distress was the inflexibility of many supply contracts between distributors and interstate pipelines. Attached to the contracts were "minimum-bill" tariffs that obligated distributors to pay for a big chunk of their contracted volumes, whether or not their own customers were interested in buying it.

Pipeline companies were not the only parties that extracted hard and fast purchase commitments from their downstream customers. For many years producers, too, had included take-or-pay clauses in their terms of sale. Some wellhead transactions called for pipelines to take, or else pay for, 90 percent or more of what the producer had to offer. With sagging markets, pipelines had to cut their gas supplies wherever contracts granted them the discretion. Unfortunately, this often meant shutting in their cheapest sources of gas—those for which they had contracted at very low prices and without minimum-bill or take-or-pay provisions prior to the shortages of the 1970s.

Big industrial customers who purchased gas directly from interstate pipelines (rather than from distributors) were in the best position to bypass the problems created by inflexible contracts. Because the pipeline had always retained the right to halt deliveries whenever supplies fell short, the purchaser was likewise free to abandon this "interruptible" service. Indeed, factory owners began to prowl the nation's gas-producing areas, searching for suppliers willing to sell shut-in gas at bargain rates. Many deals fell through, however, when recalcitrant pipelines refused to carry gas on a contract basis for customers who would otherwise have to make their purchases directly from the pipeline.

The Push for Governmental Action

By early 1983 the natural gas industry was in chaos. More gas was available, both domestically and from Canada and Mexico, than could be sold, and oil prices were going down, yet consumers saw their gas bills rising at rates exceeding 40 percent annually in many parts of the country. Distributors and state public utility commissions had little choice but to pass the increases disproportionately on to residential customers in a desperate effort to keep industrial users from switching to residual oil. Meanwhile, pipelines were shutting in some of their cheapest reserves, and industrial plants successful in lining up competitively priced supplies could not find anybody to carry the gas.

Pressure mounted on the Federal Energy Regulatory Commission (FERC), the Congress, and state public utility commissions and legislatures to do something about the gas problem and to bring some sense and stability to markets suddenly gone haywire. Specifically, these public bodies felt the need (1) to tackle the minimum-billing and take-or-pay terms in supply contracts, (2) to discourage the purchase of gas at above-market prices by ensuring that "price signals" were effectively communicated upstream from the final consumer all the way to the producer, and (3) to make the nation's pipeline network accessible to industrial users and distributors who wished to purchase gas in the field.

The push for governmental action came from all segments of the industry: consumers, distributors, pipelines, and producers. The political atmosphere, however, was terribly confusing. Not only did proposals for change differ markedly, but no segment of the industry voiced a unified position. Producers with a lot of price-controlled "old gas," for example, wanted full deregulation coupled with forced renegotiation of all gas contracts; producers that had gambled on "deep" gas (which the NGPA exempted from price controls) and had locked-in high-priced deals preferred the status quo. Interstate pipelines, too, differed in their views on whether and to what extent Congress ought to legislate generic changes in the take-or-pay terms of existing supply contracts. And consumers were becoming increasingly suspicious that *any* regulatory or institutional change would simply redivide the "upstream" revenue pie, and would bring little relief from runaway rate hikes.

There was thus no obviously correct course of action (from either a public-interest or political standpoint) for a senator with consumer leanings, a representative from a gas-producing state, or a partisan appointee on a regulatory commission. Nevertheless, the nation was searching for a legislative or regulatory solution to the industry's problems. The emphasis on a governmental "fix" for gas-market dislocations is neither new nor surprising. Gas distributors, pipelines, and producers are, after all, among the most highly regulated companies in the United States. Since the mid-1800s when municipalities and state legislatures first intervened in gas industry matters, the government's reach has expanded. It has also grown increasingly complex.

History shows us that governmental involvement in gas markets has repeatedly had a domino effect, giving rise to altogether new problems (sometimes more troublesome than their predecessors), which seem to demand even more regulation. Therefore, before one puts much faith in any new governmental fix for today's market imbalances, a careful look at the history of gas-industry regulation is in order.

The Rise of Regulation

Municipal regulation of local gas companies, which originally manufactured gas from coal, dates from the nineteenth century. Because it simply did not make sense, economic or otherwise, to have competing companies tearing up the streets to lay parallel lines, municipalities granted full or near-monopolistic franchises to local gas companies, whose rates, in turn, came under various degrees of governmental scrutiny.

Local regulation was effective for as long as *manufactured gas* (coal gas) was the commodity, because it was produced in the same town and usually by the same company that distributed it. But when *natural gas* (methane) from outside the city limits began to enter the market, municipal governments were no longer able to control gas prices. The states, however, stepped in to fill this regulatory gap. State legislatures created Public Utility Commissions and Public Service Commissions, or extended the reach of existing trade commissions into the gas transmission business. New York and Wisconsin led the nation, establishing commissions in 1907.

As technology advanced and pipelines were able to carry gas several hundred miles or more, a new regulatory gap appeared that even the states could not fill. Between 1911 and 1928, Oklahoma, West Virginia, and Missouri tried to assert jurisdiction over interstate gas transactions, but were thwarted each time by federal judges who ruled that state regulation of long distance gas transmission violated the interstate commerce clause of the U.S. Constitution.

In 1938, therefore, Congress passed the Natural Gas Act. This law imposed federal certification, or franchising, of interstate gas pipelines, coupled with oversight of rates charged by transmission companies. The Federal Power Commission (FPC, now the Federal Energy Regulatory Commission), which owed its existence to the Federal Water Power Act of 1920, became the administering agency. The reason for granting the commission broad jurisdiction was the existence of a significant *economic surplus*—the margin between the low prices producers would accept for natural gas (that would otherwise have been flared in the oil fields) and the cost of synthetic gas sold in urban areas.

The gas-transmission companies were far fewer in number than either the gas producers or local gas companies and industrial gas consumers. In the absence of rate regulation, therefore, the pipeline companies could have used their near-monopsony position on the buying end and their near-monopoly position on the selling end to take the entire surplus for themselves—buying in the fields for a pittance and selling to local gas companies at whatever price the market would bear.

And if market entry and pipeline additions had not been controlled via certification, the building of oversized and redundant lines could have inflated the unit cost of transportation and driven up consumer costs almost as surely as did the absence of rate regulation.

The Natural Gas Act clearly intended that pipelines should not be given the opportunity to pick up the entire difference between field prices and the value of gas at the city gate, but the law did not dictate how that surplus should be allocated between producers and consumers. Early on, the FPC (quite expectedly) stepped into regulating field transactions between affiliated companies. The commission reasoned that if a producer and the pipeline were related and only the pipeline was regulated, the producer affiliate could capture in an inflated wellhead price the monopoly profit that regulation sought to deny to the pipeline company. It was not until 1954, however, that a divided Supreme Court ordered the FPC to regulate prices charged by independent producers as well. Whatever the original intent of Congress, the courts and the FPC henceforth acted on the presumption that a legally "just and reasonable" wellhead price was a *cost-based* price and, therefore, that any economic surplus had to be passed on to consumers.

Price Controls and the 1970s Gas Shortages

Economists, and laypeople always have differed, and probably always will differ, about the best way to distribute economic surpluses, sometimes called economic rent. If a commodity can garner a price that far exceeds the actual cost of production, including a competitive return on investment, should the producer be allowed to reap an "unearned" windfall? Would a greater social good be accomplished if consumers were favored by means of price controls, or should the government contrive a tax that would transfer the rent to the community at large?

Despite the wide range of respectable disagreement about the *distributional* merits of price controls, production taxes, and so forth, there is a broad consensus on *efficiency* grounds that any combination of controls and taxes should not adversely impinge on the fundamental incentives to produce and market natural gas. By the late 1960s, however, producers (and some respected economists as well) warned that wellhead price controls on natural gas were indeed undermining the incentives to find and produce gas. In 1967, and for the first time since the interstate gas industry was born, the nation's reserves began to decline. The U.S. gas-reserves base continued to contract until 1981, when the producer price incentives in the 1978 Natural Gas Policy Act had begun to work their course.

The problems posed by dwindling reserves and increasing demand were exacerbated by a growing schism between inter- and intrastate gas markets. Customers in gas-producing states like Texas and Louisiana could and did offer higher prices than federal rules allowed interstate pipelines to pay. By 1975 almost half of the nation's proved reserves were committed to *intra*state markets. A more alarming statistic was the fact that producing-state customers captured virtually all new gas discoveries shoreward of the federally owned Outer Continental Shelf. (Gas found there entered interstate commerce when transported to shore).

What began as a warning in the late sixties became a national emergency by the mid-seventies. Burner-tip prices for gas began to lag seriously behind those of oil alternatives. And regulators found it impossible to adjust cost-based gas prices in a manner that could even begin to keep pace with OPEC actions. With every boost in oil prices, consumers found gas just that much more attractive, and demand surged. But because of price controls on gas, producers could not charge more for their existing supplies (thereby limiting customer demand to what was available), nor had they sufficient incentive to go out and look for more.

The combination of supply disincentives, the widening gap between end-use prices for gas and alternative fuels, and the purchasing-power advantage of customers within the producing states—all owing to cost-based wellhead price regulation— took its toll in the 1970s. The record cold snaps during the middle of the decade created an irresistible demand for action by Congress.

Government Allocation of Scarce Gas Resources

Again, economists of various schools (and politicians, for that matter) advocate markedly different strategies for dealing with an impending resource shortage. On the one hand, if gas prices are freed up and thereby can hold supply and demand in balance, a shortage is avoided altogether—but consumers have to pay more. On the other hand, if prices are held below a *market-clearing level,* some consumers do pay less, but others are shut out of the market completely, or live in fear of future curtailments. Moreover, if price is not allowed to check demand, then some governmental entity has to step in and decide who deserves to get how much of the limited supply.

Disputes revolving around these issues were the reason it took about two years for Congress to act, even after the disastrous gas shortages that racked the Midwest and the Northeast in the mid-1970s. With consumers already hard-hit by rising oil prices, it was at first unthinkable for our nation's leaders to allow our second most important fuel—and a fuel that was almost entirely domestic in origin—to have its prices dictated in effect by the workings of a foreign cartel.

While the Congress was unable to agree on an approach until 1978, state and federal regulators were using a variety of administrative programs to cope with growing shortages. They were forced to assume a role the market would otherwise have played in allocating the limited supply of gas among competing users.

First to suffer were prospective consumers who were denied gas by hookup moratoria sanctioned or imposed by state utility commissions. Later, even established customers felt the squeeze as state commissions, and then the FPC, approved curtailment schedules to systematize service cutbacks for low-priority customers (especially electric utilities and other industrial facilities that used gas to raise steam). As the shortages deepened, actual curtailments of gas reached progressively higher-priority classes, especially when severe winters coincided with diminished delivery capacity.

The movement toward comprehensive end-use regulation got a powerful boost when deliveries to even high-priority customers in the East and upper Midwest were curtailed in the winter of 1976–77. In 1978 both houses of Congress voted overwhelmingly for the Power Plant and Industrial Fuels Use Act (FUA), which contained an outright ban on construction of new gas-fired electric utility and industrial boilers, and aimed at the elimination of all use of gas as a boiler fuel by 1990.

Almost before the ink had dried, however, congressional supporters of FUA began to regret their haste. They had failed to recognize that the gas shortage was only an artifact of wellhead-price regulation. Moreover, large industrial and utility customers were anything but parasites on national gas resources. Their purchases were essential for smoothing out the temperature-sensitive seasonal ups and downs in residential gas demand. Then, too, loss of industrial load would burden remaining customers with the full complement of fixed costs, such as pipeline depreciation. Federal regulators found ways to soften these legislated restrictions until, in 1981, Congress removed the most egregious portions of the FUA. Nevertheless, in early 1983 the policy intent and many of the original provisions of the act remained intact.

The Natural Gas Policy Act of 1978

The same 1978 session of Congress that passed the Fuels Use Act also passed the Natural Gas Policy Act (NGPA). By then, the proponents of a free-market price for natural gas had gained the edge, both in the White House and Capitol Hill. There was no end in sight to the gas shortage, and national leaders feared that if struck with another winter as cold as that of 1976–1977, the situation would be intolerable. An increasing number of consumer-oriented congressmen faced up to the fact that wellhead regulation, initially put in place to protect consumers, had now gone full circle. Not only were potential users denied service altogether, but even existing customers with rights for assured service had to cope with shortages or, at best, live in fear of a cutoff. Then, too, more people were beginning to realize that relaxing wellhead price controls (thereby boosting incentives to find and produce domestic gas) could "back out" OPEC oil, which was expensive, insecure, and a drain on the U.S. balance of payments.

But proponents of decontrolling wellhead prices were able to secure the votes of a reluctant Congress only by compromising on the extent and pace of decontrol, and by accepting broader federal authority over gas allocation. Those accommodations alienated many of the key constituencies in producer states. The final versions of the NGPA and the FUA were such a hodge-podge of pro-consumer and pro-producer ideas that almost no one supported them wholeheartedly. Moreover, the legislation attracted an unlikely coalition of opponents from both consumer and producer states, which was only narrowly overcome by tough lobbying on the part of the Carter administration.

The detailed and prescriptive provisions of the 1978 legislation represented a reversal of philosophy from the 1938 Natural Gas Act, which had granted federal regulators wide discretionary authority. The NGPA instead created over a score of distinct pricing categories according to reservoir depth, location onshore or offshore, rate of gas flow, vintage (either the date a well was completed or the date on which a given gas supply was first committed to sale), and other production characteristics (see Figure 2-1). Each category had its own statutory ceiling price and schedule for price escalations.

Despite the added complexity of price regulation, the NGPA did provide for partial and phased deregulation of natural gas sold interstate. Deregulation was "phased" because, with the exception of deep gas (below 15,000 feet), which was immediately exempted from price controls, wellhead price ceilings were to be relaxed gradually until the beginning of 1985, when all price controls over gas that entered interstate commerce after 1977 would disappear. Deregulation was "partial," however, because "old" gas brought into production before the date of enactment would forever remain subject to cost-based regulation by FERC under the Natural Gas Act of 1938 or to the price ceilings and escalation schedules dictated in the NGPA.

Although the NGPA ultimately rolls back the reach of federal jurisdiction of interstate field transactions, the allocation provisions inserted to induce consumer-state support extended federal control into intrastate markets until 1985. This encroachment on state powers was rationalized and ultimately upheld in the courts on the grounds that intrastate sales did indeed affect interstate commerce.

In addition, the NGPA and the FUA together extended the reach of federal allocation policies further into end-use markets. The incremental pricing provisions of the NGPA loaded the bulk of higher gas costs onto industrial consumers. The FUA (prior to the 1981 amendments) contained a number of "off-gas" provisions to phase out boiler-fuel use of oil and gas by large industrial plants and electric utilities.

The Regulatory Web: A Retrospective

Municipal franchising and price regulation of gas distributors, starting in the nineteenth century, had thus by the early 1900s led to state oversight of intrastate transmission, which in turn prompted federal regulation of interstate transmission via the 1938 Natural Gas Act. Federal oversight of rates charged by producers selling gas to affiliated interstate pipelines followed almost immediately, with extension of field regulation to independent producers occurring much later.

Because federal regulation was indeed effective in passing the economic surplus on to consumers (especially during the rapid oil-price hikes of the OPEC era), price was not given the freedom to balance supply and demand. It was, therefore, only a matter of time before a shortage developed. When it ap-

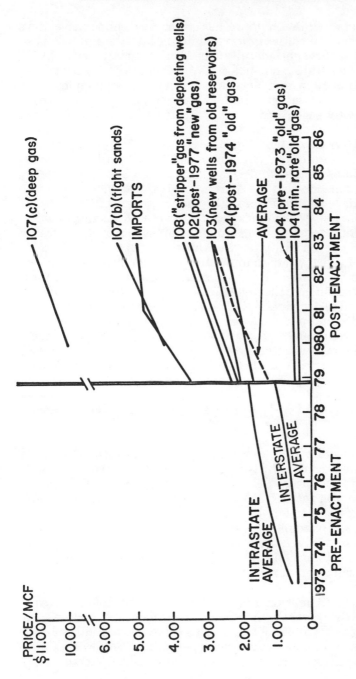

Figure 2-1. Price path of natural gas, before and after enactment of the Natural Gas Policy Act of 1978. After enactment of NGPA, intrastate gas was placed under federal control. The dashed line represents the common (interstate and intrastate) average price. The graph shows only a few of the many categories of gas established by NGPA. Source: U.S. DOE, March 9, 1983.

peared, it brought with it another government imperative: the need to regulate consumption and to allocate the limited supplies. First through federally sanctioned curtailment schedules and later through policies established in the Natural Gas Policy Act of 1978 (and more directly in its companion legislation, the Fuels Use Act), the U.S. government reached deeply into the only sector of the gas industry that had heretofore escaped such intervention.

The NGPA also extended the federal reach beyond interstate affairs and into commercial transactions within the gas-producing states. Although it was touted as "partial and phased" deregulation, the NGPA was really the keystone of a complicated edifice of governmental controls whose foundation was laid in the mid-1800s. Those controls were all fashioned with noble intent: to protect consumers against monopolistic franchises and to ensure that they fully retained the economic surplus. Each law and administrative order had its discrete and certain purpose. Nevertheless, regulation propagated further regulation, and the unforeseen impacts of complex legislation tended to overwhelm their intended effects. History shows us that attempts of government to structure markets according to some preconceived plan are likely to fail completely when, as was the case with the federal administration and Congress in 1978, the plan's authors misunderstand the underlying market forces they are attempting to channel.

Gas Market Upheavals of the Early 1980s

The transition from a supply-limited to a demand-limited environment came swiftly, catching almost everyone by surprise. End-use prices following passage of the 1978 Natural Gas Policy Act inflated at rates that mimicked the earlier oil-price upheavals. In late 1982, retail prices of natural gas in Los Angeles, for example, were 30 percent higher than in late 1981. Kansas City, Buffalo, Cleveland, Philadelphia, Washington, D.C., Minneapolis, and even Dallas suffered price hikes over the same period ranging from 30 to 44 percent.

Perhaps the most shocking statistic associated with the NGPA-induced market upheaval was the enormous spread between the highest- and lowest-cost field transactions in 1982. Some gas under old, regulated contracts was still selling for prices in the range of ten cents per million Btu, while a few deep-gas producers were pocketing almost a hundred times as much for the same volume. Transwestern Pipeline Company's purchase of gas in the Permian Basin for $10.10 per million Btu perhaps marked the zenith of the deep-gas purchasing frenzy, with industry-wide deep-gas purchases averaging $7.33 in early 1982. Meanwhile, wholesale prices for residual fuel oil (with which gas competes for industrial customers) garnered prices in the range of $3.50 to $4.50 per million Btu.

By the spring of 1982 prices in new sales of deep gas had peaked. Following protests by the public utility commissions of New York and North Carolina regarding a rate increase filed by Transcontinental Gas Pipe Line Company (Transco), the pipeline company announced a ceiling of $5.00 per million Btu on new purchases of deep gas. It also invoked the "market-out" provisions in existing contracts wherever possible, forcing producers to agree to price amendments or to peddle their gas elsewhere. Michigan-Wisconsin Pipe Line Company followed suit, and by summer, virtually all of the major interstate pipelines were doing the same.

In the first quarter of 1983 producers of even price-controlled categories of gas found it difficult to line up new customers even at the NGPA-established price ceilings. Below-ceiling sales became common and at least one pipeline, El Paso Natural Gas Company, stood by a mid-1982 announcement that it would shun new supply contracts altogether, regardless of price discounts producers might offer.

Normally it would have made sense for financially distressed gas companies to entertain new supply offers at reduced prices and to shed their higher-cost load. The problem was that they lacked (or believed that they lacked) the flexibility to do so. Gas sales are very different from oil sales. For a variety of institutional and regulatory reasons—including the fear of another shortage—pipeline companies had sought contracts that guaranteed access to gas on a long-term basis. Because of the seller's market that prevailed during the 1970s, and because federal law prevented pipelines from bidding up gas prices, the purchasing companies agreed to some very stiff take-or-pay provisions. These terms obligated the pipeline to pay for a certain percentage (often upwards of 90 percent) of the deliverable supply, whether or not the company's market outlets actually permitted that level of draw. Perhaps the biggest blunder of the U.S. gas pipeline industry was that companies almost without exception signed deep-gas contracts as late as mid-1982 that coupled stringent take-or-pay terms with extremely high prices.

By the autumn of 1982 interstate pipelines were besieged by irate customers, state utility commissions, and their congressional representatives who placed in the hopper all sorts of bills, some of which would have hit the industry like a sledgehammer. In many regions of the country, as gas prices began to exceed oil prices, the loss of industrial load to residual fuel oil was threatening company solvency as well. The price hikes and attendent gas surplus had become as big of a national headache and had fostered as much consumer unrest as had the gas shortages of recent memory.

This crisis too will pass. In an environment where price is free (at least in some areas) to move up *or down* to keep supplies and demand in balance, a persistent surplus is no more viable than a persistent shortage. This is not to say that temporary imbalances will ultimately vanish. As long as institutional and regulatory mechanisms prevent the quick and clear communication of consumer needs back to producers in the field, the gas business may be doomed to cyclical swings from surplus to shortage. These swings, however, will probably never approach the extremes engendered by price controls in the 1970s and the failure of the utility business in the early 1980s to take on the responsibilities that went hand in hand with a slackening of governmental price controls.

A Legislative "Fix"?

Although the long-term prognosis for market sanity and industry-wide health is good, business and political leaders are finding it difficult to see beyond the short-term crises now facing them. Indeed, in early 1983 many (if not most) of them still believed that a comfortable transition was impossible without further governmental intervention. In early 1983 President Reagan submitted legislation intended to ease the pains created by the 1978 Natural Gas Policy Act and its phased decontrol of gas prices. Unfortunately, that program (and virtually all concurrent proposals for legislative "fixes") would complicate the already confusing legal structure, stifle the self-correcting forces of the market, and further entrench the governmental presence in the gas industry.

Recontrol Gas Prices?

More fundamentally, there is no law that Congress can enact or rule that FERC can impose which will add anything to the incentive that consumer resistance has already given pipelines to keep their gas-purchase costs down. In this situation, there is no need for the federal government to recontrol or put a "cap" on new contracts for deep gas and other categories of "exempt" gas freed from price controls by the NGPA. Nobody is signing up for $10 (or even $5) deep gas any longer. Nor is there a need to prohibit pipelines from boosting their charges in excess of the rate of inflation; several pipelines have voluntarily sought *downward* adjustments already. The president's proposal, therefore, is unlikely to do any more than the market would (and is) doing on its own. It would, however, sorely aggravate the confusion and the governmentally induced inflexibility that is, in fact, the most insidious source of troubles in the industry today.

Decontrol Old Gas?

Old-gas deregulation is the keystone of the administration's legislative program, and an essential element in the long-term rationalization of the industry. Immediate decontrol in 1983, however, would bring us no closer to workably competitive natural gas markets except in some formalistic, ideological sense. With end-user gas prices already above market-clearing levels, the industry as a whole is incapable of extracting any more revenue from consumers. Whatever additional income deregulation might generate for old-gas producers would have to come instead out of the incomes of other classes of gas producers, the pipeline companies, or the distributors. Old-gas deregulation would *not*, therefore, ease the present market disorder but only complicate and prolong it.

Those who advocate deregulating old-gas prices immediately on the ground that doing so will get rid of the "cushion" that distorted market forces in the first place (giving rise to $8 or $10 deep-gas prices) are either foolish or disingenuous. There is no cushion left to kill. There is no room left for deep-gas producers, Canadians, or anyone else to walk away with windfalls. A handful of supplemental gas and high-cost import projects and proposals still limp along in the dreams (or nightmares) of pipeline and distribution companies, but no one is likely to propose completing any old enterprises of this kind or breaking ground for any new ones. Certainly no one will lend money to them if their viability depends on a subsidy from underpriced gas.

The proper time to consider removing price controls on old gas (and for cleaning up a lot of other regulatory debris) will come in 1985, perhaps, or 1986, *after* the transmission and distribution sectors of the industry have weathered the present crisis, and after new-gas deregulation has proved to the American people that market pricing does not necessarily mean higher prices.

Abrogate All Contracts?

There will never be a proper time, however, for the universal "market-out" or abrogation of gas-supply contracts proposed by the Reagan administration. Like the Fuels Use Act of 1978, this provision would likely turn out to be an obvious blunder as soon as the ink had time to dry.

Certainly, pipeline companies on their own will not be able to renegotiate *every* overpriced contract or successfully claim "force majeure" in abrogating all of their troublesome supply commitments. But pipelines and producers need not rewrite every contract in order to right the balance between supply and demand. Important events that can either ease or exacerbate market disorder occur, after all, *at the margin*. It is altogether likely that the institutional adjustments now taking place will have solved the market imbalance long before Congress is prepared to move legislatively.

Mandate Contract- or Common-Carrier Operations?

Several legislative proposals, including the administration's bill, would force pipelines to function wholly or in part as *contract carriers* or *common carriers*. This is an appealing program in today's market, where interstate pipelines have been loath to carry gas for third parties, thereby preventing willing sellers and willing buyers from entering into direct transactions at bargain rates.

The logic of self-help transactions, whereby pipelines would provide only transportation services rather than purchase of gas in the field and sale of gas at the city gate, is compelling in a time of market distress. (A variety of self-help arrangements, in fact, emerged during the supply shortages of the 1970s.) Over the long run, however, few buyers prefer to take gas under the delivery schedules that producers would regard as technically or financially optimum, or want to accept the risk of depending for a large part of their supply on individual producing properties or producers. The majority of producers, likewise, will not elect to supply gas on schedules dictated by the cyclical or seasonal needs of individual industrial customers or gas-distribution companies. For most transactions, therefore, it will not be cost-effective for producers, industrial consumers, or distribution companies to walk away from the special services that transmission companies have always offered as brokers, wholesalers, and custodians of natural gas inventories.

There is no way that Congress can anticipate and legislate every conceivable situation for which contract or common carrier requirements should or should not apply. And in light of today's consumer climate, any legislative action is likely to be an overreaction to largely ephemeral problems. Fortunately, a simple solution is probably at hand. FERC need only revamp its own outdated policies and accounting structures, which now actively or tacitly discourage pipelines from shipping gas owned by others.

Whither Now?

Short-Term Recommendations

There is no legislative solution, simple or complex, to the difficulties facing the natural gas industry today. But there is an enormously complex solution involving an elegantly simple policy that we would recommend. The beauties of this solution are that it is almost infinitely flexible, it can handle whatever unforeseen developments may crop up, and no single person needs to understand it completely. The solution is called the market, and the policy it depends on is for legislators to be patient, for gas industry leaders to act like normal businessmen, and for FERC to be a bit creative in encouraging those congressional and industry postures.

We urge caution in legislating a solution to the current gas market disorder, not out of an ideological faith in laissez-faire but from an even more conservative bias: skepticism about the ability of intelligent people, or government institutions led and staffed by intelligent people, to foresee or control gas market developments, and particularly to foresee the consequences of frequent radical changes in the legal and regulatory environment. The Fuels Use Act was an abomination from the beginning, NGPA has done its constructive work and only its mischief remains, while the Natural Gas Act is a dreadful anachronism. But a constant tinkering with the rules, the legislative attempt to chase gas prices and supply conditions back and forth across the business cycle, and efforts to fix transitory problems with grand structural reforms are likely to catapult the industry from one crisis to another. These actions cannot hold consumer prices down, they cannot guarantee supply additions to cope with a booming economy or a cold winter, and they will not even ward off pipeline and distribution-company bankruptcies.

In our judgment, the most acute danger our nation faces is that Congress, egged on by an administration and FERC officials who should know better, will overreact to the present disorder in ways that will make the system even less flexible than it is today. We must beware of the tendency to mandate sweeping structural changes for the sake of dealing with symptoms of market imbalance that are both marginal and ephemeral. We should not overlook the marvelous capacity for adjustment to changing conditions that is inherent in the profit-seeking and loss-minimizing activity of thousands of producers, dozens of pipeline companies, and hundreds of gas-distribution companies, nor should we belittle the remarkable adaptations this allegedly timid and hogtied industry has already made.

Finally, we "experts" should not quickly forget our vividly demonstrated fallibilities as gas market forecasters and policy planners. Errors of foresight or policy judgment can be locally or episodically painful when they occur randomly within the gas production, transmission, or distribution sectors, among the state regulatory commissions, or among consumers. Imposed by the federal government as national policy, however, bad forecasts and decisions can and do metastasize into costly nationwide dislocations and crises.

If anything, recent history of the gas industry and its regulatory institutions shows that we should proceed from a posture of great humility in the demands we make for federal action. If Congress acts at all during the surplus and price fly-up crises of the early 1980s, experience suggests that it should act only in small and tentative ways.

Long-Term Recommendations

Our recommendations for the long run are, however, markedly different. Once the present market crises have been settled, we believe the system merits a complete overhaul. *We recommend dismantling the federal system of economic regulation of the gas industry—scrapping the Fuels Use Act, the Natural Gas Policy Act, and even the 1938 Natural Gas Act.*

The original rationale for regulation, after all, has now been stood on its head. The regulatory system was created and nurtured in the hope of compelling local gas distributors, interstate pipelines, and gas producers to pass the economic surplus on to consumers. Some consumers did indeed reap the intended benefits, but other would-be consumers were shut out of the market altogether, and almost everybody faced the risk of curtailment during the height of the gas shortages.

The gap between regulated prices and market value, moreover, encouraged regulated utilities to eat into the surplus and bridge the supply-and-demand gap by acquiring or manufacturing gas that cost more than it was worth. And, from a consumer perspective, the "partial" deregulation of the Natural Gas Policy Act was tantamount to complete deregulation, with end-use prices rising to whatever the market would bear. The difference between partial and complete deregulation was that implementation of the former spawned a whole new generation of supply and production inefficiencies, culminating in the headlong quest for deep gas and imported gas, whatever the cost.

The creep of regulation from local gas distributors to intrastate transmission to interstate transmission to wellhead transactions and finally to end-users, had its place in history, and in hindsight it seems that perhaps all of these actions were justifiable. But today's web of regulation and dependent industry is terribly confusing as a result, and is beset with all sorts of perverse incentives, inefficiencies, and unintended inequities—most of which would fall by the wayside if economic regulation were withdrawn.

And why should it not be withdrawn? After all, there is no way that producers, pipelines, or anybody else, regulated or unregulated, can extract additional revenues from gas consumers who simply turn down the thermostat or switch to alternative fuels. *The prospective economic surplus has vanished,* and with it has vanished the fundamental condition that prompted much of the regulatory edifice in the first place. Indeed, instead of extracting unearned profits in the early and mid-1980s, producers and pipelines face the prospect of adjusting their expectations to a revenue flow that cannot fulfill what their present contracts and commission-approved tariffs would seem to promise.

In addition to the disappearance of the economic surplus heretofore created by producer-price regulation, and the growth in market power wielded by consumers (who react to price increases by conserving gas or switching to alternate fuels), one other condition has evolved that was not present when producer and pipeline regulation was first implemented. Today, virtually all gas is produced, and virtually all gas is consumed, in areas served by more than one pipeline company. Granted, there will always be examples of an isolated producer facing a pipeline monopsony (that is, a producer will have practical access to only a single pipeline buyer and will be faced with meeting the pipeline's terms of sale, or not selling at all), and some isolated communities will always face a pipeline monopoly. These situations are not, however, unique to the gas business. They are part of the cost of living and doing business in certain places, and they no more call for a special regulatory regime governing the purchase or sale of natural gas than does, for example, the buying of fish in a small harbor or the sale of beer at an isolated resort.

The potential abuses and stresses generated by this and many other problems that now preoccupy government and industry in their deliberations over deregulation tend to have commonplace and flexible remedies in the Uniform Commercial Code, in the antitrust laws, and in ordinary business practice. Then, too, the existence of take-or-pay contracts at uneconomic prices or with unlimited escalator clauses may not be nearly the problem that so many industry officials and regulators now believe it to be. Such inflexible contracts may in some cases be a drag on economic efficiency, and they could limit the ability of gas utilities to buy lower-cost gas as it becomes available. But as the companies that have simply walked away from these commitments have found, a seemingly iron-clad agreement is tenuous in an economic setting in which it no longer makes a shred of sense. Whether the pipeline purchaser legally invokes "force majeure" or claims an "inability to perform under changed circumstance" (or simply banks on the fact that an aggrieved supplier not only must take him to court but also must win, and win again and again—and then *collect*), there is a strong suggestion that purchasing companies have a lot to gain by taking inordinately aggressive action.

Regulation of pipeline construction and rates may have been socially useful during the industry's infancy, but within a mature pipeline network, certification and rate-review accomplishes little that could not be ensured via generic laws, standards, and business practices. For example, with the economic surplus gone, the risk of overbuilding or duplication of facilities in an environment of unregulated entry will be borne by investors, while the benefits of increased pipeline competition will be shared by producers and consumers.

Likewise, there will no longer be any reason to continue federal supervision of the *operational status* of gas pipelines. Because competitive field markets and competitive city- and factory-gate markets encourage competition in the gas-transportation business (unless the government continues to suppress it), pipelines will choose to act as *private carriers* buying and selling gas as they do now, or as *contract* or *common carriers* selling only transportation services, depending on the preferences of their suppliers and downstream customers. And because of competition, the role the pipelines actually assume will probably have little effect on the net price of their services.

Although an unraveling of the regulatory web that engulfed the gas industry and culminated in the 1978 Natural Gas Policy Act and the Fuels Use Act is perhaps as inevitable as it is desirable, the most important arena for action now lies within the existing legal framework. All companies engaged in producing, moving, and distributing natural gas must reacquaint themselves with the fundamental laws of free-market economics. And regulators and legislators would be wise to do everything in their power to let all segments of the industry know that they are on their own, and to ensure that the financial and institutional incentives are in place to encourage a productive and creative response.

3

EMERGENCY MANAGEMENT

Benjamin Zycher

Summary

This chapter discusses appropriate public policy with respect to each of the two major effects of significant oil supply disruptions. These effects are, first, *relative price changes in the economy*, and second, *wealth transfers among individuals*. Efficient adjustment to the allocational effects of oil price increases can be achieved only through the operation of market forces using price changes as the vehicle for such adjustment. Experience during the 1970s regulatory period is wholly consistent with this argument. Furthermore, market analysis implies that reduced dependence on foreign oil has little to do with U.S. "vulnerability" to the effects of oil supply disruptions, and that the embargo threat is and always has been empty. The chapter argues further that price and allocation controls cannot make the poor better off generally, and that punitive taxes imposed on politically disfavored groups (or political minorities), such as "big oil," constitute poor social policy, particularly in a constitutional context. Alternative redistribution policies and criteria for choice are presented; such redistribution can be viewed as an investment in the political viability of market-based (efficient) allocation policies. The policy of the Reagan administration, then, is summarized and considered; this includes international emergency management policy as embodied in the International Energy Agency sharing arrangement. Finally, some conclusions and policy recommendations are offered.

Introduction

Because the theme of this volume is energy policy in a free market environment, the title "Emergency Management" may be misleading, given that the argument to be presented below is, in short, that the market works. Hence, "management" per se is to be avoided. Muttering and grumbling from various quarters aside, reliance on market adjustments as a general policy in the event of energy supply disruptions does not imply an "absence of policy," "complacency," or a "failure to plan in advance." A central theme of this chapter is that past policies such as those in force during the 1970s—and a resulting expectation that such policies may emerge again from the democratic process—have the ironic effect of *inducing* complacency and of providing *disincentives* to prepare for emergencies. A policy of market reliance is no less "a policy" than one of government intervention; what the popular discussion of "emergency management" reflects is the lack of systematically developed criteria for policy decisions and careful analysis of alternative policies in light of the relevant criteria. One goal of this chapter is to provide such a framework.

Effects of a Supply Disruption

Oil supply disruptions may be accompanied by significant increases in the market price of oil, which have two main classes of effects. *First*, the oil price increase changes *relative prices* in the economy; the prices of oil, oil substitutes, and oil-intensive goods rise relative to the prices of other goods. (The relative prices of imported and import-substitute goods may change also if the exchange rate is affected by increased domestic spending on foreign oil.) The overall change in relative prices has the aggregate effect of inducing perhaps substantial resource shifts across industries, sectors, and regions, as labor, capital, and other resources "search" for the most profitable (i.e., productive) employment in light of the changed price relationships in the economy. This search process is not instantaneous; therefore, some unemployment of resources, including labor, is an inevitable result of a significant increase in the price of oil, particularly if the increase is expected to persist. Expansion of the money supply can delay but cannot prevent this outcome, because changes in the money stock, however measured, cannot undo the real relative price changes in the economy.

Second, the oil price increase has the effect of inducing a *transfer of wealth* among individuals or, more loosely, among groups. In particular, owners of oil, oil substitutes, and oil-producing inputs (such as some labor) enjoy an increase in wealth, while consumers of oil, owners of oil-consumption complements (such as large cars), and owners of oil-complement producing inputs (such as some automobile industry workers) suffer a reduction in wealth as a result of the oil price increase. There may take place also a general transfer of wealth from the domestic economy to overseas economies (particularly those of oil-exporting nations).

The wealth transfers among (groups of) individuals may constitute a cause for some concern if the oil price increase makes the attainment of a minimum standard of living difficult for some individuals. This may generate within the political system a demand (i.e., pressure) for increased income redistribution.[1] Such redistribution for purposes of providing a minimum living standard to those in need is an example of what economists call a "collective good." Each individual (by assumption, every nonpoor member of the society) would have this preference for redistribution satisfied whether he or someone else actually provided the resources (i.e., funds) necessary. Therefore, each individual has an incentive to wait for others to provide the transfer payment, just as on a windy day each homeowner has an incentive to wait for someone else to gather the newspapers blowing around the neighborhood (or, alternatively, to wait for the refuse to blow farther down the street). In short, each individual would like to obtain a "free ride." Standard analysis implies therefore that the market without some sort of intervention by individuals acting collectively (i.e., the government) would supply a quantity of transfers smaller than that actually demanded by all potential donors. If this standard analysis is correct,[2] incorporating a demand for redistribution by donors, government transfer programs may increase the total amount of redistribution toward a more nearly optimal level.

Chapter Outline

As the emphasis is on the behavior of and reliance on markets, the chapter begins with a nontechnical discussion of how markets operate and adjust to the effects of exogenous events, such as oil supply disruptions. This leads directly into a discussion of the standard "dependence" and "vulnerability" arguments, and of the unimportance of targeted "embargoes," such as that of 1973. The case for reliance on market adjustments as the best policy for dealing with future disruptions is then presented, along with a history of the regulatory evolution of the 1970s.

A Brief Primer on Markets

Maximization of Social Wealth

The appropriate goal of economic policy, both generally and specifically, is *to maximize total social wealth*.[3] This is consistent with income distribution goals in principle because a larger economic pie yields more for all members of the society, and also because a shrinking pie retards upward mobility, particularly across generations, and is likely to be borne at least in part by the poor, however defined. Social wealth is maximized when scarce resources are allocated "efficiently"— that is, when the total value of all output produced from available inputs is maximized. Resources and final products must be allocated to their most valuable (productive) uses; inputs must be combined in such a way as to minimize waste. In most cases, only decentralized (free) markets, responding to price signals emerging from a multitude of individual deci sions, lead to efficient outcomes and hence to greater total wealth.

Prices as Economic Coordinators

Prices can be viewed analytically as a set of signals coordinating consumption and production decisions by consumers and producers. Prices measure both the value (i.e., the amount of other goods that consumers are willing to bid, or forgo, for any particular good) and also the value of resources needed (i.e., the value of other production forgone) to produce any particular good. In short, prices induce consumers to forgo consumption of goods for which production costs (forgone alternative production) exceed the (marginal) value of the good. Similarly, prices spur producers to produce goods the value of which exceed production costs. For the final units, production costs are equated with the value of consumption.

For any given production level, prices serve also to induce individuals to allocate supplies toward highest-valued (i.e., most productive) uses. Any individual valuing a good at less than its market price is induced to sell (or relinquish or forgo) it. Hence, prices constitute a set of coordinating devices that serve to maximize both the total value in use of goods being produced and the total value (productivity) of resources used in the production of goods. Note that this "efficient" allocation of resources and goods depends partially but critically on the distribution of wealth or endowments; different income distributions may result in different "efficient" outcomes.

In short, prices provide information or signals about the economic environment, constitute a set of constraints shaping individual and economy-wide behavior, and themselves emerge from the decision-making processes of many consumers and producers. Since centralized governmental decision-making processes cannot have the informational efficiency of decentralized markets, prices established by regulatory fiat generally are unable to provide information or signals leading to efficient outcomes. Bureaucrats cannot know, and have few incentives to discover, market-clearing prices (i.e., prices that equate quantities demanded and supplied) and the efficient allocation of inputs and goods. Price ceilings, in particular, induce producers to produce too little and consumers to demand too much. Because reallocation of supplies is a process that itself consumes resources (only one of which is time), price controls reduce also the ability of the economy to allocate available supplies to highest-valued uses.

An Aside on Import Dependence, Vulnerability, and Embargoes

An argument that seems to have had a large following for a considerable period holds that the degree of "dependence" on imported oil—that is, the volume or percentage of domestic consumption supplied from foreign sources, particularly "unstable" ones—is a direct, or even the primary, determinant of U.S. "vulnerability" to the effects of foreign oil supply disruptions. For example, one prominent economist argues as follows:

> [With a tariff on imported oil] the United States would then be less dependent on any one foreign supplier and could choose to curb purchases from nations more likely to embargo shipments for political or ideological reasons.[4]

In short, the degree of dependence is held to determine the degree of vulnerability. However, market analysis, summarized above, suggests strongly that this argument is in error. The seeming simplicity and self-evident nature of the argument stem from the ambiguity inherent in the term "vulnerability." Vulnerability to what? It must mean vulnerability to the effects of significant oil price increases. Does the degree of dependence affect the size of the oil price increase resulting from any given supply disruption? The answer is no. When international oil prices rise (or fall), domestic prices change by an identical amount *regardless of whether we import all or none of our oil*.[5] If they did not, oil would be reallocated among international and domestic markets so that price equality (net of transport costs) is achieved. Therefore, efforts to reduce "vulnerability" by reducing "dependence" cannot have the advertised effects; they can only impose costs on the domestic economy through increased use of more expensive oil.

Because reallocation of oil supplies in world markets is not difficult, price changes affect all oil-consuming nations. Another important implication of economic analysis is that "embargoes" targeted at a particular nation cannot have important effects on the targeted nation in terms of its own access to (i.e., cost of obtaining) oil. If an oil producer refuses to sell to a particular nation, that nation can obtain oil that otherwise would have gone to some third party; the producer imposing the embargo will sell to some other buyer. Only by reducing production (and thereby raising prices) can a producer affect a given buyer, but in so doing must affect *all* buyers for the reasons discussed above. The evidence is wholly consistent with this argument: during the 1973 embargo, the United States and the Netherlands, despite their status as the intended targets of the embargo, faced no greater difficulty obtaining oil—that is, faced the same oil prices—than did all other oil-consuming nations. Prices rose sharply, not because of the embargo against the two nations, but because production was reduced. The price increase affected all nations equally: producers, consumers, and those with both high and low dependence levels. Gasoline lines in the United States were caused by neither the embargo nor the production cutback, but instead by the regulatory apparatus, a subject to which we will return below.

The General Case for Market Reliance in Emergency Management

The goal of "economic efficiency" may seem somewhat sterile; but if the satisfaction of human wants is the goal,[6] the quest for "efficiency" is nothing more than a search for ways to satisfy those wants most fully in a world in which not all wants can be satisfied simultaneously. In short, choices must be made. The policy issue then becomes one of choosing institutions and modes of economic organization that most effectively will lead the system to behave efficiently in terms of allocational choices. In particular, given that an energy supply disruption has occurred, policy should seek to minimize the aggregate adverse effects of the disruption by inducing allocation of available supplies to their most productive uses, and by encouraging production of substitute fuels. Which adjustment mechanism—price adjustments emerging through the operation of decentralized markets, or centralized regulatory fiat— has in the past and is more likely in the future to achieve these goals?

Since the early 1950s thirteen oil supply disruptions have occurred. These disruptions were of varying magnitude and duration, and were characterized by differing institutional arrangements and U.S. policy contexts.[7] Gasoline lines and other manifestations of "chaos" accompanied only those occurring with a centralized price and allocation regulatory system in place.[8] As discussed above, markets have strong incentives in the face of price adjustments to reallocate available fuels and to spur the production of substitutes. The democratic process on the other hand has strong incentives to satisfy the demands of important interest groups;[9] such redistribution policy can only shift some of the adverse effects of the disruption away from politically favored groups, thus making matters worse for other groups. Greater efficiency, however, accrues to the benefit of the whole decentralized economy. It is difficult to bring pressure on Congress to develop efficient policies due to the small stake that any individual has in efficiency per se.[10]

The Crisis of the 1970s

The impossibility of adjusting efficiently to the effects of oil supply disruptions through the use of a price and allocation regulatory apparatus is illustrated clearly by the 1970s experience. The bureaucracy does not have the incentives, nor can it have the information necessary to achieve efficient outcomes. Because of this, the regulatory apparatus tends over time to become increasingly complex, ad hoc, and receptive to goals and pressures having little to do with the pricing and allocation of fuel. Because the complexities and absurdities of the regulatory system generally are not understood, it is useful to describe in some detail the evolution of the regulatory system in 1973–74, along with some additional aspects in full bloom in 1979. The important point to bear in mind is that the irrationality of the system is a *necessary* by-product of the very existence of the regulations. It is necessary because the regulations are put in place to serve inconsistent ends, and because the complexities of the market force the regulators to patch each successive round of rules with further rules, which then in turn require further patching. There can be no such thing as a price and allocation system that avoids these problems, particularly because the world changes over time while the rules remain static.

The Economic Stabilization Program

The Nixon administration's Economic Stabilization Program (ESP) general price controls were initiated on August 15, 1971; they were divided into several phases:

Phase	Dates
Initial Freeze	August 15–November 13, 1971
II	November 14, 1971–January 10, 1973
III	January 11–June 12, 1973
Second Freeze	June 13–August 11, 1973
IV	August 12, 1973–April 30, 1974

Source: *Historical Working Papers of the Economic Stabilization Program*, U.S. Department of the Treasury, 1974.

The petroleum regulations were part of the larger program, and like the overall ESP, had two basic objectives which were substantially inconsistent: reduction of the growth rate of prices, and avoidance of shortages. These conflicting goals led to increasing complexity in the regulations, to a growing ad hoc system of response to emerging problems, and to creation of a large and cumbersome bureaucracy.

Phase II

Oil prices during Phase II were not considered a special problem, and so were subject only to the general provisions of the ESP. The Price Commission (the administrative arm of the ESP) believed, erroneously, that price controls at the refinery level would reduce costs, and hence prices, at the retail level.[11] Price increases by refiners were permitted to reflect cost increases borne after November 14, 1971, only. Under the Phase II controls, major refiners accumulated large backlogs of cost increases that could not be passed through because of the price ceilings; at the same time, most independent retailers and wholesalers generally were free from controls because of the Small Business Exemption. The major problem generated by the Phase II controls was a potentially serious shortage of home heating oil; the controls fixed relative prices at their August 15, 1971, level, reflecting the normal seasonal high for gasoline and the normal seasonal low for distillates. (Any control program must use base dates, which introduce substantial rigidity into the system.) As a result, home heating oil (distillate) stocks were substantially below normal levels at the outset of the 1972–1973 heating season. The problem was exacerbated by an unusually cold winter, but most demands were met, in part through a drawdown of inventories.

Phase III

The Phase II requirement that large firms obtain prior approval for price increases was removed by Phase III. This resulted in a surge in the reported GNP deflator. Sharp cost increases were being experienced in the oil sector, and political pressure mounted for action by the Cost of Living Council (CLC), the policymaking arm of the ESP. The response was predictably ad hoc: the CLC issued Special Rule Number 1, which imposed mandatory controls on the sale of crude oil and refined products by firms with annual sales of $250 million or more, of which there were twenty-four. The regulations limited product price increases to a weighted average of 1 percent above the base period price for the year beginning January 11, 1973. Price increases between 1 percent and 1.5 percent required documentation of corresponding cost increases after March 6, 1973, while increases above 1.5 percent were subject to profit margin limits and prenotification.

The controllers apparently believed that the twenty-four largest firms "controlled the industry," and hence the price constraint applied to them would constrain all industry prices automatically. Thus, Special Rule Number 1 exempted the large number of remaining firms in the industry. The effect was to create two classes of sellers, controlled and uncontrolled, with differing abilities to compete for (marginal) supplies of crude oil and refined products. It became profitable for these uncontrolled sellers to stockpile certain products, particularly propane, in anticipation of higher prices; furthermore, many sellers began passing through cost increases accumulated during Phase II. Crude oil cost increases added to the pressure, and the result was a sharp increase in refined product prices during the first half of 1973. Price changes in the market would have been much smoother in the absence of the controls.

The large firms subject to Special Rule Number 1 were

reluctant to bid up crude oil prices because doing so would require consumption of part of their pricing authority under the 1.5 percent rule, and because the prenotification process consumed at least thirty days and could result in a denial by the CLC. The majors thus had clear incentives to ship their foreign crude oil to their overseas refineries, which were not subject to controls. In order to avoid this, the CLC produced another ad hoc response: increases in crude oil or product costs could be passed through without affecting the percentage price increase constraint. Phase III then ended, however, largely because of the rapid growth in the reported price index.

Phase IV

The ESP was extended for an additional year to April 30, 1974, and a second freeze of sixty days was instituted to allow for preparation of Phase IV. After a massive tug-of-war exercise between independents, majors, producers, refiners, resellers, and retailers, the Phase IV oil industry regulations (soon incorporated into the 1973 Emergency Petroleum Allocation Act [EPAA]) were announced.[12] The central figure of the Phase IV-EPAA controls was a two-tier system of crude oil price controls. In essence, monthly crude oil output from a property up to its (monthly) 1972 base period level was defined as "old" oil subject to price controls. Other "new" oil and imported oil were not controlled. Efforts by refiners to obtain "cheap" old oil forced the government to impose crude oil allocation regulations essentially freezing buyer–seller relationships as of 1972. In short, the government entered the business of effectively running the daily operations of the oil industry.

The crude oil allocation regulations, based on historical transaction patterns, created serious problems because some refiners had access to proportionately greater amounts of crude oil, relative to refining capacity, than did others. Furthermore, some refiners had access to proportionately greater amounts of "cheap" crude oil, thus bestowing on these refiners proportionately lower costs. The regulatory response to the first problem was the "buy-sell" program under which refiners with access to proportionately above-average amounts of cheaper crude oil were required to sell (at controlled prices) crude oil to other refiners. This program became unmanageable administratively, so that after a few months the buy-sell requirements were limited to the fifteen largest integrated refiners (who were required to sell) and small refiners (who were allowed to buy).

The problem of different average crude oil costs caused by differential access to crude oil controlled under the price regulations became increasingly serious over time. The regulatory answer to this was the entitlements program, a system of staggering complexity, the purpose of which was to equalize crude oil costs across domestic refiners. In essence, those with above-average access to "cheap" crude oil were forced to buy rights to this crude oil from those with below-average access. The effect was to subsidize the importation of foreign crude oil, because anyone importing an additional barrel, in effect, became entitled to an additional payment from some other member of the industry. The entitlements program was used also to subsidize imports of distillate and residual fuel oil, to subsidize small refiners, and for other purposes not closely related to the initial cost-equalization goal. The confusion and irrationality of the system is illustrated by the fact that to this day disputes and lawsuits emanating from ambiguities and inconsistencies in the entitlements regulations still are being filed, pursued, and litigated.

The two-tier pricing structure for domestic crude oil provided incentives to shift month-to-month production patterns and to abandon old reservoirs on new properties. Two regulations were implemented to avoid these problems. First, in order to count the additional production from an oil field in a given month as new oil, cumulative production at the end of that month had to exceed the cumulative production at the end of the same month in 1972 by an amount equal to the added production. If it did not, then only the addition to cumulative production could be counted as new oil. Second, a "released oil" regulation was written to encourage producers to work existing fields with secondary and tertiary recovery techniques. For each barrel of new oil produced on an old property, a barrel of old oil was released from price controls. Whatever the effect on production behavior from old properties, it is clear that the released oil regulations added to the complexity of the system and to the problem of enforcement. In a larger sense, the two-tier system was intended to restrain price increases while providing incentives for added production; to whatever degree these objectives were achieved, it is clear that the two-tier system added considerably to the complexity of the regulations and to the distortions caused by them. The two-tier system is an example of the growing complexity caused by the strains and pressures inherent in a price-control program.

This growing complexity was illustrated further by the cost

allocation regulations written to control the relative prices of refined products. It was felt at the outset of Phase IV that certain products—gasoline, heating oil, and diesel fuel—were sufficiently inelastic in demand to require tighter controls. The regulations therefore divided refining yields into two categories: the above three "special products" and "other covered products." Only the prices of special products were controlled at the retail level; furthermore, increases in domestic crude oil costs could not be passed through into special product prices. The upshot of these cost-allocation regulations was that a disproportionate share of increased costs was forced onto the other products. An almost immediate effect was the crisis in propane prices. Propane, produced mainly from natural gas liquids, was in tight supply at the outset of Phase IV because of declining natural gas production and sharply increased demand for natural gas and propane. This upward price pressure induced refiners to allocate disproportionate cost shares into propane prices, since propane was an "other covered product." In response to the propane crisis, the Federal Energy Office, which by then had come into being, gave modified special product status to propane. The propane episode illustrated once again the need for growing complexity in the regulations, and the increasingly ad hoc nature of the responses to emerging problems.

Phase IV brought with it a new crisis in home heating oil, since relative prices were frozen at May 15, 1973, levels. A new set of regulations, the Refinery Balance Incentive Program, was instituted in order to deal with this problem. In essence, the plan allowed refiners to raise distillate prices by two cents per gallon, but required an equal reduction in gasoline prices. Further sliding-scale incentives were provided for additional distillate production. A more ad hoc response to the problem hardly can be imagined.

Further Comments on the 1970s Regulations

Most provisions of the EPAA were scheduled for expiration in February 1975. Another furious political battle resulted from this imminent demise of EPAA, culminating with enactment of the 1975 Energy Policy and Conservation Act (EPCA). Aside from abolition of the released oil regulations, the main feature of the EPCA was transformation of the two-tier crude oil pricing system into a three-tier system, along with removal of price and allocation controls on most refined petroleum products. Needless to say, the complexity of the regulations and of emerging problems bedeviled both the industry and the regulators, and resulted in disputes and controversy lasting to this day.

One striking feature of the evolution of the control program was the tendency for new objectives to arise over time. At first, restraint of price and promotion of production were the main—but inconsistent—objectives. Then the Small Business Exemption was implemented in an attempt to further "competition."[13] The allocation regulations were promoted as a tool for the furtherance of "equity," particularly during the embargo. Finally, the specter of "windfall profits" was raised as an additional justification for the controls. Nowhere was this clearer than with respect to the controversy over pricing of state royalty crude oil produced from land owned by state governments. At first, the CLC exempted such production from controls; this exemption was viewed as an extension of the revenue-sharing concept. Further analysis revealed that small refiners who purchased royalty oil would be hurt considerably by the exemption, and it was argued that the exemption should be removed for purposes of promoting "competition" in refining.[14] Finally, the CLC decided to remove the exemption on the grounds that it was not providing sufficient stimulus to added production. Throughout the price-control period, the choices among conflicting objectives were largely arbitrary.

Further problems were created because the refined product allocation regulations were based in part on historical geographic use patterns, which became increasingly irrelevant as time went on. This deficiency, which for practical purposes is unavoidable, became glaring during the 1979 episode: newly developed, urban, and other areas characterized by growing gasoline use suffered disproportionate shortages while other areas had only minor problems. William Simon, the Nixon administration energy "czar," wrote as follows with respect to the price and allocation control system:

> As for the centralized allocation process itself, the kindest thing I can say about it is that it was a disaster. Even with a stack of sensible-sounding plans for evenhanded allocation all over the country, the system kept falling apart, and chunks of the populace suddenly found themselves without gas. In Palm Beach suddenly there was no gas, while ten miles away gas was plentiful. Parts of New Jersey suddenly went dry, while other parts of New Jersey were well supplied. Every day, in different parts of the country, people waited in line for gasoline for two, three, and four hours. The normal market distribution system is so complex, yet so smooth

that no government mechanism could simulate it. All we were actually doing with our so-called bureaucratic efficiency was damaging the existent distribution system. As the shortages grew more erratic and unpredictable, people began to "top off" their tanks. Instead of waiting, as is customary, to refill the tank when it is about one-quarter full, all over the country people started buying fifty cents' worth of gas, a dollar's worth of gas, using every opportunity to keep their tanks full at all times. And that fiercely compounded the shortages and expanded the queues. The psychology of hysteria took over.

Essentially, the allocation plan had failed because there had been a ludicrous reliance on a legion of government lawyers, who drafted their regulations in indecipherable language, and bureaucratic technocrats, who imagined that they could simulate the complex free-market processes by pushing computer buttons. In fact, they couldn't.[15]

In short, price ceilings cannot reduce true prices, which are the sum of the pecuniary and nonpecuniary cost of obtaining supplies. They cannot alleviate but must in fact create and exacerbate shortages, and in practice they have the inevitable result of seriously hindering smooth adjustment to the effects of supply disruptions. Experience is wholly consistent with these observations. Only decentralized (free) markets operating through price adjustments can adjust smoothly to emergencies, allocating available fuels efficiently and promoting the production of substitute fuels. As discussed below, expectations that the market will not be allowed to operate freely have the further effect of discouraging advance private sector preparation for emergencies.

Can Price and Allocation Controls Help the Poor?

As discussed above, the wealth transfer effects emanating from oil supply disruptions may produce a demand within the political system for increased redistribution. If the traditional argument is correct that the market supplies too little charity because of the "free rider" problem discussed above, then increased public redistribution *may* lead to a more optimal level of such transfers, whether in kind or in pecuniary form.

The issue to be addressed here is not whether such increased redistribution is justified. Instead, the focus is on whether price controls can be expected to better the lot of the poor generally, and, briefly, on alternative redistribution policies and criteria for choice among them.

Four Reasons Why Price Controls Cannot Help the Poor Generally

A standard argument made by those favoring government intervention as a response to supply disruptions is that rationing by price (i.e., free-market allocation) would be unfair to the poor. By holding prices down through government fiat, so goes the argument, the poor can be spared the adverse economic consequences of rising fuel prices. In short, price rationing purportedly hurts the poor generally. This argument is false for four reasons. It is important to note also that price controls do not reduce *true* prices, defined as the marginal cost of obtaining supplies; only reported or legal prices are affected.[16]

The four reasons that price controls cannot help the poor generally are as follows.

1. *Price effects in uncontrolled markets.* Price controls on, say, gasoline, constitute a constraint on price rationing of gasoline, that is, on the free-market allocation process. In other words, controls constrain the ability of individuals to compete for gasoline in terms of dollars, so that competition must take other forms. To the extent that the nonpoor are prevented by such constraints from consuming additional units of gasoline, they will consume additional units of other goods instead, driving up the prices of those goods. In short, the oil controls may reduce the ability of the poor to compete for goods in non-oil markets. Nonprice rationing of petroleum thus may induce a demand shift process, making the poor worse off with respect to the entire consumption basket. Furthermore, this shift in relative prices, caused by the artificial constraints imposed by the oil price controls, will induce a series of windfall gains and losses for a wide variety of individuals and groups, both consumers and producers. The net distributional effect of these windfalls cannot be predicted in advance, but some poor individuals undoubtedly would be losers.

2. *Effects of alternative rationing mechanisms.* Constraints on market allocation (i.e., price rationing) cannot make the poor better off unless the alternative allocation mechanism itself favors this group. Surely the queue as an allocational device does not obviously favor the poor

either in general or with respect to any particular good. This problem *cannot* be avoided; once market allocation is constrained, there *must* be an alternative rationing mechanism for any good the market value of which is greater than zero. In order for the poor to be made better off by queuing, they must have greater willingness than others to spend time in the queue. The poor do not obviously have lower marginal time values than the nonpoor if their schedules are less flexible. Even if the marginal time values of the poor are lower than those of the nonpoor, queuing still may favor the nonpoor if the relative value of the good to the nonpoor is sufficiently high relative to its value to the poor. Nor is allocation by government regulation more likely to achieve the stated "equity" aim: What regulatory process can be expected with confidence to favor the poor? If price and allocation controls are advocated in the name of equity, it is mandatory that the alternative allocation mechanism be specified and shown to be more consistent with the interests of the poor.

3. *Burden of price ceilings on producers.* Price ceilings reduce prices received by producers, and increase true prices paid by consumers.[17] Therefore, ceilings are analytically equivalent to a tax, the incidence of which is likely to fall on both (some) consumers and the owners of inputs in the production of the good in question. Are the producers of "luxury" goods—or of gasoline—themselves wealthy? Are pension funds and other owners of oil company stocks "wealthy"? An answer of No is wholly plausible for at least some of these owners, in which case the group of losers might well include some poor individuals.

4. *Effects of controls on economic growth.* A major vehicle for improvement in the position of the poor—particularly in terms of upward mobility across generations—is economic growth. The interest of the poor lies in a growing economy with expanding opportunities for employment and the acquisition of wealth. Policies that hinder allocative efficiency—price and allocation controls—must have the effect of reducing aggregate wealth, because of both direct and disincentive effects. It is implausible that at least part of this burden of reduced wealth would not fall on the poor. In addition, reduced aggregate wealth may reduce the amount of transfers supplied to the poor by the nonpoor.

Conclusion: Price Controls and the Poor

For these reasons, the net effect of price and allocation controls on the poor is uncertain at best. What is likely to be true is that such controls, like most policies, make some members of each group better off, and others worse off.

Issues and Options in Redistribution Policy

An Aside: Redistribution and Windfall Profits Taxes on "Big Oil"

Clearly, wealthy and civilized societies cannot allow those in need to go hungry and cold. The definition of "need," and discrimination against individuals according to this definition, are tricky problems not of direct interest here. If individuals have demands for the welfare of others, then I would argue that it is incumbent on such individuals, *speaking through their elected representatives,* to provide the funds necessary. Cries of "Tax the oil companies!" are somewhat questionable, since those making such arguments, in effect, proclaim their great concern for the needy and then point their fingers at someone else when the time comes to ante up—a strange manifestation of charity indeed.

Such hypocrisy aside, there is in a normative sense a more fundamental reason that the political system should discipline itself and refrain from imposing taxes on political minorities (e.g., "big oil") unless such taxes can be shown to be efficient in terms of the particular minority's demand for public services. A basic purpose of constitutional (i.e., contractual) government is to constrain the behavior of political majorities, particularly in terms of confiscatory behavior directed at opposition political groups. Members of the majority, implicitly, may agree to such constraints as a means of acquiring insurance against the confiscatory behavior of others when members of the current majority become members of the minority. At the constitutional level, any individual, not knowing whether he will be in the majority or minority, may refrain from such confiscatory behavior, if in the majority, in order to be free of similar confiscation if in the minority. The constitutional contract, therefore, is analogous to a common (unanimous) commitment to refrain from stealing. This leaves open the issue of how the contract is enforced over time. One hypothesis is that uncertainty and risk aversion induce each successive majority to honor the contract.

Once the contract is broken—say, through the imposition of punitive taxes by the current majority—the value and cost of majority and minority status, respectively, must grow. Unless a new agreement can be reached, implicitly, between the majority and minority—presenting enforcement problems at least as great as those of the constitution itself—both the majority and the minority will be driven by the increased stakes to invest additional real resources in efforts to retain or obtain majority status. These resource investments are a pure social waste because they produce only a changed level or direction of politically mandated wealth transfers among groups. They do not increase overall wealth. In addition, the increased resulting taxation is likely to impose additional deadweight inefficiency and losses on the economy. Breakdown of the constitutional contract thus leads to an increase in the social costs imposed by the tax/transfer negative sum game emerging from competition within the democratic process. In other words, individuals are induced to shift some investment away from production of increased wealth and toward efforts to steal from others or to prevent others from stealing from them. By shunning punitive taxation of politically disfavored groups—such as oil and gas producers—society can reduce incentives for such socially wasteful investments.

Alternative Redistribution Policies and Criteria

The redistribution issue is important not only because redistribution, as discussed above, may be efficient social policy, but also because the wealth transfer effects of oil price increases may reduce sharply the political viability of reliance on the market as policy for dealing with the price effects of supply disruptions. In other words, redistribution may constitute a socially efficent investment in efficient *allocational* policies. It is important, therefore, to continue the process of examining alternative redistribution policies in the light of relevant criteria.

What potential policies are available? What criteria would be useful for purposes of guiding choices among them? There is not space here for a detailed examination of these questions. A brief list of potential policies and criteria may be useful, however, for purposes of illustrating the tradeoffs and analytic issues lurking in this area.

Potential redistribution policies can be summarized as follows:

1. Price and allocation controls, possibly combined with coupon or other formal types of rationing, which in turn could be combined with "white market" arrangements for the legal exchange of ration rights.
2. Revenue "recycling" in which (assumed) increased tax revenues from the oil sector substitute for reduced revenues raised by other tax instruments. These revenues could be earmarked for redistribution or other purposes, and could be combined with intergovernmental grants from federal to state or local governments.
3. Standard tax/transfer redistribution programs in which revenues are raised as general revenue and then budgeted for particular redistribution programs. As with revenue "recycling," this could be combined with intergovernmental grants.
4. Nonprice (administrative) allocation of Strategic Petroleum Reserve (SPR) crude oil.
5. Market allocation of SPR crude oil, possibly combined with earmarking of sales revenue for particular redistribution programs or for intergovernmental grants.

Again, this is merely a *listing* of general options; no inference should be drawn that any or all of them are being *advocated* here. Some criteria for purposes of discrimination among the options can be described as follows:

1. What are the differing allocational and efficiency effects of the options?
2. What are the comparative effects of the options in terms of optimal tax policy? In other words, which options minimize distortions in private sector consumption, production, and investment decisions? Which minimize distortions in signals received by government with respect to the size and composition of the government budget demanded by the taxpayers as a group?
3. What are the comparative effects of the options with respect to optimal energy policy? Which policy minimizes distortions in fuel mix choices? Which optimizes incentives for private sector preparation for possible future supply disruptions? Which minimizes distortions in investment decisions?

4. What are the implications of the options for federalism and for the allocation of functions and responsibility across federal, state, and local governments? What are the public sector efficiency implications of these effects?
5. What are the likely comparative effects of the options in terms of the advertised goal—specifically, redistribution to the needy during periods of energy price increases? To what extent can other governmental units (e.g., the states) offset the objectives of each option through substitution in public spending?
6. What differential incentives does each option provide the government in terms of optimal use of the SPR?

Clearly, the redistribution policy issue is complex, far more so than would permit even minimally adequate discussion in this chapter. The central point is that price and allocation controls are likely to be seriously defective in terms of the criteria summarized above, are unlikely to benefit the poor generally, and present massive administrative difficulties. Further research in the redistribution policy issues noted above would not be a waste of time.

Energy Emergency Policy Under the Reagan Administration

Two brief statements summarize nicely the emergency management policy of the Reagan Administration. *The Annual Report of the Council of Economic Advisers,* submitted to Congress in February 1982, states the following: "The policy of this Administration is . . . to facilitate the operation of market forces as the guiding and disciplining constraints shaping investment, production, and consumption decisions in the energy sector."

The president's veto message (March 1982) on the Standby Petroleum Allocation Act states: "What I do not have, do not want, and do not need is general power to reimpose on all Americans another web of price controls and mandatory allocations."

Furthermore, the administration policy with respect to allocation of SPR crude oil complements the market policy for allocation decisions described in the two quotations above.[18] Specifically, SPR crude oil will be sold at auction to the highest bidder, and the range of bidders is not limited.[19] In other words, the Reagan administration policy will facilitate efficient pricing and allocation of private sector oil, and will do the same for federal oil held in the SPR. For all of the reasons discussed above, this policy will minimize the adverse effects of future supply disruptions and will contribute toward the goal of maximizing social wealth.

Two issues of some importance now require attention from the administration. First, the SPR Drawdown Plan does not specify when or at what rate[20] the SPR will be drawn down. Instead, the drawndown "trigger" will be kept flexible in light of uncertainty surrounding the nature of future of oil supply disruptions. A flexible trigger allows adjustment of SPR use as the actual circumstances of a future disruption become known. It therefore avoids some of the rigidities inherent in a policy fixed and announced in advance.

On the other hand, while it is true that the characteristics of future disruptions cannot be known in advance, a predetermined trigger offers an important advantage in that it allows the private sector to proceed with greater certainty in its own preparations for emergencies. This is consistent with the administration's policy of primary reliance on the private sector for preparation in advance of any adjustment to oil supply disruptions. The relevant comparison to be made is between an imperfect policy announced in advance and an imperfect policy formulated when substantial price increases are being borne by important interest groups. Therefore, it is not obvious that flexibility is the correct policy; further analysis of this issue is advisable. What kinds of triggers could be announced in advance? How would they operate during disruptions? How would the private sector adjust its behavior in light of them? What is the effect of increased uncertainty under the flexible trigger? These are the kinds of questions that ought to be addressed seriously.

Second, income distribution issues have been paramount during past interruptions. As discussed above, substantial price increases cause important wealth redistribution effects. Equity arguments have played in the past and are likely to play again in the future an important role in the policy-formulation process. Again, the market policy may carry with it distributional implications with important political drawbacks, thus threatening the viability of the market policy. Therefore, the analysis of alternative redistribution policies as outlined above is important. A simple pattern of negative responses to the proposals of others may not be enough to defend the policy of market reliance. I suggest that these are the analytic issues now paramount in the emergency management area.

A Further Aside on the SPR

Why should there be a governmental oil stockpile? Surely the market can foresee the possibility of future supply disruptions; can it not prepare for them? Do we not believe that private sector preparation is likely to be more efficient than that of the government? The answer is "Yes, but . . ." The "but" arises because past policies—price and allocation controls during the 1970s—had the effect of penalizing those who had prepared for disruptions of price increases, and of subsidizing those who had not. Specifically, those who had invested in "speculative" inventories saw some of the value of those stocks confiscated by price controls, while others received "cheap" oil—a free lunch—because of the price and allocation regulations. The fact that such policies were implemented in the past must give rise to an expectation that similar policies may emerge again from the political process. This expectation acts as a disincentive to private sector preparation: both those who expect to see their oil confiscated and those who expect to be bailed out have reduced incentives to prepare.

This perverse effect of past government policy suggests that private sector investment in preparation will be lower than socially optimal; as with public goods, a subsidy for stockpiling of oil may be efficient. The SPR is one possible form of such a subsidy.

But is it the efficient form of the subsidy? The government could, as just one example, merely subsidize the acquisition of (marginal) private stocks, a program that would involve some, but not overwhelming, administrative difficulty. One reason that the SPR may be the efficient form of the subsidy is that it is something that actually can be pointed to when pressures begin to arise for price and allocation controls on oil. To the extent that the existence of the SPR facilitates resistance to such pressures, its social value goes far beyond the mere value of the oil that it contains. To the extent that the existence of the SPR reduces the perceived likelihood that controls will be imposed in the event of a disruption, the SPR may enhance private sector preparation as well.

The International Energy Agency Sharing Arrangement

It was noted above that price increases caused by disruptions affect all oil-consuming nations equally, regardless of whether they import all or none of their oil. Because free trade and the absence of international price controls (which would be impossible to enforce in any case) will reallocate oil supplies during any future disruption, there is good reason to review the International Energy Agency (IEA) sharing arrangement, under which IEA members have agreed to share oil according to a predetermined formula. This formula is merely a nonprice rationing scheme; it cannot increase the efficiency of world oil allocation during a disruption. Furthermore, in the inevitable absence of international price controls, the sharing formula cannot affect even the final allocation of oil, since there is nothing to prevent those receiving allocations from reselling their oil. Hence, the sharing arrangement will have the effect merely of redistributing wealth if oil is shared at below-market prices, and of imposing some additional transport and transaction costs on the system during a future disruption.

Free trade in oil will facilitate international adjustment to the effects of a future disruption, and will serve to minimize the adverse effects of the disruption on the world economy. Export controls imposed by some nations in an effort to shield themselves from some of these effects would reduce this beneficial effect of world market activity. Therefore, notwithstanding the basic innocuousness of the IEA sharing arrangement, it does have an important value: by committing the United States and others to "share" oil, it implicitly precludes the imposition of export controls during disruptions. Laws and treaties that have the (unintended) effect of preventing damaging intrusion by government should be treasured.

Conclusions and Recommendations

The overriding goal of government energy policy must be to facilitate and defend the operation of market forces. This is true particularly for emergency policy because only the market can adjust smoothly in the face of price shocks and other important exogenous phenomena. Regulatory policy cannot do it and causes substantial damage when it tries.

Therefore, now that the market policy is in place in the United States, it is important both to defend it against inevitable attacks and attempts to dilute it, and to enhance credibility and confidence that the market policy will withstand pressures for its abandonment when the next supply disruption occurs. This means that the Administration must be willing to spend some of its political capital to defend market policies, not only in the energy area, but also more generally.

A good place to begin is with the current ban on export of Alaska crude oil. This policy is inconsistent with free trade policy, with free market energy policy, and more generally with efforts to increase the efficiency of resource allocation. No one denies that the sole purpose of the export ban is to subsidize domestic maritime interests (see Chapter 5). If the U.S. government cannot or will not stand up to these groups, the credibility of its commitment to defend the market policy must be reduced. During a serious supply disruption, far more than the mere maritime industry will stampede in favor of controls; if the government is not willing to invest the political capital necessary to defeat the maritime unions and shipping interests, who can believe that it will do so when both the political and economic stakes are far higher? Every compromise leads to an expectation of further compromise. Failure now to stand on principle makes future defense of principle more difficult. This is one reason that the administration's effort to achieve full decontrol of natural gas prices has importance beyond issues of natural gas or energy policy. The administration's effort constitutes evidence that it is willing to take some lumps in order to defend the operation of free markets.

Second, the SPR is more than mere oil in storage; it is a bulwark against pressures for price and allocation controls during future disruptions. Investments in oil for the SPR, therefore, are also investments in part in the enhanced political viability of reliance on markets; this will have the beneficial side effect of enhancing private sector preparation as well. The Reagan administration has done a good job in building the SPR; at only about 100 million barrels in early 1981, the SPR in May 1983 contained over 320 million barrels. Decisions about the current fill rate and ultimate size of the SPR should consider this added dimension of social value of SPR crude oil.

Third, the administration must consider on a broader scope the distributional and the political implications of reliance on the market; it must fashion a distributional policy designed both to deal with the real needs of the poor and to protect the political viability of the market policy. The sort of analytic effort described earlier would be a good start.

Fourth, the issue of SPR trigger and drawdown policy needs to be addressed further. The administration policy of market allocation of SPR crude oil is the correct policy. The flexible trigger *may* be the correct policy, but it should be the subject of further analytic effort. Again the research outlined above would constitute a useful way to further SPR policy development.

Finally, defense of the market policy requires that there not exist in Washington a concentrated interest group, supported by tax dollars, with an institutional interest in a return to regulatory fiat as a policy for dealing with supply disruptions. That is what much of the Department of Energy is, however muted it may be during the Reagan administration. The bureaucracy has an interest in maximization of its work load, authority, and influence. The present market policy serves none of those ends; a price and allocation control apparatus does—beautifully. The Energy Department serves also as a central receiving station for the demands and appeals of all of the special interests lobbying for regulatory policies favoring them. It is important to remove institutional arrangements that facilitate such political pressure. The Energy Department is such an institution. In short, abolition of the Department of Energy is important in its own right and also in order to defend the policy of a free market.

4

RESTRICTIONS ON OIL IMPORTS?

S. Fred Singer

Summary

General agreement seems to exist that oil imports should be reduced. Some of the reasons make sense; some do not. After all, the reason for importing oil, or anything else, is that imports are cheaper than the corresponding domestic item. But oil is special, in the sense that the fear of an embargo still exists. There are many misconceptions involved here. For example, while a directed embargo cannot hurt us, production cutbacks—for any reason, anywhere—can raise the price of world oil, even to countries that do not import oil. However, our dependence on imports and our fear of interruptions have forced us to incur extra costs for protection, specifically the establishment and maintenance of a Strategic Petroleum Reserve (SPR).

The usual reasons given for restricting imports are: (1) emergency situations and economic damage produced by supply interruptions; (2) the high cost of preparing for such interruptions, including the cost of the SPR; (3) the outflow of dollars for oil imports, and its effect on the trade balance; (4) national security, which was invoked as the reason for the U.S. oil import quota program from 1957 to 1973. None of these reasons is completely valid. Reliance on market forces is in most cases equally effective and much less costly.

On the other hand, a real concern must be collusive price actions by oil producers, and especially by OPEC. Import restrictions—in the form of countervailing tariffs or, equivalently, variable import fees—can be used to counteract predatory pricing. Such fees can also neutralize the effects of possible price wars among producers on domestic projects of conservation and energy production by maintaining domestic price stability even if the world price collapses.

Price stability is much to be desired—by both consumers and

producers—although the price *level* would be under contention.
A combination of variable import fees by consumers and stockpiles by *producers* could achieve more stability.

Why Restrict Oil Imports?

Several arguments have been advanced for reducing or restricting oil imports.

The Oil Crisis Issue

During the past ten years, Americans have become painfully
aware of the importance of oil. The world price rose from $3 a
barrel in 1973 to $30 in1983, reaching a high of about $34 in
1981. Because of price controls in the United States, many had
to contend with "shortages," especially in 1974, and again in
1979, when millions of Americans waited in line at gas stations.[1]

The public also witnessed a tremendous preoccupation with
energy issues at the highest level of the federal government,
especially during the Nixon and Carter administrations.
Jimmy Carter, in March 1977, declared the energy problem to
be the "moral equivalent of war." But once President Reagan
had decontrolled the price of oil, in January 1981, the furor
surrounding the oil problem seemed to abate. In a free market,
prices can allocate the supplies of gasoline and other products.
A crisis based on bureaucratic misallocations, as in 1974 and
1979, is no longer possible (see Chapter 3).

Nevertheless, we are left with some semipermanent re-
minders of the oil crisis era. We now have a Department of
Energy (set up in 1977), a Synthetic Fuels Corporation, and a
growing Strategic Petroleum Reserve (SPR). The word OPEC
has entered our vocabulary. The Middle East and everything
going on there have become the stuff of which headlines are
made.

The National Security Issue

The much publicized Arab oil embargo of 1973 never succeeded in keeping oil away from the United States, nor can any
embargo be effective; oil is a fungible substance that can be
purchased through various channels throughout the world.[2]
Nevertheless, explicit and implicit embargo threats directed
against the United States (together with pressure to adjust our
foreign policy to match Arab objectives), the high price of oil,
and particularly the price jumps, have all reinforced in the

public mind the fear of being dependent on insecure Middle East oil.

As a consequence, under President Nixon we had Project Independence—which did not make us energy independent. We had efforts under the Carter administration to set up a huge, wildly expensive, government-subsidized synthetic fuels industry—which fortunately did not come about. And we are accumulating a stockpile of oil, a Strategic Petroleum Reserve, whose goal was originally 500 million barrels, later doubled to 1 billion by President Carter, and now targeted for 750 million barrels. (While our oil import projections have been shrinking, the stockpile goal has grown.)

Even though a directed embargo does not work, there is a national security threat from a possible oil cutoff—whether intentional, accidental, or caused by third parties (e.g., terrorists). In spite of its insurance feature, the ability of a stockpile to moderate price jumps is limited; there can be severe economic impacts due to higher oil prices, if the cutoffs are of long duration. We must also consider the impact—economic, but particularly psychological—on our allies, many of whom may not have large stockpiles. For this reason, sharing arrangements are contemplated under the terms of the International Energy Agency (IEA) agreement.[3]

But a stockpile does not protect against a physical shortfall. Rather, release of stockpiled oil serves to dampen a price increase—for *all* oil consumers in the world. Therefore, restricting imports would not protect us from a price increase caused by a supply interruption in world oil. Even if we imported no oil, we would still experience a price increase.[4]

The provisions of the SPR will cost the nation over $3 billion annually—and more, if interest rates rise again. The maintenance of military forces to intervene in case of serious problems in the Persian Gulf is also a costly undertaking. Finally, there are the indirect security costs that come from restrictions on our foreign policy because of fears of a possible cutoff, whether such fears are justified or not.

Money Outflow

Another reason often cited for reducing oil imports has to do with the large outflow of money. At the peak of 1980, this outflow from the United States came to more than $80 billion per year. Even though most of this money was returned in the form of investments or purchases of U.S. goods and services, it represented a large sum in relation to our foreign trade balance. During 1980, oil constituted about a third of our imports in dollar terms.

Of course, this outflow could be reduced if we were willing to develop more domestic oil and especially synthetic oil, but this latter option is quite costly. Indeed, if there were no national security considerations, we would be better off purchasing imported oil, even at the high price of $30 per barrel, rather than developing high-cost oil or manufacturing synthetic oil domestically at a resource cost, which might be as much as $50 per barrel.

Fortunately, there is an alternative to imported oil, namely more medium-cost domestic crude oil. Although expensive, it is still cheaper than synthetic oil—at least for the foreseeable future.

Is Oil Special?

Arguments have been offered at various times that oil is special and that imports should be restricted. During the 1950s and 1960s, the United States imposed oil import quotas to prevent cheap foreign oil from flooding the market and crowding out domestic production.[5] (The program was poorly handled and caused many inequities.) What can we say about oil today?

- Oil is pervasive; it is more widely used throughout our economy than any other imported material. For example, we may import 100 percent of cobalt, diamonds, and other minerals, but none of these imports represents as large a percentage of GNP as does oil.
- There are no viable substitutes for oil in the short run, although more and more boilers are now set to burn either coal or gas, in addition to oil. (Gas can be substituted readily, but it takes time to develop the means to substitute coal. Nuclear energy requires lead times on the order of ten years or more.)
- However, the main feature of oil is that its price has been and continues to be affected by the often capricious production decisions of OPEC, which are outside the control of oil consumers.

Problem of Unstable Prices

Import restrictions on oil are receiving consideration at this time because of the rapid decline of the price of oil, the first such decline since 1970 when OPEC countries first gained significant control over production decisions by taking over the concessions from the multinational oil companies.

Consider these facts:

1. Middle East oil, although high-priced, is considerably cheaper to produce than U.S. oil. Middle East oil can be produced at less than $5 per barrel—most of it, actually at less than $1 per barrel. U.S. oil, on the other hand, has an average production cost of probably over $20, although there is still some "old" oil in existence whose marginal production cost is less than $10 (i.e., the cost of producing an extra barrel is less than $10). The marginal cost of developing a new oil field in the United States would certainly be more than $20 per barrel.

2. The price rises of world oil, and particularly the 1979 jump to above $30, have stimulated large investments in high-cost energy throughout the world, and especially in the United States. This investment here has been mainly in high-cost oil, in high-cost gas, in conservation efforts involving equipment with high capital cost, and in alternative energy sources such as solar and geothermal—all justified on the basis of high oil prices and the expectation of even higher prices in the future.

With the world price slipping toward $20 and perhaps even lower, these investments are endangered. The real fear is not that oil prices will decline and stay low; that would simply make oil cheap and benefit the consumer ultimately. Rather it is that the price is likely to be unstable for the next decade or longer. A decline to below $20 would wipe out much of the current effort on conservation and renewable-energy development. Followed by a revival of OPEC market power, restrictions on production could drive the price up to $30 or more. (In the short run, not much could be done about this higher price except to release oil from the SPR. For a variety of reasons, this may not take place.) The consumer will then be stuck with much higher oil prices, but the alternatives that were developed earlier will no longer be available. They will have been abandoned as the price went below $20.

To summarize, a united OPEC would have the power to drive the price of oil both down and up. The fear is that the OPEC nations could wipe out investments in the consuming nations for alternate oil and other energy sources, cripple the conservation effort, and make the West vulnerable to higher oil prices in the future. (Even if OPEC does not adopt this strategy in a conscious manner, a price war among revenue-hungry producers may result in a temporary price decline deep enough and long enough to accomplish the same purpose.)

This scenario is, of course, similar to the classic anti-dumping situation to which much legislative effort has been devoted. Here a foreign manufacturer underprices his goods in order to destroy competitive manufacturers in the United States; once this goal has been accomplished, he raises the price to a higher level. The standard remedy against such predatory pricing is a countervailing tariff.

Price stability is much to be desired; everyone would agree to that. The major problem is who should get the rents, that is, the difference between the price and the actual production cost of oil, which may be considerably less than the price. From our point of view, we would like rents now going to OPEC to go instead to consumers or, alternatively, to the U.S. Treasury as the proxy for U.S. taxpayers.

After a review of past U.S. efforts to set up import quotas and tariffs for oil, we will describe various proposals that have been made and discuss their merits in terms of the criterion which has just been set up.

Past Restrictions on Oil Supply and Demand

The United States has a complicated history of restricting oil demand and oil supplies, domestic and foreign, involving a variety of nonmarket approaches. By and large, these interventions have not been in the long-term national interest.

Demand Restrictions

Demand restrictions have been rare: coupon rationing during World War II, and allocations (another form of rationing) during the so-called oil crises of 1974 and 1979. Both efforts can be judged as unsuccessful. Rationing led to a black market in coupons. Allocations led to massive lines at gasoline stations and other distortions. Fortunately, the restrictions were shortlived, thereby avoiding a major economic disaster.

It is generally agreed, at least by economists, that the efficient (and equitable) way to handle "shortfalls" is to use the market system and free prices. Unfortunately, in spite of all of the evidence, there is still a clamor to use methods other than price-rationing (see Chapter 3).

Domestic Supply Restrictions

The outstanding example is the market-demand prorationing program of the Texas Railroad Commission to limit the output of wells in Texas; other states adopted similar programs in the early 1930s. "Prorationing" refers to a proportionate reduction of the output of each well; other restrictions included limiting the number of wells per unit area in an oil field. The states' programs were backed up by the federal government through the Connally "Hot Oil" Act.[6]

Critics of the program claimed that it created a cartel in order to raise the price of oil. Proponents argued that the program was necessary in order to obtain conservation. Both claims are to some extent true. The price of oil would have been much lower without prorationing. But while some see the Texas Railroad Commission as a prototype of OPEC, prorationing did not cause price jumps. Rather, it avoided price wars among producers during periods of oversupply. It produced a stable and slowly declining price while allowing a fair profit for producers—a worthy goal for OPEC.

Prorationing also aided conservation by preventing overproduction. Under the "rule of capture" a leaseholder can pump as fast as he wishes from as many wells as he cares to drill on his lease. But with small lease acreages (i.e., compared to the size of the oil reservoir) this procedure would remove oil from underneath adjacent leases, thereby providing an incentive for everyone to pump beyond the limit of maximum long-term yield, thereby reducing the total amount of oil recovered from the reservoir.

Import Restrictions

Methods to restrict imports fall into two general categories: (1) quotas provide a precise way to limit the amount of imports, but thereby raise the domestic price; (2) import duties, whether called fees or tariffs, raise the price in a precise way but do not control directly the amount imported. There is thus a rough equivalence (or complementarity) between the two methods, although they differ considerably when used as an instrument of policy.[6]

Duties on imports (called a "fee") were first included in the Internal Revenue Act of 1932. They were substantial in terms of the price of oil at the time, but became inconsequential after 1974 when nominal oil prices were ten times higher. In 1975 President Ford imposed much greater import duties and proposed a permanent $2 per barrel tariff. But the special duties were removed by the Energy Policy and Conservation Act (December 1975), which did not include a tariff proposal.

The import quota program has been the main effort for limiting oil imports and has become a classic in discussions of government involvement in energy markets. We quote here from the Introduction in the well-known analysis of Bohi and Russell of Resources for the Future:

> There was no formal U.S. position on controlling oil imports until 1957, when a voluntary control program was implemented to hold back an anticipated surge in imports. Predictably, the voluntary program foundered as soon as it seriously affected private interests, and the Mandatory Oil Import Program was implemented in 1959. That program came to limit crude oil and petroleum product imports to approximately 12.2 percent of domestic production. It was designed in such a way that its costs were somewhat camouflaged, and its effects could be shifted among interest groups. It was administered with flexibility and political skill. Consequently, the basic elements and structure of this program survived great changes in its economic effects, rationale, and political impact. During its life, the quota brought about substantial changes in the structure of petroleum production, refining, and consumption; effects that will far outlive its formal abandonment in 1973.
>
> The mandatory quota program came under increasing attack in the late 1960s on two fronts: it was regarded by oil producers as badly administered because of an increasing number of special quota allocations, both inside and outside the 12.2 percent rule; it was offensive to consumer groups because its cost had become higher, more obvious, and the inefficiencies of its discriminatory application were better recognized. A Cabinet Task Force on Oil Import Control was formed in 1969 to study oil impact policy and make appropriate recommendations for changes. The task force recommended in 1970 that the quota program be abolished in favor of a less restrictive and more neutral import tariff.
>
> The quota instrument was not abandoned, but quantitative restrictions were eased to allow more imports during the period from 1970 to 1973. Relaxation of controls owed less to a change in import policy than it did to the Nixon administration's willingness to abandon restraint in order to relieve the upward pressure on prices. On April 18, 1973, the mandatory quota program was suspended and a license-fee system was imposed—but at rates which actually lowered the total levy below the previous tariff. Oil imports returned to the formally uncontrolled status that they had known before 1957.
>
> The Arab oil embargo of October 1973 revived general support for a policy of limiting U.S. dependence on imports—or in any event, for a policy which would avoid the vulnerability which had so forcefully been exposed. President Nixon responded with the

Project Independence Program, promoted initially as a series of actions which would eliminate U.S. dependence on foreign energy supplies in 1980. It was soon clear that the cost of any program would be far higher than the public would tolerate, and, moreover, far higher than the present expected value of any benefits it might yield.

Denied easy solutions, the nation agonized over how much energy security it was worth buying, how efficiently it was to be provided, who was going to pay for it, and in what way. The issues were not posed in this way, of course, but in such terms as: demand reduction versus supply enhancement; market decision-making versus government allocations; strategic storage and standby emergency preparedness versus "drain America first" or "strength through exhaustion"; and unjust enrichment of energy companies through exploitation of the poor versus subsidization of wasteful energy consumption. While laws were passed, research budgets increased, slogans adopted, international conference called, solemn declarations signed, and developments in the oil market closely watched, the level of oil imports rose, the ratio of imported oil to domestic oil grew, the proportion of imported oil used in total energy consumption increased, and dependence on those nations which sought to restrict oil supplies in 1973 was escalated.[6]

Proposed Policies For Import Restrictions

This section presents some representative proposals for restricting imports that are currently under public discussion. They would all result in higher prices to the consumer (and therefore in reduced oil demand to match the reduced supply). But this fact in itself should not be an obstacle: various refunding or recycling schemes are available that would get an equivalent amount of money back to U.S. residents (but not directly to oil consumers); one might even subsidize the lowest-income group. For example, an automatic (and rapid) recycling scheme would: (1) reduce withholding for income tax (or for Social Security) in the weekly paycheck; and (2) increase transfer payments such as welfare benefits, unemployment insurance payments, or Social Security payments to non–wage earners.

Who Gets the Surplus?

The surplus—that is, the extra amount paid by the oil consumer as a result of import restrictions—must end up in the pockets of one (or more) of the following: (1) a foreign oil producer, (2) a domestic oil producer, or (3) the U.S. government (which has imposed the import fee, duty, tax, or quota). It should be noted that the government can always take away the surplus from the domestic producer through an excise tax or the existing Windfall Profits Tax (see Chapter 6). (Even without a special tax, income taxes and ad valorem severance taxes will remove a part of the surplus.) The important policy objective then is to transfer the surplus away from the foreign producer.

Import Quotas

In the quota program of the 1960s, refiners received import license tickets free of charge, entitling them to imported oil at the (lower) world price. Favored refiners, especially smaller ones who got more tickets, thus received a clear subsidy, since they could sell oil products at the going domestic price or effectively "sell" their valuable quota tickets, which the government had given them.

Any future quota program would likely incorporate an auction for quota tickets, in which refiners would bid against one another. The proceeds of the auction would transfer part of the surplus to the federal government. But since the quota would restrict imports, the domestic price would rise by a certain amount (which can be estimated from past economic data). Domestic producers would therefore also receive a surplus, roughly equal to the auction bid.

An important variation on the auctioning of quota tickets has been suggested by M. A. Adelman. Sealed bids should be instituted; this might give oil-exporting countries a means of selling more oil to the United States by secretly subsidizing the bid of an oil importer. At the same time, of course, some of the exporter's profits would be transferred to the U.S. Treasury. Adelman saw this scheme also as a means to encourage OPEC members to break cartel discipline by selling larger amounts of oil at an effectively lower price.

Tariffs

The idea has been generally accepted that oil imports involve some externalities, so that the incremental barrel of oil, which must always be imported, is more costly than the average (given by the price). Earlier work, under conditions of domestic price control, gave a quite clear-cut result.[7] The incremental barrel cost more because: (1) the world price exceeded the average domestic price by about $5 (in 1977); and

(2) the Strategic Petroleum Reserve was configured on the basis of the rate of imports, so that the security cost per imported barrel was about $2. But, (1) price controls no longer exist, and (2) the SPR size need not (or should not) be related to the import rate. Also, the cost of the SPR is an imperfect "proxy" for the security cost incurred by oil imports.

Optimal Tariff

More recent work has been concerned with the concept of an "optimal" tariff.[8] One line of research has concentrated on the economic damage of an oil supply interruption, arguing that the greater the rate of imports the larger the economic disruption. But this argument is only partly true. Most of the disruption is due to the world price increase following a supply interruption; since oil is a fungible substance, this price increase affects *all* oil consumers, even those who do not import, in the absence of price controls and trade restrictions. A separate argument has to do with whether increased imports increase the *risk* of a disruption occurring.

A second approach argues that a decrease in demand, brought about by a tariff, will lower the world price—just because the supply curve exhibits a slope. (This argument, of course, applies not only to oil but to any commodity.) In fact, the argument can be made even stronger when applied to OPEC: a higher price will also bring forth more domestic oil (and other energy sources) and reduce demand for OPEC oil even further. OPEC, and more particularly the Arabian residual producers, face a high demand elasticity.

In principle, there are several kinds of possible tariffs. All of them would act as a source of revenue to the Treasury, although any reduction in demand would reduce imported oil (on a barrel-for-barrel basis) so that Treasury revenues would also be somewhat reduced.

Tariffs will also raise the price of domestic oil to the level of (world oil price plus tariff) as refiners compete for (now) cheaper oil and bid up the price. As mentioned earlier, these gains to domestic producers can be taxed away and refunded to consumers. But since imposition of a tariff may decrease the world price, the domestic price may not rise by the full amount of the tariff.

Fixed Tariff

A fixed tariff is simplest—so many dollars a barrel, independent of the price of oil.

Ad Valorem Tariff

This tariff is a certain percentage of the world price. It thus presents a higher demand elasticity to OPEC, especially the residual producers.

Increasing Percentage Tariff

This form of tariff would cut OPEC oil demand most severely in case of a price increase and presumably discourage such increases. Unfortunately, it would also harshly penalize the consumer unless an elaborate system of recycling were immediately available.

Consumption Taxes

Taxes at the point of consumption would apply to both imported and domestic oil, as opposed to tariffs, which are placed only on imported oil. Consumption taxes are mainly designed to raise revenues but would indirectly affect imports as well. It is important to note that they do not encourage domestic firms to find and produce more oil (as would be the case with import quotas or tariffs).

Gasoline Tax

This widespread consumption tax is especially popular in Europe and Japan. It is primarily a revenue-raising measure and has had some of the aspects of a luxury tax. (In the United States, however, the automobile has never been considered a luxury.)

In the United States the federal tax hardly acts as a disincentive for the use of gasoline but can be thought of as a user's fee to pay for highway construction and maintenance essential for automobile operation and safety. The tax on gasoline was, until April 1, 1983, four cents per gallon, and now stands at nine cents per gallon (see Chapter 6). State taxes fulfill a similar purpose.

An argument has often been made for a higher federal tax. President Carter, and more recently presidential candidate John Anderson, have suggested fifty cents per gallon. Such a tax might be economically efficient—in other words, about the right amount to pay for various externalities produced by automobiles, especially in the urban environment: noise, crowding, dust, pollution, hazards to life and limb, and similar disamenities.

Incidentally, such a higher tax would also reduce oil imports, especially in the United States where presently close to 60 percent of oil is used for transportation.

Oil Tax

On the one hand, taxes on all oil products would act as an incentive for more fuel switching, thereby reducing oil demand and imports. On the other hand, they may just drive up prices for coal and natural gas.

Energy Tax

An across-the-board Btu tax on all forms of energy could decrease fuel switching, and under some conditions might even increase oil imports.

Variable Import Fee

As opposed to other forms of import duties, the variable import fee (VIF) is specifically designed to deal with the problem of maintaining domestic price stability in the face of unstable world prices. With the current price probably higher than the world oil market can support, sudden collapses, even though shortlived, are a real possibility, especially in view of competition for market shares among OPEC producers and the likelihood of a price war (see Prologue).

The basic idea behind a VIF is quite simple. Administratively it is handled like a tariff, except that the amount of VIF is set so that adding it to the world price would give a desired target price:

$$WP + VIF = TP$$

VIF goes into effect only if and when WP (world price) drops below TP (target price). Note that if the world price of oil should go above the target price, then there would be no import fee. No simple policy measure by the United States can avoid the possibility of sudden price rises caused by, say, major interruptions in the world oil supply. But for this purpose we have a strategic stockpile that can limit these price rises by releasing oil from storage.

Purpose

A VIF serves several purposes:

1. Keeping the domestic price always at the target price achieves price stability. This protects the ongoing conser-

vation effort as well as investments already made (in conservation, domestic oil, and other energy resources) with the expectation of constant or higher future prices. There is a large constituency of energy-related firms (including thousands of small oil producers and coal operators) as well as banks that desire stability.

2. Knowledge of the future price floor allows more certain planning. Investments can be made without fear that "the rug will be pulled out from under." It prevents oil exporters from discouraging investments that would provide substitutes for imported oil.

3. If the domestic price is kept constant, the demand for oil will not rise, even if the world price should drop to low values, which is likely for short periods only. This prevents oil exporters from manipulating our consumption of oil.

4. Finally, the VIF will produce a (variable) stream of revenues for the Treasury, which can of course be recycled or used to reduce general taxes. With domestic prices stable, the Treasury and the states will continue to collect a variety of existing taxes from the energy industry. If the price were to drop by $5, such revenues would decline by about $15 billion per year.

Target Price (TP)

There is no good way to set the best target price except by the political method. It should be no higher than $34 per barrel, the 1982 price, and probably not lower than $25, a price that would keep coal, gas, and nuclear energy competitive with oil. (Because of the Windfall Profits Tax and other taxes, many domestic producers will not be too sensitive to the exact value of TP.)

A decision will have to be made whether to escalate the nominal TP every year to keep step with inflation. Without escalation, or with partial escalation, TP will drop in real terms; after a number of years, then, the VIF program could be effectively phased out.

World Price (WP)

World Price would be defined in terms of a current (or recent) world average, according to some accepted formula. Allowance would have to be made for price differentials for crudes of different quality. The actual price paid for the imported oil would not be relevant; in this way one would encourage importers to bargain with foreign oil suppliers so as to obtain the lowest possible price.

International Issues

A country that reduces its oil imports, by whatever means, reduces the exports from the swing producer (usually OPEC, or Arabian producers within OPEC). This creates more excess production capacity in the world and puts downward pressure on the world price, and thus benefits all oil consumers, including those that do not import oil.

(In a country that neither imports nor exports oil, the consumers will gain but the producers will lose in case of a decrease in world price. But the net effect for such a country is zero.)

Thus if the United States restricts imports, it will create a public good for all of the world's oil consumers—with the highest per capita importers (like Japan) benefiting most. The same reasoning applies if the United States releases some of its SPR oil during a supply interruption.[4]

The existence of these interconnections, by way of a single world oil market and essentially single world price, argues for international coordination among oil importers. The International Energy Agency (IEA) could and should supply such coordination.

Instead of worrying about supply interruptions, which will be handled automatically through higher prices that allocate the available supply worldwide, the IEA might usefully consider how to meet the real possibility of price collapses. Joint action— for example, by a VIF system—would put all industrialized countries on a similar basis, remove the competitive threat of cheaper oil for some countries' industries, and put joint pressure on OPEC to lower the price even further.

Conclusions and Recommendations

We need to distinguish carefully between two extreme situations: (1) import restrictions under conditions where the price of oil is expected to be stable; and (2) import restrictions under conditions where the price is unstable.

The Case of Stable Oil Prices

In the case of a steady price (or a price that varies slowly, say, by less than 5 percent per year), the two major arguments for restricting imports are national security and externalities (i.e., various costs to society not borne by those who import oil).

Clearly, if one can reduce the amount of imported oil, the security situation improves. However, it is not at all certain that a tariff is required in order to achieve lower oil imports. Development of domestic oil and other substitute energy resources, as well as conservation, may be a cheaper and more effective means of limiting imports. One has to guard, of course, against the development of energy alternatives that are more costly than imported oil and that would result in a substantial loss of economic welfare.

An argument can be made that importing oil imposes externalities. Certainly there is the cost of maintaining the SPR as well as certain military and other security expenditures. If we postulate a stockpile of 750 million barrels, then the annual maintenance cost of the stockpile works out to be $3 billion— about $2 per imported barrel for a current import rate of under 4 million barrels per day. (These figures could double if the interest rate goes to about 20 percent.) It would seem fair and economically efficient that this security cost be paid by those who cause the security problem, namely the unstable suppliers. But since oil is a fungible substance, there does not seem to be a simple way of putting the cost burden on only those countries that have declared embargoes against us in the past. One can put a security tax on all imported oil and perhaps exempt oil imported from the Western hemisphere, but any such arrangement would be awkward and would likely lead to abuses.

The trouble with a tariff, even one that is defined as an optimal tariff, is that it also increases the price of domestic oil. This "windfall" for domestic oil can always be removed by an appropriate tax and even recycled to the consumer, but such schemes are complicated and likely to be abused.

Of course, there are those who would opt for an import tariff as a revenue-raising measure. This invites comparison with an oil consumption tax or value-added tax on manufacturing processes that use oil. The advantage of a tariff is that it encourages more domestic production and also takes care of some of the externalities of imported oil.

Others have recommended that a consumption tax be put on oil products, particularly on gasoline. However, as discussed earlier, this is not so much energy policy as transportation policy. Clearly, there are externalities—in other words, costs imposed on society by the automobile, such as road construction and maintenance, highway safety expenditures, and the like—that should properly be borne by those who use the highways.

Import Restrictions under Unstable Price Conditions

Unstable oil prices, especially large, unpredictable fluctuations, can raise havoc with our domestic energy industry and with our economy generally. A stable price is much to be preferred. The price of crude oil in constant dollars has been reasonably stable since 1880, and quite stable from 1950 to 1970, when it actually declined very slowly. Since 1973 the price has been unstable, however, first increasing in sudden jumps and now likely to decrease, perhaps smoothly or perhaps with collapses followed by recoveries (see Prologue).

There is, of course, the possibility that oil producers will resort to predatory pricing. They could try to extract extra revenues through price rises brought about by supply restrictions. For example, if OPEC members jointly decide to cut their production (or if some of the major members cut their production), the world price will rise; but because of the inelastic response in the short term, demand cannot fall, and large amounts of revenues will then be transferred to the producers. Over the long term, of course, conservation and substitutions would reduce the demand, as is happening now in response to the most recent price rise of 1979 to 1980; but at that point the cartel can drop the price by increasing production, and thereby destroy both conservation efforts and the investments in alternatives to imported oil.

The standard response to this classic pattern of predatory pricing is a countervailing tariff, covered by existing legislation. However, the situation just described may develop even without any specific intent on the part of the OPEC cartel— just by the fact that the response to price changes tends to be rather inelastic. What should be our response to this kind of a "price yo-yo," which could cause great damage to our economy? We will discuss five possibilities:

First, in principle, price fluctuations can be smoothed out if people would buy oil when the price is low, store it, and then sell it when the price is high. In practice, however, such schemes do not work very well, even for innocuous commodities, such as copper and timber. For oil, the situation would be worse because of the high visibility and the political problems of oil.

Some positive encouragement would be needed to induce speculators to buy and sell oil and to profit from price swings. They can do this now in futures markets on a quite short time scale, on the order of weeks or months, but they are effectively discouraged from doing this on a longer time scale by two factors. One is the existence of the federal SPR, which overhangs the market. Since it is not known how the SPR will be used, it is possible that the release of SPR oil can limit the price rise and, therefore, the potential profits of those who have bought oil for speculation. The other factor has to do with whether Congress will allow speculators to profit from price rises during an "emergency." There is always the likelihood that price controls or price limits will be imposed and that "hoarders" will be penalized for attempting to take profits. The straightforward free-market solution to price stability is, therefore, not feasible.

Second, the obvious alternative would be for the government itself to utilize the SPR as a price-stabilization device, by selling from the SPR when the price increases and buying for the SPR when the price drops. Although this scheme may make sense in theory, it seems difficult to put into practice. The generally accepted view, within both the administration and the Congress, is that the SPR should be saved for the proverbial rainy day and should not be used for "price manipulation." Yet under a prudent operation of the SPR—that is, as a private owner would operate it—oil should be sold right now, by auction or otherwise, in anticipation of lower prices in the future, at which point stockpile purchases would be resumed. There is an additional advantage: selling from the stockpile now would help to drive down the world price by putting more pressure on the OPEC cartel.

Third, the hybrid solution would be to privatize the stockpile—in other words, to leave it in government hands but to permit and encourage private individuals to purchase shares and to sell oil to the stockpile and buy oil from the stockpile. Based on the SPR, one might even build a more extensive oil futures operation.

Fourth, a sensible alternative to the schemes discussed so far would be for the oil producers themselves to stabilize the world price. This, in fact, is a preferred solution. For example, Saudi Arabia, by using its considerable excess capacity, could set up a rather large stockpile very cheaply so as to meet almost any emergency situation that might drive the price up too far.[9]

It was widely believed, at least by some of us, that Saudi Arabia had implicitly agreed to perform this balancing operation by acting as the residual producer. Together with the Gulf sheikhdoms, the Saudis maintained a reasonably constant price from 1974 to 1978. However, in 1979 they became intrigued by the idea of letting the price go up, and thus increasing their revenues. They should, of course, have stemmed the price rise by producing at maximum capacity and by announcing their intent to increase production even further, thereby aborting the panic buying that drove the price up. Following the price tripling, from about $12 to $36 per barrel, they then failed to bring the price back down again, choosing instead to defend the higher price, which could not be supported by the market, by cutting their production from 10.5 mbd down to below 4 mbd. Apparently, they did not realize that it was in their interest to maintain the lower price level. Yet, it was this high price, sustained for more than two years, that really spurred on consumers to replace oil with cheaper alternatives such as coal, gas and nuclear energy.

Judging from this experience, therefore, it appears unlikely that the producers would be willing to stabilize the price—except, of course, at an extremely high level that cannot be supported by the market and is, therefore, unrealistic.

Fifth, a number of conservative economists have independently suggested the imposition of some kind of import fee or tariff.[10] The president has authority to apply such a fee under the Trade Expansion Act of 1962. A direct method for stabilizing the price on the down side is the variable import fee, as discussed earlier. This can be done only by the government of a consuming country, or preferably by a number of governments in coordination.

A variable import fee, used as a stabilization measure, would: (1) protect our investments in conservation and domestic energy development; (2) deprive OPEC of increases in oil demand that might follow a price decrease; (3) transfer the losses of OPEC into gains to consuming countries and to their citizens, by means of recycling through the treasuries.

A variable import fee is, of course, analogous to a countervailing tariff imposed in response to dumping or predatory pricing. Even though there may have been no intent on the part of OPEC to carry out predatory pricing, the effect on our industry and economy of a large price drop or a price collapse would be the same as if it had been planned. Hence, the response could well be the same within our present laws.

5
EXPORT OF ALASKAN OIL AND GAS

Stephen D. Eule
and
S. Fred Singer

Summary

A huge resource of oil and gas in Alaska is locked up by federal legislation that prohibits its free commercial export to overseas users. As a result, Alaskan oil is currently creating a glut and discouraging oil production in California. Half of Alaska's oil production has to be shipped to the East Coast and Gulf Coast at considerable cost, ultimately borne by American consumers.

Removing export restrictions would gain the federal treasury about $1 billion per year and also increase Alaska's revenues substantially. It would reduce the nation's deficit trade balance with Japan ($18 billion in 1982) and the rest of the Far East by up to $20 billion in potential oil and gas exports. In addition, it would render unnecessary the construction of a $2 billion pipeline from Puget Sound to the Midwest and eliminate the current costly and wasteful tanker traffic to the East Coast.

Most important, it would stimulate Alaskan producers to develop more oil for export, probably from 0.5 to 1 million barrels per day (mbd), worth about $5 to $10 billion per year. And it would blaze the way for exports of natural gas in the form of liquid natural gas (LNG) or as raw materials for fertilizer, with great benefits to the economic development of Alaska. Gas exports of about 1 TCF (trillion cubic feet) would be worth about $5 billion per year and would add the equivalent of another 0.5 mbd to the world oil market.

Changing the restrictions that ban overseas sales of Alaskan oil will take political effort. That ban has a powerful constituency in the maritime unions. Under a 1920 law, the Jones Act, all shipments between American ports must be made in American-flag ships manned by American crews. All the oil that leaves the southern Alaska port of Valdez for terminals on the West and Gulf coasts falls under the Jones Act. Even though only part of the 1.6 million barrels of oil that run through the Alaska pipeline each day might be involved in export to Japan, the maritime unions would fight to keep the export ban from being dropped.

But on balance, more would be gained than lost if exports were permitted. Moreover, export of Alaskan hydrocarbons poses no threat whatsoever to U.S. security. On the contrary, putting another 1 mbd (or more) of non-OPEC hydrocarbons on the world market would enable the consumer nations to import less oil from unreliable OPEC producers.

Introduction: History and Background

Alaska Strikes Oil

In January 1968, significant reserves of heavy, high-sulfur oil, and associated gas, were discovered at Prudhoe Bay on Alaska's North Slope. In 1968–69, Alaska sold the basic leases for the approximately 19 billion barrels of oil and 26 trillion cubic feet (TCF) of natural gas for around $950 million, with a 12.5 percent royalty interest attached.

Estimates of the oil and gas reserves on the North Slope, which includes the Kuparuk River field as well as Prudhoe Bay, vary. In August 1980, the state of Alaska estimated its "most likely" discovered reserves of oil at 10.2 billion barrels. National Petroleum Council estimates suggested in December 1981 that there could be anywhere from 13 to 55 billion barrels of oil yet to be discovered, their mean estimate being 24 billion. These estimates are for the North Slope and Bering Sea, and incorporate averages from other studies. Richard Nehring projects that by the year 2000, total recovery of crude in Alaska will be between 22.1 and 31.5 billion barrels. For gas, the picture is equally bright. Estimates by Alaska place its discovered North Slope gas reserves at 35.4 TCF. Under stricter definitions, the American Gas Association estimated in January 1982 proven reserves to be 26 TCF for the North Slope, and 31.9 TCF for all of Alaska. The same study estimated potential gas reserves for Alaska at 145 TCF; and the National Petroleum Council estimates undiscovered reserves at 109 TCF, with a high estimate of 246 TCF.

While oil and natural gas have been produced at Cook Inlet for thirty years—with some of the gas successfully being shipped to Japan in liquefied form by Phillips-Marathon—it was the opening of the Trans-Alaska Pipeline System (TAPS) to the Prudhoe Bay field in 1977 that turned Alaska into a major energy supplier. For 1982, Alaska averaged over 1.7 million barrels per day (mbd) of crude oil production, with close to 95 percent (1.6 mbd) of this coming from the North Slope. Another 90,000 barrels per day were added to the TAPS throughput with production from the recently developed Kuparuk River field just west of Prudhoe Bay. By the mid 1980s, the $3 billion Waterflood Project will keep up Prudhoe Bay production by maintaining reservoir pressure by means of seawater injection. At the same time, production in the Kuparuk River field is expected to be at least 0.2 mbd. With the capacity of TAPS being 2.0 mbd, it is clear that production from the North Slope will approach this limit during the decade, even if there are no new oil discoveries.

The Trans-Alaska Pipeline

Coupled with the development of the North Slope oil reserves came plans for transporting the crude to market. Two realistic choices were open for consideration: (1) the first was the construction of an oil pipeline through Canada to refineries in the U.S. Midwest (the Trans-Canada pipeline); (2) the second option called for the construction of a pipeline to tidewater in southern Alaska, where crude could be loaded on tankers for shipment to the West Coast and the Pacific. Although the Trans-Canada pipeline to the Midwest seemed attractive from a strictly economic point of view,[1] the disadvantages of constructing and operating an international pipeline (i.e., the lengthy negotiations and increased regulation) were sufficient to discourage such an enterprise. Instead, the oil companies felt that the Department of Interior would quickly grant a right-of-way from the North Slope across Alaska to Valdez, and they applied for permits in June 1969. But because this involved pipeline construction on federal land, Congress became involved in the decision-making process.

Objections to the proposed TAPS came from many quarters, and construction was not allowed to proceed until four years later. Environmentalists, spurred by the passage of the National Environmental Policy Act of 1969, wanted to block construction because of (unfounded) fears that the pipeline would cause serious ecological degradation of the delicate tundra. Others (correctly) argued that the West Coast would be unable to absorb all of the Alaskan oil. And there were also charges by consumer groups and midwestern congressmen that the ultimate purpose of the proposed TAPS was the oil companies' desire to increase profits by shipping North Slope oil to Japan. Another of their concerns was that declining exports of Canadian crude would result in regional supply disruptions. Maritime interests, however, were all in favor of the TAPS bill.

After much congressional testimony, debate and hand-wringing, construction of the pipeline was finally allowed to begin in 1973 with the passage of the Trans-Alaska Pipeline Authorization Act, which included the requirement that any exports to noncontiguous countries, such as Japan, receive presidential approval. Although this provision had the effect of partially placating midwestern interests (they still did not get their Trans-Canada pipeline, however), it arose from heightened concern about domestic energy security: the TAPS bill was passed two weeks after the Arab oil embargo was imposed.

Restrictions on Crude Exports

The Trans-Alaska Pipeline Authorization Act established two broad criteria to determine whether exports are to be permitted: (1) the president must conclude that allowing exports would "not diminish the total quantity or quality of petroleum refined within, stored within, or legally committed to be transported to and sold within the United States" and that it would be in "the national interest"; and (2) after making such a determination, the president is required to report the findings to Congress which, after review, can overturn the presidential initiative by passing a joint resolution within sixty days.

In 1977 and 1979, amendments to the Export Administration Act (EAA) placed additional restrictions on the export of oil to noncontiguous nations. The 1979 changes required that any export proposal from the president be confirmed by both houses of Congress.

One particularly odious stipulation, found in Section 7(d) of the EAA, which must be met before exports are permitted, states that "within three months following the initiation of such exports or exchanges," a situation must obtain with: "(I) acquisition costs to the refiners which purchase the imported crude oil being lower than the acquisition costs such refiners would have to pay for domestically produced oil in the absence of such an export or exchange[2] and (II) not less than 75 percent of such savings in costs being reflected in wholesale and retail prices of products refined from such imported crude oil." Given the improbability that domestic import-price differentials can be predicted with confidence and the lag time involved for acquisition price changes to take effect and reach consumers at the pump, such legislative legerdemain makes certifying exports difficult at best. Although proponents of export restrictions voice their disagreement, the restrictions embodied in the TAPS bill, the Energy Policy Conservation Act, the Naval Petroleum Reserves Production Act, and the EAA have become so tight that it is accurate to speak of a *ban* on exports to noncontiguous countries.

The primary reason for the increasing severity of export restrictions has been the continuing perception that the North Slope oil would be consumed in California and that exports would undermine national security. This assumption proved incorrect; demand for oil has not increased as much as anticipated by government and industry because oil prices increased from only $2.50 per barrel in 1972 to over $30.00 per barrel. In fact, one can show that energy security will *increase* with the lifting of the ban.

The Ban, Maritime Interests, and the Jones Act

Protection from supply disruptions was not, however the only driving force behind the increasingly stringent restrictions. Because the Merchant Marine Act of 1920—the Jones Act—mandated that waterborne commerce between U.S. ports be carried on U.S. built, owned, registered, and manned ships, the maritime industry has had a continuing interest in upholding the ban. As early as 1969, seamen's unions and domestic shipbuilders skillfully lobbied in support of the Trans-Alaska route. Although they opposed the Trans-Canada pipeline proposal, they nonetheless strategically paid lip service to the problems posed by possible supply disruptions, damaging any notion of allowing Alaskan exports. Since passage of the TAPS bill, the American Maritime Association, Maritime Trades Council, National Maritime Union, American Bulk Ship Owners Committee, and others connected with the maritime industry have been active in increasing the restrictions on ANS exports. Essentially, the maritime interests have ensured a demand for tankers and the elimination of foreign competi-

tion through their support of the TAPS bill and export restrictions.
It is no secret that, in the absence of a protected market, the U.S. maritime industry would be unable to compete against foreign tankers hauling Alaska crude, as the representative tanker rates listed in Table 5-1 clearly indicate. Instead, mari-

Table 5-1
Transportation Costs for
Crude Shipped From the Persian Gulf
and Alaska to Selected Markets
(Dollars per Barrel)

Origin	Destination	Costs Foreign Flag	U.S. Flag
Valdez, Alaska	West Coast	0.60ᶜ	1.40
Valdez, Alaska	Gulf Coast	1.20ᶜ	4.20ᵇ
Valdez, Alaska	Japan	0.50	0.90ᵃ
Persian Gulf	West Coast	1.60	—
Persian Gulf	Gulf Coast	1.60	—
Persian Gulf	Japan	0.75	—

Sources: Except where noted, figures are from *The Export of Alaska Crude Oil* (Cambridge, Mass.: Putnam, Hayes & Bartlett, 1983), Figure 1.
    a. M. Hoyler, "Statement on Alaskan Exports" (before the Budget and Audit Committee, Alaskan Legislature, April 23, 1983).
    b. State of Alaska.
    c. Estimates.

time interests receive an implicit subsidy* to support inefficient shipyards and provide generous salaries. For instance, the average second mate on a U.S.-flag vessel makes $60,550 for six months' work, a master $119,000—three times the salaries paid to seamen from other developed countries.[3]
The level of dependence of the maritime industry on the implicit subsidy embodied in the Jones Act, the TAPS bill, and the export ban is startling. Of the 11 million deadweight tons (DWT) of American-flag shipping capacity, about 10 million DWT (91 percent) are involved in the North Slope crude trade.

---

* This subsidy is supported by unwitting consumers who are not aware that they are providing it. An *explicit* subsidy, e.g., one from the federal government, would instead transfer costs to the taxpayer. As all subsidies involve social costs, the explicit subsidy is the more appropriate form because the costs are visible and can be accounted for, whereas with an implicit subsidy these costs are effectively hidden.

It is therefore not surprising that the extent of lobbying and the amount of political contributions by maritime interests is well out of proportion to their number. Indeed, what should ostensibly be strictly a debate on national security has instead become a classic example of a highly visible, vocal, and powerful interest group protecting benefits it receives at the expense of the politically invisible and unorganized—the American consumer and taxpayer.

Alaskan Oil Trade

West Coast Glut and Responses

Soon after completion of the TAPS in 1977, a surplus of heavy, high-sulfur crude developed on the West Coast. Historically, West Coast refiners depended on lighter crudes (such as Indonesian) to meet the large demand for motor fuel products. Refinery capacity equipped to process the "sour" ANS crude represented a small portion at the total. With the ban on exports in place, ANS producers were forced to expand the market for their crude in the United States, as their natural market (the Pacific Rim) was closed to them. This they accomplished in two ways: (1) refineries on the West Coast were retrofitted in order to handle the heavy ANS crude; and (2) because combined Alaska-California production exceeded West Coast refinery demand,[4] excess crude was shipped to refineries on the Gulf and Atlantic coasts, and in the Caribbean.

Of the three largest Alaskan North Slope producers, Standard Oil Company of Ohio (Sohio), Atlantic Richfield Company (ARCO), and Exxon[5] all ship crude to the eastern United States. ARCO, which produces about 0.34 mbd of North Slope crude and has a West Coast refinery capacity of 0.28 mbd, ships its excess to the Gulf in its own tankers. Exxon, too, meets its West Coast capacity first and subsequently ships the remainder to the Gulf (it owns 30 percent of the ships it assigns to the Gulf trade). Sohio, the largest ANS producer, provides crude to its eastern refineries through exchange agreements with companies that have West Coast refining capacity but little ANS production (notably Chevron). The remainder of Sohio's ANS crude is shipped to the Gulf in contracted tankers. For 1982, of the over 1.4 million barrels of ANS crude that left Valdez each day, more than 0.8 mbd found its way to the eastern United States.[6] Figure 5-1 illustrates the flow of Alaska crude for 1982, and the associated shipping costs.

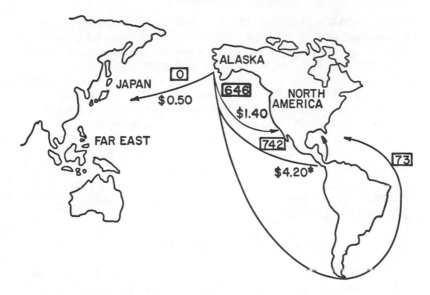

Figure 5-1. Flow of Alaskan Crude Oil in 1982 (in 1,000 barrels per day) and Associated Shipping Costs. The cost figure to the East Coast provided by Putnam, Hayes & Bartlett ($5.25) appears to be high. The $4.20 figure shown instead is the average transportation cost reported to the State of Alaska for royalty purposes in December 1982. Source: *The Export of Alaska Crude Oil* (Cambridge, Mass.: Putnam, Hayes & Bartlett, 1983), Figs. 1 and 10.

Alaskan producers, however, recognized that transporting crude by U.S. tankers was not the most efficient mode. There emerged in response a number of proposals involving more efficient, large-diameter pipelines to assist movement from the West Coast to the eastern United States. The most prominent among these was the Northern Tier Pipeline, a proposed trunk from Port Angeles, Washington, to Clearbrook, Minnesota. The Northern Tier Pipeline Company estimated that the construction of the 42-inch pipeline would cost $1.9 billion (1981 dollars). The original project was vetoed by Washington Governor John Spellman for environmental reasons, but a new proposal would carry the oil around, rather than across, Puget Sound. Sohio planned converting to oil transport the existing natural gas pipelines between Texas and California, which would involve reversing the flow. Others have suggested reversing the flow of existing oil pipelines in Alberta connected with the U.S. system (Trans-Mountain System).

Of the many proposals put forward, the Northville Pipeline is the only project that has been completed thus far (in 1982). Spanning Panama, the pipeline provides an efficient means of avoiding the bottleneck that results from the transfer of oil from large vessels to the smaller ones capable of using the canal. Of the 700,000 barrels of oil that pass through the pipeline each day, about 350,000 barrels are under contract for three years (150,000 barrels of this coming from Sohio).[7]

It is important to note here that all these proposals have one major objective: finding a substitute for transporting crude in expensive U.S. tankers.[8] Naturally, maritime interests looked on these proposals with undisguised horror, as they would stand to lose a good portion of their market, which is now protected by the Jones Act.

The Security Issue

The maritime industry aside, the principal objection to Alaskan exports stems from security concerns. In case of an embargo or oil cutoff, the argument goes, the United States must be guaranteed sufficient Alaskan oil to meet American needs. This argument was born in the period of the first Arab oil embargo. It is no longer valid, if it ever was. The export of Alaskan oil would in no way compromise U.S. security. Indeed, it could enhance it, for the following reasons.

The Inconsistency of Oil Export Restrictions

There are no prohibitions regarding the export of oil products, such as gasoline and fuel oil. It seems strange, therefore, that there should be a prohibition against exporting crude oil. There also are no restrictions on exporting oil during emergencies to U.S. partners in the International Energy Agency. In fact, the United States has an agreement concerning the sharing of oil supplies during emergencies (see Chapter 3). Under the IEA agreement, an oil-sharing system between its twenty-one member countries[9] would become activated in an emergency affecting 7 percent of expected supplies. (The system has never been implemented, but all IEA members are bound to honor it.) If the United States is prepared to export Alaskan oil during an emergency, why prohibit its export during nonemergency periods?

Ineffective Embargoes

There are two kinds of potential embargoes: (1) an embargo declared against the United States without a production cutback; and (2) an embargo coupled with a production cutback. The level of production is the critical factor; the simple declaration of an embargo would make little difference to the United States except for psychological pressure.

Any selective embargo against the United States cannot be effective—and has never been effective. Oil imported from overseas comes from a number of different sources. If any one of these, or even a combination of them, should put an embargo on oil to the United States, one or both of the following scenarios might develop: (1) the oil companies would sell the oil to another customer, say France, but oil destined to France from, say, Africa, would be diverted and shipped to the United States; (2) oil from the countries involved in the boycott would come into a transshipping terminal, such as Rotterdam, and then be shifted to the United States under a swap arrangement. The point is that oil is a fungible substance; its source matters little.

An embargo would be effective in one instance: if an adversary imposed a naval blockade against the United States along both coasts. Such action would be difficult for any power to mount. But if it were successful, it would also interfere with the traffic from Alaska to California and certainly to the East Coast. Short of military actions by opponents, however, the United States is immune to any simple embargo.

Production Cuts and the Market Price

But what if the embargo were coupled with production cutbacks in such a way that simple swapping procedures would not be possible? In that case the market could take over and adjust the available supply—now reduced—to the demand. Any production cutback thus would raise the world price, whether the production cutbacks were coupled with an embargo, or caused by an accident or by third parties, such as through a war or sabotage. Everyone would have to pay the higher price in these circumstances, not just the United States. Indeed, the Alaskan oil exported would also command the higher price (as would all domestic oil, in the absence of price controls).

There is often talk about countries "outbidding" one another during a supply crisis, but in a free market this would not be the case. As the price went up, those persons (not countries) wishing to buy the oil would have to pay the higher price, and oil use by others would fall.This redistribution of oil would be entirely automatic, in response to normal market forces, not to government allocation efforts.

Some time could elapse before the new supply relationships were established following an oil cutoff. During this time, there could be dislocations and shortages just as there are shortages in retail outlets when the inventory is low. To soften such short-term disruption, the United States and other industrialized countries have provided for strategic reserves of petroleum. The release of oil from the U.S. stockpile (or from the stockpiles of other industrialized countries) would limit any price increase due to sudden interruptions in production levels. If the supply interruption persisted, the oil market would reach equilibrium at a higher price; if it were only temporary, there would be no long-term change in price—although, of course, stockpiles would be partly depleted.

Two Case Studies—1973 and 1979

What happened during past embargoes? In October 1973, producers on the Arabian Peninsula declared an embargo against the United States and the Netherlands, and later cut back their production. The declaration itself did little but scare people. The cutback in production, however, increased the price of oil, which eventually soared from about $3 to $12 per barrel.

There was considerable market disruption in the United States in the spring of 1974, characterized chiefly by long lines at gasoline stations. These lines were caused by the exaggerated reaction of the federal government, which sought to allocate gasoline and other oil products to achieve a "fair distribution." Yet federal bureaucrats had no more success than any other planners in trying to simulate the workings of the market, and misallocation inevitably followed. "Shortages" occurred widely in 1974 because well-meaning government interference with the market process was compounded by price controls on domestic oil. Without free movement of prices, there was no reason for demand to fall to the new, reduced level of supply—other than by the forced decline in consumption because of waiting in line.[10] But nothing was learned. In 1979, the Department of Energy again put into effect an allocation system—with predictable results: long lines at gasoline stations.

Further proof that embargoes do not work is found in the events of November 1979. When the U.S. embassy in Teheran was occupied, President Carter declared that the United States would no longer buy Iranian oil. The action was, in effect, a self-imposed embargo—a boycott. Of course, nothing happened. The Iranian oil went elsewhere, and the United States bought oil from other sources. There was no psychological impact either—perhaps because the word "embargo" was never used.

One of the first acts of the Reagan administration was to remove price controls on oil. Congress still believed that an allocation system had to be instituted during emergencies and tried to force the White House to agree to such a system. In vetoing the bill, President Reagan explained why the market allocates more successfully than any bureaucrat or combination of bureaucrats. The U.S. Senate upheld the presidential veto.

It should seem clear that embargoes and production cutbacks do not work when oil prices are decontrolled and a large strategic stockpile keeps prices from moving too high. An embargo threat is little more than a psychological tool that is effective only if the victim thinks it might be harmful.[11]

The Consumer Issue

Ensuring energy security is only half the battle. The other statutory condition for permitting Alaskan exports is that they must confer almost immediate (within three months) benefits to the U.S. consumer. As Marshall Hoyler notes: "By setting this unrealistic condition, Section 7(d) [of the EAA] effectively prohibits exports and denies their benefits to consumers. More importantly, it prevents long-run benefits for consumers while *appearing* to provide consumer protection."[3]

Permitting exports would benefit consumers:

1. By wasting less resources, thus lowering the costs to the economy. The majority of these savings would go to federal and state governments.
2. By encouraging increased production, both in Alaska and in California, which would put downward pressure on the world price.

Exports Will Reduce Oil Prices in the Long Run

As we have seen, the effective ban on exports has led to an established market of Alaskan North Slope crude on the West and Gulf coasts. Presently, about 44 percent of North Slope production is used on the West Coast; the remainder is shipped to the eastern seaboard on U.S.-flag tankers.[12] The export ban, the Jones Act, and the absence of a west-to-east oil pipeline in the United States means there is no other marketing option—except not selling oil at all. Eliminating the export ban would open up other, more economic markets for the surplus now being shipped east.

Export would be more efficient economically simply because the transportation costs of North Slope crude to the Far East are lower, even if the oil is shipped on U.S.-flag vessels. Figure 5-1 provides relevant tanker rates, and these are listed in Table 5-1. As the figures reveal, Jones Act requirements set U.S. tanker rates well above world tanker rates.

For an estimate of the scale of the export potential, market prices may be approximated using the price of Persian Gulf (i.e., Saudi Arabian) oil, plus the costs of its transportation to each market. The wellhead price "netback" that producers receive for their crude oil is the market price (say, in Houston) minus transportation costs. These costs vary with shipping distances, tanker size, and other factors.

The different rates mean that North Slope producers receive different wellhead prices for their oil, depending on its destination. Using market prices established by Persian Gulf oil, North Slope producers would net back $2.80 per barrel more for their West Coast shipments than for their Gulf Coast shipments. They could use such a price advantage to drive down the West Coast price *at the refinery* and expand their market share by discounting. In fact, there is already increasing, albeit incomplete, evidence of some West Coast refinery "discounting."[13]

If the ban on exports were lifted, Alaskan North Slope producers could increase the wellhead prices of their currently Gulf Coast–bound shipments (before state and federal taxes) by $2.85 (i.e., $4.20* − 1.60 + 0.75 − 0.50) by changing the destination to Japan and taking advantage of the lower shipping costs. At the same time, a change in the destination of surplus North Slope crude would reduce the glut in the West Coast market, and West Coast refinery prices could rise by as much as $2.00 per barrel. There would thus also be an increase in the netback received by California producers who were previously forced to lower their refinery prices to match North Slope competition.

---

* Putnam, Hayes & Bartlett, Inc., report an Alaska–Gulf Coast transport cost of $5.25, which would raise the netback to $3.89. We believe this is too high, but if accurate it would further support our contentions.

Another factor to be considered is the prospect of reduced shipping costs associated with the Alaska–West Coast route. With the ban lifted, the reduction in demand for U.S. tankers because of the reduced Alaska to West Coast trade and the absence of Jones Act requirements on exported oil would mean more competition along the American coastline—lowering transport costs and further raising the North Slope netback. Although larger transportation cost decreases have been forecast, we will assume a decline of $0.42 per barrel, from $1.40 to $0.98. A summary of the total possible increases in wellhead prices resulting from these factors is given in Table 5-2.

Alaskan North Slope oil shipments to the eastern United States amounted to 0.815 mbd in 1982. Gross wellhead revenue increases were calculated using this figure minus the 0.35 mbd of oil contracted for passage through the Northville Pipeline for the next three years. A 1.3 mbd figure for California production was used. The estimated wellhead increases *before taxes* would be between $582 million per year and $1,048 million per year for North Slope production for the first three years after lifting the ban, and up to $949 million per year for California production. With the addition of the 0.35 mbd now passing through Panama, the North Slope before-tax figures could increase to between $946 million and $1,412 million per year. Analysis suggests that *after taxes,* North Slope producers would be left with only about 8.3 percent of the increases, or between $48 million and $87 million per year for the first three years, and $78 million and $117 million a year thereafter, the other 91.7 percent of the increases going to the federal and state governments.

If exports were permitted, oil companies would certainly increase production and put more oil onto the world market to take advantage of the increased netbacks. A conservative estimate is that the additional output could amount to 0.5 mbd (in addition to the 1.6 mbd now being supplied through the pipeline), with more optimistic estimates exceeding 1 mbd. (If production rose above the 2 mbd TAPS capacity, the pipeline could easily be upgraded to handle the increased output at relatively little cost.) The increased oil in the world market would reduce the amount that the OPEC cartel could sell and thus put downward pressure on the world price.

If U.S. oil exports to Japan were 1 mbd, for instance, Japan could reduce its imports from other, less secure sources, and the United States could replace 1 mbd of Middle East imports with more Mexican oil.[14] Thus, not only would U.S. security be

Table 5-2
Estimated Increases of Crude Oil Wellhead Prices
After the Lifting of the Oil Export Ban
(Dollars per Barrel)

Market Conditions (prevailing under the export ban)	Originating Port	With Export Ban		Without Export Ban (or Discounting)		Wellhead Increases[a]
		Destination Port	Wellhead Price	Destination Port	Wellhead Price	
No Discounting[b]	California	West Coast	PG[c] + 1.50	West Coast	PG + 1.50	0.00
	Alaska	West Coast	PG + 0.20	West Coast	PG + 0.62	0.42
	Alaska	Gulf Coast	PG − 2.60	Japan	PG + 0.25	2.85
Full Discounting[b]	California	West Coast	PG − .50	West Coast	PG + 1.50	2.00
	Alaska	West Coast	PG − 1.80	West Coast	PG + 0.62	2.42
	Alaska	Gulf Coast	PG − 2.60	Japan	PG + 0.25	2.85

a. With export ban removed
b. By Alaskan producers in California
c. PG = Persian Gulf price

enhanced, and not only would financially strapped Mexico gain through lower transportation costs, but the world price of oil could conceivably be lowered by about 5 percent. Since OPEC is currently earning about $200 billion a year in revenues, this would reduce the oil bill of the importing countries, including the United States, by approximately $10 billion a year.

Exports Will Increase Federal and State of Alaska Revenues

The biggest gainers from removing the ban are not, however, the oil producers, but rather the federal government and the Alaska and California state governments, which together would receive over 90 percent of the wellhead increases. Alaskan royalty oil and severance and income taxes would take over 32 percent of North Slope increases. The federal government would take 7 percent in corporate income taxes and 52 percent in windfall profits taxes for most current production. California oil producers would also pay more in taxes as a result of wellhead increases. The distribution of potential savings resulting from lifting the ban are given in Table 5-3.

Our calculations show that, if the ban were lifted now, the federal deficit in the first three years could be lowered by as much as $603 million per year, and $1.044 billion per year thereafter. Admittedly these increases in revenues are comparatively small when the size of the present deficit is considered, but because the government would have to borrow that much less, more capital would be available to consumers and entrepreneurs, and thousands of new jobs could be created.

Opponents of the ban claim that lifting the ban would cause a loss in federal revenue through defaults on federally guaranteed shipbuilding loans and taxes paid by shipowners and seamen. The Maritime Administration puts these one-time Title XI loan losses at around $594 million. Our figures show that this loss could be negated by the gains made in the first year alone. Of the ships put out of business by lifting the ban, those needed for defense could, according to the Maritime Administration, be purchased for about $200 million, well within the projected increase in federal revenues.[3] And even if the $1 billion to $1.7 billion figure that export opponents cite were valid, this could easily be made up within the first few years of lifting the ban—and this does not even consider increased revenue from Outer Continental Shelf lease sale bids, estimated by the Department of Interior at over $80 million in fiscal year 1984 alone.

Table 5-3

Estimates of Potential Gross Revenue Increases from Alaska
and California Oil Production and Their Distribution,
With Export Ban Removed
(Millions of Dollars Per Year)

Market Conditions Prevailing Under the Export Ban	Gross Wellhead Revenue		To Producers		To State Revenues		To Federal Revenues	
	First Three Years	Thereafter	First Three Years	Thereafter	First Three Years	Thereafter	First Three Years	Thereafter
No Discounting[a]								
From Alaskan Oil	582	946	48	78	190	308	344	559
From California Oil	199	199	155	155	—	—	44	44
							388	603
Full Discounting[a]								
From Alaskan Oil	1,048	1,412	87	117	342	451	619	834
From California Oil	949	949	739	739	—	—	210	210
	1,997	2,361					829	1,044

a. By Alaskan producers in California

Hoyler's analysis suggests that the total increase in federal revenues over the life of the Alaskan oil fields would be between $5 billion and $8 billion. In addition to increased federal and state revenues, our balance of trade would improve by about $15 billion per year, and the economy of Alaska would improve perceptibly.[15]

Alaska Gas Transportation Options

Through the middle 1970s, the development of Alaska's hydrocarbon resources focused primarily on the state's enormous oil reserves. For nearly thirty years, a small amount of natural gas has been produced in the southern portion of the state for export to Japan in the form of LNG; but the huge gas reserves of the Alaskan North Slope remain untapped. The opening of the TAPS gave the gas reserves associated with that oil a new importance. The gas may be reinjected into the formation from which it is drawn, but such reinjection provides only a temporary solution. In fact, over time this practice results in a reduction in oil field pressure and thus in a reduction in the ultimate amount of recovered oil. And because reinjection consumes up to one-third of the gas, the cost in wasted energy increases. Still, in the absence of a means of transporting the gas, the only other option is to burn it off, or "flare" it.

The Alaskan Natural Gas Transportation Act of 1976 also contains export limitations. Section 12 states that "the President must make and publish an express finding that such exports will not diminish the total quantity or quality, nor increase the total price of energy available to the United States."

The situation with respect to natural gas is somewhat similar to the oil situation. The Prudhoe Bay field contains the largest discovered gas reserves on the North American continent; it represents 10 percent of proven reserves and more than a year's supply for U.S. consumers.[16] Several companies studied ways to move the natural gas to markets and proposals were filed with the Federal Power Commission (now the Federal Energy Regulatory Commission) in 1974. Of the various proposals, the one finally selected, the Alaska Natural Gas Transportation System, would move gas by pipeline from the North Slope to the Midwest through Canada. However, the very high cost of the pipeline, now estimated to be $40.9 billion, has made the proposal impractical. With higher wellhead prices for natural gas, and with a limited deregulation approaching in 1985, a great deal of gas has been developed in the lower forty-eight states. The various provisions of the act can do nothing to make Alaskan gas competitive in price with gas from the lower forty-eight.

The Alaska Natural Gas Transportation System (ANGTS) faces problems with financing, cost overruns, and doubts over the marketability of the relatively expensive Alaskan gas in the lower forty-eight states, which are supplied with less expensive conventional gas. As a result, Alaskans have begun to reexamine the alternatives available to them to determine if some other approach to the problem of marketing their gas might be more sensible. The principal options currently under consideration include:[18]

1. To continue to pursue financing for the ANGTS project, in hopes that the use of innovative rate structuring and the decline of interest expense might make Alaskan gas more competitive at some future date.
2. To select an alternative means of transporting North Slope gas in hopes that it will prove less expensive, again making the gas more competitive in the lower forty-eight states.
3. To determine whether Alaskan producers should abandon the notion of marketing the gas in the lower forty-eight and instead focus on the export market.
4. To examine ways of using the gas within the state to establish some sort of manufacturing base.

One of the proposals submitted to the Federal Power Commission was by the El Paso Company. The company proposed to transport natural gas from Prudhoe Bay through approximately 800 miles of 42-inch pipeline, to a gas liquefaction plant and terminal located on Prince William Sound at Point Gravina, Alaska. There the gas would be converted to LNG and shipped via cryogenic tankers to Point Conception near Santa Barbara, California.[17] However, the LNG could be shipped just as easily to Japan, Korea, Taiwan, and other users in the Pacific Ocean basin—but more cheaply from the Kenai Peninsula than from Point Gravina. The amount would be on the order of 2.8 billion cubic feet per day or approximately 1.0 TCF per year, worth approximately $5 billion.

Determining the best solution for the North Slope gas is doubly difficult because the oil and gas market, both in the

United States and internationally, is undergoing a period of rapid and dramatic change. As the patterns of this change become clearer, it is evident that the traditional view of the gas market is no longer valid. The policymakers currently examining Alaska's options must thoroughly understand the evolution that is taking place in order to make sound economic decisions.

The Changing Natural Gas Market

"Shortage" into Surplus

It is easy to forget that, as recently as five years ago, the conventional wisdom held that the United States would soon run out of natural gas. Throughout the first half of the 1970s, interruptions in natural gas deliveries on the interstate market increased, and gas reserves committed to that market diminished. By the winter of 1976–77, the situation had reached crisis proportions, as regions of the Northeast and Midwest faced massive gas shortages that threatened economic chaos (see Chapter 2). Policymakers were quick to point to these shortages as evidence that the exhaustion of America's natural gas reserves was imminent. This view was embraced with particular enthusiasm by officials of the Carter administration, many of whom were convinced that all of the world's resources were on the verge of exhaustion.[19]

Against this background, Alaska's enormous North Slope gas reserves were very tempting to policymakers who believed the United States faced the prospect of running out of oil and gas. The high cost of utilizing these reserves seemed of little consequence.

As early as 1979, however, evidence began to appear that the dire assessment of gas reserves, widely taken as axiomatic, was grossly overstated. The first sign was the appearance of a so-called gas "bubble"—a large volume of gas that "found" its way into the market. According to the prevailing view of reserves, it should not have appeared. Analysts tried to explain it as merely a temporary "market anomaly" that would soon be absorbed, leaving the United States once again with the shortage. The bubble, however, did not disappear; the shortage did. In fact, in 1981, for the first time in more than a decade, the United States added more new natural gas to its reserve base than it used. In 1982, instead of a shortage, there was a surplus of natural gas estimated at 15 percent. Currently the surplus is so great that gas companies, which once could not serve all of their existing customers, are now seeking new ones. But more important, the unexpected availability of natural gas has taken place at prices far below those needed to make North Slope gas economic. Should natural gas prices be decontrolled, even greater volumes of gas priced below an economic level for Alaskan production under current circumstances are expected to find their way into the market.

Growing Competition

Competition from natural gas produced in the lower forty-eight states is not the only factor limiting the marketability of Alaskan gas in the United States. The importation of large volumes of natural gas from Canada and Mexico will also provide stiff competition. Both Mexico and Canada are experiencing great economic pressure to move their gas into the U.S. market. Until recently, both countries priced gas at levels that limited its attractiveness to U.S. consumers. But these pricing policies—which seemed strangely similar—were simply the product of the seller's market for energy existing in the middle to late 1970s. With the crumbling of OPEC, the steady decline of world oil prices, and energy conservation, both Canada and Mexico have had to rethink their policies. As a result, both nations are now willing to make price concessions; for instance, Canadian gas sells in the United States at only 65 percent of its authorized price. Even given the price reduction, however, the volume of gas taken is down from just a few years ago.

For Mexico, whose gas reserves far outstrip those of either the United States or Canada, increased sales of both oil and natural gas are critically important. The country's near financial collapse was only a warning signal. The need to feed and find employment for its burgeoning population makes it imperative for Mexico to expand sales of its oil and gas. The United States is its most logical market, and so competition from Mexico seems likely to be an even greater barrier to the marketing of Alaskan gas in the lower forty-eight states than is competition from domestic or Canadian gas producers.

Competition from conventional sources of natural gas, whether domestic or foreign, is not the only factor affecting the marketability of Alaskan gas to consumers in the lower forty-eight states. Of equal importance will be competition from other fuels, and especially from residual fuel oil, or "resid." Since the largest share of natural gas is consumed in the industrial boiler market, industrial consumers effectively determine the price at which gas is sold. Part of their ability to influence gas prices stems from the fact that most industrial boilers were modified to accommodate a variety of fuels during the 1970s when natural gas supplies were subject to federal regulations. Many of these boilers can burn either natural gas or resid. As a result, the price of resid effectively caps the price at which natural gas can be sold. At present, resid sells for roughly the equivalent of gas, priced at between $4.00 and $4.50 per thousand cubic feet (MCF). But residual fuel oil prices are expected to decline further in the future because of oversupply.

Given the intense competition and the probable future price trends in the natural gas market of the lower forty-eight states, it seems unlikely that North Slope natural gas will be competitive in the near future. Therefore, the current price structure must be modified, or an alternative market sought, if Alaska's gas resources are to be utilized and further developed.

Reshaping the ANGTS Project

One of the reasons Alaskan gas will be so expensive in the first few years after ANGTS comes into service is that loans made for its construction must be repaid. If the repayment schedule can be renegotiated to stretch the payments over a longer period, the selling price of the gas might be reduced. The effectiveness of this approach will hinge on two major factors.

The first factor is the interest rate. Since most plans to restructure the pipeline's financing call for the payment of interest, the interest rate and capital repayment schedule (even if deferred) will have to be such that the final price of Alaska gas is competitive.

The second factor, of course, is the prevailing price in the lower forty-eight gas market. Just what this might be in the future is hard to say, but one thing is certain: if Alaskan gas expects to compete, its current projected cost of $10 per MCF (in 1982 dollars, equivalent to $60 per barrel of oil) must be reduced. Recent attempts to market deep gas at a similar price have failed. In fact, several pipeline companies recently informed a group of deep gas producers that the lines would pay no more than $5.00 to $6.00 per MCF for deep gas. This seems to be compelling evidence that Alaskan gas will have to sell in the $5.00 range if it is to compete with alternative sources of gas.

Possible Alternative Routes

One possible solution to the North Slope gas dilemma would be an alternate means of transportation. The best alternative to the ANGTS appears to be the so-called All-Alaska Pipeline System (AAPS), proposed several years ago by the El Paso Company, a system that would be built parallel to the existing oil pipeline. At the time the proposal was first put forward, estimates of its cost included funds to build a California LNG terminal and purchase eleven LNG tankers. Adjusted to current 1982 dollars, the original cost estimate for the AAPS was $11.1 billion, which compares favorably with the $24.2 billion capital cost estimate for the Alaskan segment of the ANGTS pipeline.[20] More recently, U.S. Government Accounting Office consultants placed the capital cost of the AAPS at $26.8 billion.[21]

At the request of the governor of Alaska, a committee was appointed to examine the feasibility of constructing a gas pipeline to South Alaska. In January 1983, the Governor's Economic Committee on North Slope Natural Gas reported its findings,[22] and recommended the construction of an 820-mile, 36-inch-diameter gas pipeline from Prudhoe Bay to the Kenai Peninsula,[23] where the gas would be exported in LNG tankers. Total capital costs for the proposed Trans-Alaska Gas System (TAGS) are estimated to be $25.5 billion (which includes pipeline, conditioning plant, and liquefaction plant), $15.4 billion less than the anticipated capital cost of the proposed ANGTS. The TAGS estimate does not, however, consider the cost of compression facilities on the North Slope (estimated at $4 billion) or LNG tankers (overestimated at $3.3 billion).[24] Actually, lower interest rates and a shorter construction time frame would lower the total cost appreciably. But even given these estimates, TAGS is still a bargain compared with the alternative, ANGTS. Besides its lower cost, TAGS has another, possibly even more important, advantage over ANGTS: it opens the prospect of exporting North Slope gas. If the problems encountered in financing the project can be solved expeditiously,[25] and if North Slope gas is marketed aggressively, Alaska could become an important exporter of LNG.

Alaskan Natural Gas and the Export Market

A worldwide trend toward greater use of natural gas has been well established. The most logical export markets for

Alaskan gas are the nations of the Pacific Rim, especially Japan. The Japanese already import small amounts of LNG from Alaska. Significantly, Japan is moving aggressively to make use of LNG, and recently contracted with Indonesia for major purchases of the fuel. As a result of this policy, Japan has the necessary LNG terminals in place, and already owns LNG tankers. Hence a pipeline, processing facilities, and liquefaction plant would be the only U.S. infrastructure necessary to market LNG to Japan. Japan would probably even be willing to help finance the project. However, the decision would have to be made quickly; otherwise Japan will move to find supplies elsewhere.

The pessimism of the GAO report aside, the Japanese would probably purchase as much Alaskan gas as they could for the same reason they would purchase North Slope crude: increased energy security. Present, scheduled, and potential sources of LNG to Japan include Indonesia, Malaysia, Thailand, and the Soviet Union, so it is easy to see why the Japanese would welcome U.S. gas; they could even use the increased supply to back out OPEC oil. With the implementation of the Alaskan gas pipeline, we would also expect that South Korea and Taiwan would initiate development of LNG terminals and regasification plants, further expanding the market.

A number of economic advantages, beyond the obvious revenues, would be associated with the export of Alaskan gas to Japan. First, such trade would go a long way toward reducing the current U.S.–Japan trade imbalance. Second, it would reduce Japan's dependence on fuel imported from the politically unstable Persian Gulf, and thereby greatly enhance the world's energy security. Most important, by directly reducing the world's oil consumption, Alaskan gas exports could also help to keep world oil prices down.

It would seem, therefore, that exporting Alaskan gas to foreign markets would be advantageous—for Alaska and for the world in general. These advantages would not materialize if the gas were marketed only within the United States. Alaskan gas sold in the lower forty-eight would not displace foreign oil; domestic usage would have no effect on the U.S. balance of payments; and building a pipeline to transport gas domestically would be a far more expensive proposition than building a pipeline to transport gas for foreign markets. Exporting Alaskan gas would therefore appear to be the optimum solution to the North Slope gas dilemma, from a national standpoint.

Conclusion: Alaskan Oil And Gas—Don't Fence Me In

Removing the ban on exports of Alaskan oil and gas will (1) enhance America's (and Japan's) energy security, (2) increase revenues to the federal government and to Alaska, and, in the longer run, (3) encourage increased oil production and thereby lower world oil prices. Originally a response to the fear of oil embargoes, it is now evident that the export ban does not protect us from price hikes associated with supply shortfalls. Price disruptions will be felt whether Alaskan oil is exported or not. As oil is a fungible commodity that cannot be effectively embargoed against the United States, any supply disruption will be felt worldwide and will cause an increase in world prices. Alaskan oil would thus sell at the higher price, whether in Houston or in Tokyo. On the contrary, increased North Slope production as a result of exports and free trade will reduce the real need for unreliable OPEC oil. Together with the Strategic Petroleum Reserve, an export policy will soften the price effects of any future supply disruption.

Blocking the export of Alaskan oil imposes great costs, ultimately borne by the American taxpayer. Shipping North Slope crude to the eastern United States in expensive U.S.-flag tankers leads to great economic waste, as would the construction of a pipeline to the lower forty-eight states. With the removal of the ban, increased production, as a response to increased netbacks and an expanded market, would benefit the consumer by: (1) putting downward pressure on the world price of oil; (2) increasing federal revenues and thus reducing deficits; (3) increasing the revenues to Alaska and California; (4) reducing the balance-of-trade deficit with Japan and improving U.S.–Japan relations; and (5) providing a boost to Alaska's economy.

Granted, under free trade, the maritime industry would suffer, and seamen and shipyard jobs would be lost. But because the overall effect of removing the ban would be a more efficient economy, in aggregate there would be more jobs created than lost.

This analysis thus concludes that the total removal of the export ban is in the best interests of the United States. However, because we recognize that any proposal for full removal of the ban would come under intense political pressure from maritime interests, we note three options that would ease the transition problems to the maritime industry. They are:

1. Permitting only the export of incremental North Slope production.
2. Mandating that Alaskan oil exports be carried in U.S.-flag tankers.
3. Encouraging the export of Alaskan LNG in U.S. tankers.

Although each of these alternatives would confer obvious advantages over the status quo, they fall far short of the benefits gained by a complete removal of the present export ban.

6

TAXATION OF ENERGY PRODUCERS AND CONSUMERS

James W. Wetzler

Summary

Taxes and subsidies (in the form of tax exemptions and tax credits) influence energy consumption and energy production in the United States strongly but not always consistently. For example, tax policy may no longer encourage domestic oil production relative to many other kinds of investments, but coal producers are granted considerable benefits.

The issue is: should tax policies discourage consumption and encourage production, especially of oil. If so, how can such policies be made consistent, equitable and reliable?

1. Consistent—so that different policies are not at cross purposes with each other;
2. Equitable—for different income groups and different regions of the nation;
3. Reliable—so that expectations of frequent change will not negate the purpose of the policy.

Introduction

Much of the nation's energy policy resides in an unlikely place—the Internal Revenue Code. To some extent, the tax law reflects conscious congressional decisions to alter energy production and consumption, but much of its impact appears not to have been intended. This chapter attempts an overall assessment of the impact of the tax system on energy policy and analyzes some suggestions to rationalize that impact.

The chapter addresses the effect of the tax system, first, on energy consumption, then, on energy production. The focus is limited to the United States, although there also has been considerable controversy about the tax treatment of U.S. corporations that produce energy abroad. The chapter describes existing energy-related tax laws, summarizes the principal proposals for change, and analyzes the major issues involved in choosing among those proposals.

Taxes On Energy Consumers

Present Tax Laws

The present tax law, as of mid-1983, contains two principal types of provisions that affect energy consumers. First, there are a variety of excise taxes on specific uses of energy. One of these, the "gas-guzzler tax," was enacted with the intention of influencing energy use. The others were enacted as revenue sources for various trust funds not directly connected with energy policy, although energy policy considerations have in fluenced their structure. Second, the income tax contains several tax credits specifically intended to promote conservation of oil and natural gas. These "tax expenditures" are much larger than direct spending by the federal government on energy conservation.

Taxes On Energy Consumption

By far the most significant tax on energy consumption is the nine-cent-per-gallon federal tax on gasoline and other motor fuels, the revenues from which flow mainly into the Highway Trust Fund.[1] (The tax rate was four cents until April 1, 1983.) State taxes on motor fuels average an additional ten cents per gallon. Other energy-related user fees are the taxes on fuel used for noncommercial aviation and for commercial cargo transportation in certain inland waterways, which provide important conservation incentives for the industries involved but have little impact on overall fuel consumption.[2]

Motor-Fuels Taxes. There is little controversy over the proposition that fuel consumption is an adequate measure of the extent to which people use the highways and, therefore, a fair way to finance highway-related spending at all levels of government. In accordance with the theory that the motor-fuels taxes are really fees for use of the highways, the present federal taxes contain exemptions for farming and for business-related off-highway use of motor fuels. These exemptions would probably be questioned by anyone who wanted to analyze the motor-fuels taxes as instruments to discourage energy consumption, instead of as highway user fees.

Congress has provided a number of other exemptions from the motor-fuels taxes with the express intention of affecting energy consumption. These include complete exemptions for buses and for fuels that are 85 percent or more alcohol. There are also special tax rates of four cents per gallon for gasohol and five cents per gallon for taxicabs.

The special tax rate for gasohol applies to alcohol-gasoline mixtures that are 10 percent or more alcohol. It amounts to a subsidy of fifty cents per gallon of alcohol, since one gallon of alcohol exempts ten gallons of gasohol from a tax of five cents per gallon. This sizable subsidy to alcohol consumption (as motor fuel), an estimated $83 million in fiscal year 1984, is available without regard to the amount of oil used to fuel the distillery, despite evidence that there may be little net oil saving when alcohol fuel is produced in an oil-powered distillery.[3] The excise tax exemption, however, is limited to alcohol produced from feedstocks other than oil, natural gas, or coal. A separate subsidy for alcohol fuel derived from coal was enacted in 1982 through an exemption for "near-neat" alcohol fuels; that is, those that are 85 percent or more alcohol.

Many states have also attempted to subsidize gasohol through exemptions from state gasoline taxes, but these have had the effect of distorting gasohol use by causing the available gasohol to migrate to those states with the largest tax exemptions. Given the high level of the federal subsidy, it is probably appropriate for states to reduce or eliminate their excise tax exemptions.

The exemption for buses was originally designed (in 1978) as a subsidy to this form of public transportation in light of its energy-saving features. Even though this exemption can be justified on the grounds that it equalizes the treatment of government-owned and private transit systems, since government use of motor fuel is exempt under a separate provision of the law, it is useful to evaluate the exemption under its original rationale as energy policy because doing so illustrates some of the problems with using subsidies to encourage conservation.

Exempting buses encourages conservation to the extent that it shifts people from autos to buses with lower fuel use per passenger-mile; however, it also discourages conservation to the extent that it provides an incentive to buy less fuel-efficient buses and encourages trips by lightly traveled buses whose fuel use per passenger-mile is high. Which of these effects predominates and whether the present exemption actually reduces energy use is unclear as a theoretical matter. This sort of indeterminacy arises again and again in evaluation of energy-related tax incentives.

Gas-guzzler tax. The gas-guzzler tax is being phased in through 1986. For that model year, the tax will range between $500 for 22-mile-per-gallon cars to a maximum of $3,850 on cars whose fuel economy is below 12.5 miles per gallon. Currently, the tax applies only to a handful of luxury cars and raises little money, but a beefed-up tax could have an impact in the future, particularly if gasoline prices decline further.

Unfortunately, here, too, it is impossible to prove a priori that a larger tax would significantly reduce fuel consumption. Such a tax would give consumers who plan to buy a car an incentive to buy a more fuel-efficient model, but other consumers would respond perversely to a higher tax, either by keeping old gas guzzlers on the road for a longer period of time or by buying a less fuel-efficient pickup truck (since trucks with gross vehicle weight of 6,000 pounds or more are exempt). Also, any improvement in fuel economy resulting from the gas-guzzler tax may simply duplicate what would have occurred anyway through efforts by the auto manufacturers to meet federally mandated fuel economy standards. As with the fuels tax exemptions and special rates, the gas-guzzler tax appears at first glance to be a plausible way to encourage conservation, but on analysis the beneficial effects of such a tax become much more problematical.

Income Tax Incentives

Since 1977, Congress has enacted an array of new income tax incentives designed to encourage energy conservation and conversion from oil and gas to alternative energy sources. In fiscal year 1984, these energy conservation "tax expenditures" will reduce revenues by $800 million. In contrast, the Reagan administration has requested budget authority of $74 million for fiscal year 1984 for spending on energy conservation.[4]

Residential credit. The residential energy credit consists of two parts: a 15 percent credit for the first $2,000 of expenditures on home insulation and other specified energy-conserving items and a credit of 40 percent of the first $10,000 of expenditures for solar energy and other renewable energy sources. The insulation credit will reduce 1984 revenues by $305 million, and the solar credit by $450 million. The credit was quite controversial when enacted in 1978, but since then the main issue has been whether to expand the original list of items eligible for the credit.

One of the problems with encouraging energy conservation through subsidies is that it puts energy-conserving products that have not made it onto the list of eligible items at a competitive disadvantage and thereby distorts consumer choices. The principal omission in the solar credit is the lack of any incentive for passive solar design techniques, which use the design of the building itself to reduce energy use. Many people believe that passive solar techniques offer greater potential for energy conservation than does active solar equipment such as solar reflectors and related equipment, for which the 40 percent credit is provided; but so far the inability to draft a workable definition of passive solar design has prevented the enactment of a tax incentive for it.[5] The question of whether to include wood-burning stoves and heat pumps has been a principal area of controversy for the insulation credit; concern over the environmental impact and safety of wood stoves and an ongoing debate about the extent to which heat pumps really conserve energy have so far prevented the inclusion of these products.[6]

Business credits. An array of extra investment tax credits for business energy conservation expired at the end of 1982. The energy credits that remain in the law until their scheduled expiration after 1985 include a 15 percent credit for investments in solar, wind, and geothermal energy, a 10 percent credit for equipment used to burn fuel from biomass, and a 10 percent credit for intercity buses. These credits are added on top of the regular 10 percent investment tax credit and ordinary depreciation deductions. The subsidy provided by the energy credits was reduced by about one-fifth in 1982 by the requirement that the basis of property on which depreciation deductions are computed be reduced by one-half of investment tax credits, including the energy credit.[7] The business energy conservation credits involve an estimated revenue loss of only about $40 million per year.

In the fiscal year 1983 budget, the Reagan administration proposed the repeal of all business energy tax incentives on the grounds that they interfered with free-market allocation of resources and were unnecessary as a result of the deregulation of oil prices. This was the only one of the administration's "revenue enhancement" proposals not acted on by Congress in 1982, and it was dropped without comment from the 1984 budget.

Evaluation of Present Law

The tax law contains an array of incentives to reduce oil and natural gas consumption. Most important, the motor fuels taxes, state and federal, raise the price of motor fuel by about 20 percent. Although not very much by European standards, this is still enough to have a significant impact on consumption in the intermediate or long run.

The energy conservation tax incentives—both the credits and the various excise tax exemptions—are equivalent to a subsidy program of well over a billion dollars per year to various investments that are believed to conserve energy. Two questions arise: (1) Should this large a budgetary commitment be made to subsidizing particular ways of conserving energy? (2) Should the subsidy be moved from the tax to the outlay side of the budget?

Since enactment of the energy tax incentives, oil prices have been decontrolled, natural gas prices have risen close to market levels, the worldwide oil shortage has ended, and Americans have made substantial investments in ways to conserve oil and gas. These considerations all point toward reducing subsidies for energy conservation. The case for subsidies is strongest when they encourage experimentation with new technologies, like solar energy, because such experimentation yields information of value to the nation as a whole; and it is weakest when there is doubt that the subsidies actually work in the right direction or when the subsidies introduce distortions by favoring certain ways to conserve energy over others. Furthermore, the subsidy approach forgoes a potentially important form of energy conservation—shifting consumer demands away from energy-intensive products and toward other goods and services. Indeed, to the extent that subsidizing energy conservation by energy-intensive industries lowers the price of products produced by those industries, the energy-saving impact of the subsidy will be diluted and the subsidies could act perversely to increase energy consumption. Similarly, subsidies for energy-conserving investments by homeowners can backfire by encouraging such energy-using activities as turning up thermostats or adding new rooms to the house.

The standard criticisms of designing subsidies as tax incentives apply to the energy conservation tax incentives. They are entitlements not subject to the annual appropriations process (as most direct outlay programs would be),[8] are not available to taxpayers without tax liability, and contribute to public perception that the tax system is complex and inequitable. However, for a program with several million clients, like the residential energy credit, the income tax may be an administratively efficient vehicle for delivering the subsidy.

Alternative Approaches

A widely discussed alternative to the present approach is to discourage energy consumption by taxing it. If designed properly, an energy consumption tax can introduce fewer distortions than the present array of taxes and tax incentives can, because businesses and consumers would be free to respond to higher energy prices with what they consider the most cost-effective ways to conserve energy. In recent years, there have been numerous proposals for broad-based taxes on energy consumption, and discussion of these ideas can be expected to continue as long as the federal government faces a significant revenue shortfall, if for no other reason than that our 71 quadrillion Btu of annual energy consumption offers an enormous tax base. The principal alternatives are (1) a tax on oil consumption, (2) an increase in the tax on gasoline and other motor fuels, (3) an oil import fee, and (4) a broader-based tax on energy consumption.

Oil Consumption Tax

In his 1984 budget, President Reagan proposed a $5-per-barrel tax on consumption of oil, either domestically produced or imported. Such a tax would raise about $21 billion per year. The president's proposal is for a temporary, three-year tax to be triggered only if certain budgetary and economic contingencies come to pass. Such a temporary, contingent tax would provide a much smaller incentive for longer-term investments in oil conservation than would a permanent, or at least a longer-lived, tax.

Motor-Fuels Tax

President Carter, in 1977, proposed a 50-cent-per-gallon gasoline tax, which was to be phased in at a rate of 5 cents per year if national gasoline consumption exceeded stated targets.[9] A similar proposal for a 23-cent tax had been reported by the House Ways and Means Committee in 1975, but it met with a humiliating defeat on the House floor. As with the Reagan proposal, the contingent nature of these taxes would have reduced the incentives provided. Undaunted by the failure of his original proposal, President Carter proposed a gasoline tax of 10 cents per gallon in 1980, but this, too, was defeated.[10]

Presumably, since it would not be a highway user fee, an additional gasoline tax should not include the present exemptions for farming and off-highway uses, only whatever exemptions or special rates are deemed appropriate as energy-conservation incentives.

Oil Import Fee

An oil import fee would raise the price paid by consumers in essentially the same manner as would a tax on oil consumption. The principal difference is that it would also raise the price received by domestic producers. About one-half of this price increase to producers would be recaptured as additional windfall profit tax. Thus, relative to a tax on oil consumption, the import fee would raise less revenue per dollar but would provide a greater incentive for domestic oil production.

Broader-Based Energy Consumption Taxes

In 1975, President Ford proposed a $2-per-barrel tax on oil consumption and a companion tax on consumption of natural gas equal to the Btu-equivalent of $2 per barrel (37 cents per thousand cubic feet). This proposal was not seriously considered by Congress, but recently there have been proposals for a tax on energy consumption that is more broadly based than a tax limited to oil. Tax rates could be a fixed amount per Btu or a percentage of the value of the energy (termed an ad valorem tax).

A per-Btu tax on energy consumption would be fairly straightforward, although some decisions must be made about just what energy sources ought to be taxed. Presumably, exhaustible energy sources like oil, gas, coal, and nuclear energy and renewable sources whose supply is limited near present levels of demand, like hydroelectric energy, would all get taxed. Virtually unlimited, inexhaustible sources like solar and wind energy would be exempt. Renewable energy sources whose supply is potentially limited but at levels far above present demand, like geothermal or biomass energy, should probably be exempt, at least initially. An important feature of the per-Btu tax is that, if the tax is expressed as a percentage of the price of the energy, the tax appears to place a disproportionately large burden on those forms of energy, especially coal, with low value per Btu.

The ad valorem approach solves this problem but raises the issue of exactly where in the chain of energy production and distribution to impose the tax. The closer to the consumer the tax is imposed, the higher would be the tax per Btu. For administrative reasons, the tax should be imposed at the point in the production and distribution process at which there are fewest taxpayers (e.g., the oil refinery), but since that point will differ among the different energy sources, this approach would cause the effective tax rates to differ if each energy source has the same nominal percentage tax rate.

An ad valorem tax on major energy sources might be designed as follows: the tax would apply to oil, gas, coal, and electricity; the oil tax would be a percentage of the value of petroleum products as they leave the refinery (or as they enter the country as imported products).[11] The natural gas tax would be imposed on the sale of the gas by pipelines to either final consumers or local gas distribution systems; that is, the point in the distribution chain where there are relatively few taxpayers. Excluding the value added by the local gas-distribution system is roughly equivalent to excluding value added after the oil refinery. For coal, there is no convenient place to impose the tax, and the most promising possibility appears to be to tax the sale to the ultimate consumer and reduce the tax rate on coal to compensate for the fact that all of its value to the consumer will be taxed while only part of the value of oil and gas will be taxed. For electricity, there could be a tax on the sale to the customer, again with a reduction in the tax rate to compensate for the taxation of all value added. In order to prevent double taxation, the tax would have to be refunded to anyone who uses energy to produce taxable energy. To limit the tax to domestic consumption, it would have to be refunded to exporters.

Issues

Three issues seem relevant in analyzing the present tax structure and proposed changes: (1) Should the federal government have a policy to achieve a particular amount or pattern of energy consumption? (2) How can this goal be achieved in a manner that minimizes unintended distortions caused by government policies? (3) To what extent do energy-related tax changes redistribute income and wealth, including redistribution between different regions of the country. A fourth question, which often dominates public discussion of energy tax issues, has to do with the effect of policy on the overall level of revenues. In this discussion, it is assumed that other taxes are altered to achieve whatever level of aggregate revenue is deemed appropriate.

Why Discourage Energy Consumption?

There are essentially two reasons why the United States might want to discourage energy consumption below what consumers would be willing to buy in a free market—the "optimal tariff" argument and the "externality" argument. (See Chapter 4.) Furthermore, tax changes might be desirable to raise prices to the market level to offset other policies that interfere with market pricing of energy.

Optimal tariff. A nation that imports a particular product and consumes so much of it that changes in its imports exert an appreciable impact on the world price of the product may be able to increase its real income by implementing policies that reduce its imports. In the economic literature, a tariff that achieves this result is called an optimal tariff, although the argument also extends to other import-reducing policies. By reducing imports, the country in question can lower the world price of imported products and thus effect a transfer of income from the exporting country to itself. The cost of such a policy is that the tariff, or other policy to reduce imports, creates inefficiencies of its own; and the optimal tariff is the tariff at which the benefits of reduced import prices just balance the costs of the related inefficiencies. Some commentators have argued that the United States is in a position to impose an optimal tariff on oil.[12]

Some hypothetical calculations illustrate the potential benefits from energy conservation. Presently, the United States consumes 30 percent of world oil production. Thus, a 10 percent reduction in U.S. oil consumption would reduce world oil demand by about 3 percent, enough to have an appreciable impact on prices. U.S. oil imports are about 5 million barrels per day, or 1.8 billion barrels per year. Hence, if a 3 percent drop in worldwide oil demand would reduce oil prices by 3 percent (by about $1 a barrel), the transfer of dollars from oil exporters to U.S. consumers and the U.S. Treasury would be about $1.8 billion. (There would also be a transfer of several billion dollars per year to other oil importers.) A higher level of oil imports (like the 1977 level of 8.6 million barrels per day) or a greater response of oil prices to changes in demand would mean that the nation would derive still greater benefits from reductions in oil consumption. To evaluate any particular government actions to encourage conservation, these benefits from reduced oil prices must be weighed against whatever distortions are introduced by the government policies.

Externalities. To the economist, an "externality" occurs whenever one person's actions create benefits or costs for other persons that are not properly reflected in market prices. Two important externalities are frequently associated with energy imports—their impact on national security and their effect on world political and economic stability.

The national security argument for reducing oil consumption is that imports of oil pose a threat to national security and that consumers should be forced to take this "cost" into account in the price they pay for oil. At the present (1983) level of oil imports, given the volume of imports available from Mexico, Canada, and Venezuela, the direct security threat to the United States seems too modest to justify major action. However, there is still a significant national security threat to Western Europe and Japan, a large fraction of whose oil comes from the Persian Gulf, and this becomes an American problem to the extent that we agree to supply our allies with oil or that they would gain access to our relatively secure supplies during an emergency.

A related issue is the threat to world economic and political stability posed by the concentration of financial wealth in the hands of a few oil-exporting countries. This cost, too, should be borne by oil consumers. However, it is a difficult cost to evaluate since oil exporters vary widely in their use of their incremental income.

A third set of externalities involves the effects of energy use on the environment. To the extent that use of high-sulfur oil, coal, or nuclear power harms the environment, consumers should bear these costs through higher prices for these products.

Offsetting other distortions. A third argument for making energy conservation a matter of national policy is that price controls encourage excessive consumption of energy and that these should be offset by countervailing policies to discourage consumption. With the decontrol of oil prices, users of oil now pay the market price, as do users of coal. However, users of natural gas and electricity do not necessarily pay the market price, because prices charged by these industries are regulated, and there may be a case for taxing these uses of energy to raise the price consumers pay for these products up to the marginal cost of producing them. Electric utility rate regulation, based on average costs, lowers consumer prices well below marginal cost in most jurisdictions. However, it is no longer clear that the same can be said about regulation of natural gas prices.

What kinds of energy should be conserved? These three arguments for adopting a national policy of encouraging energy conservation have different implications for the question of just what kinds of energy consumption should be curtailed. The optimal tariff argument applies mainly to oil, since natural gas imports are relatively small and the United States is a coal exporter. The same is true of the arguments regarding national security and world political and economic stability. These considerations all suggest limiting energy-conservation efforts to oil, except to the extent that regulation has lowered the prices of electricity or natural gas below marginal cost, or environmental problems justify attempts to reduce the consumption of coal and nuclear power.

Note that many of the existing energy tax incentives do not distinguish between oil and other sources of energy. The home insulation credit, for example, applies equally to all homes, regardless of how they are heated or cooled.

Efficient Energy-Conservation Policies

Once it is decided that there should be a policy to encourage conservation of oil or of other forms of energy, the question arises as to how this can be done so as to minimize unintended distortions created by the policies. The present tax policy is to tax the use of energy in transportation and to subsidize various investments that are believed to save energy: home insulation, active solar systems, equipment used to burn alternative sources of fuel, buses, and so forth. The distortions caused by this ad hoc approach have been alluded to earlier: it is not clear that some of the energy tax subsidies operate to encourage conservation, and distinctions between items that are on or off the list of favored expenditures distort consumer choices. Other tax-induced distortions arise from the fact that the same subsidy may be provided for investments that save widely varying amounts of energy; that is, solar energy systems installed where it rains most of the year get the same subsidy as systems installed in sunny areas.

Far fewer distortions would result with an excise tax on those types of energy whose consumption it is the policy to discourage. With the tax, consumers and businesses would make their own judgments about how best to respond to higher prices by reducing energy consumption or converting from taxed to untaxed sources of energy; and their information about what is the best response is likely to be far better than that of the government.[13] Furthermore, from the standpoint of minimizing distortions, the tax should apply to all uses of the types of energy whose consumption is to be discouraged. Achieving a certain amount of oil conservation through a tax limited to certain uses of oil, such as motor fuel, will be more painful than doing so with a tax on all uses of oil, because an across-the-board tax enables users to decide which uses of oil are least necessary, rather than having the tax-writers make that decision.

Equity

Tax policy inevitably raises questions of equity. One consequence of a decision to conduct energy policy through the tax system is that equity sometimes gets more attention than it would in other forums. Excise taxes on energy consumption raise three equity issues: (1) How is the tax burden distributed by income class? (2) Does the tax create windfall gains and losses? (3) How is it distributed by region?

Distribution by income class. One issue that arises with any tax on energy consumption is that the percentage of income devoted to energy consumption declines as income rises, so that an energy-consumption tax burdens lower- and middle-income groups more than would an equivalent amount of revenue raised through an increase in most other taxes. Thus, depending on one's preferences about the distribution of the tax burden, it may be desirable to combine an energy-consumption tax with modifications to the income tax and to income maintenance programs designed to offset any adverse distributional effects. (Note that some income maintenance programs adjust automatically to higher energy prices because benefits are indexed to consumer prices.)

Just how much of an offset would be appropriate depends, to a large extent, on whether one wants to view the income distribution issue from a year-by-year or a longer perspective. Many lower-income families who devote a large fraction of their income to energy consumption in a particular year do so because their income is temporarily low and they have made commitments to a life-style requiring a rate of energy consumption that would not be sustainable for someone whose income was permanently that low. Similarly, middle- and upper-income people who save a portion of their income for future consumption will be burdened by an energy tax in the future. An alternative measure of the relative burden of an energy consumption tax is the variation in energy consumption as a percentage of total consumption (rather than of income) as income rises. By this measure, taxes on natural gas and heating oil would still be regressive, although less so than would appear from simply considering energy consumption as a fraction of income, but a tax on gasoline would be approximately proportional. Of course, people who are unemployed or have already retired are likely to find the year-by-year perspective more congenial than the longer-run perspective of a tenured economics professor.

Windfall gains and losses. An energy tax would create significant windfall gains and losses for owners of homes and businesses. A tax on oil, for example, would lower the value of oil-heated homes, especially poorly insulated ones, and may raise that of homes heated by other fuels. It would lower the value of equipment that uses oil as a fuel or feedstock. To the extent that it would raise the price of natural gas or coal, producers of those fuels would receive windfall gains and consumers would incur windfall losses. One argument for a tax on gasoline as an alternative to a tax on oil is that, because few fuels compete directly with gasoline and the equipment that uses gasoline is relatively short-lived, the gasoline tax would create smaller windfall gains and losses than would the other types of energy taxes. Similarly, a tax on all energy sources would probably create smaller windfalls than an oil tax. One way to reduce the extent of windfall gains and losses would be to phase in any large increases in energy taxes.

These windfalls are an important political consideration. Those who experience windfall losses tend to express their displeasure to Congress with considerably more vigor than the winners express their gratitude, and Congress often responds to these concerns with exemptions from the tax or with special tax rates that reduce the windfall losses but introduce new distortions. For example, when Congress has considered new taxes on energy consumption, there has been considerable interest in exemptions for both home heating oil and petrochemical feedstocks.

Regional impacts. Patterns of energy use vary widely in different regions of the country. The Northeast consumes a lot of home heating oil, the West a lot of gasoline, the Pacific Northwest a lot of hydroelectric power. How a particular legislator responds to an energy tax proposal is likely to be strongly influenced by the proposal's impact on the region he represents.

While data exist on energy consumption by state, these data are hard to interpret because it is not clear who bears the burden of a tax on business use of energy. A disproportionate amount of business use of energy is concentrated in energy-producing states, and these states appear to be bearing a larger share of the burden of an energy-consumption tax if business use is allocated to the state where the consumption occurs than if it is allocated to consumers nationwide. Moreover, to the extent that any energy tax changes the price of energy received by energy producers, it has regional impacts not reflected in data on energy consumption. On balance, however, it appears that some combination of a tax on oil consumption and an additional tax on gasoline can be made relatively even across regions.

Evaluation of Alternatives

There appears to be a good case for a federal policy to encourage energy conservation, but this case is subject to the important qualifications that the policies not introduce significant unintended distortions, and that action be taken to counterbalance significant inequities that may result. From the standpoint of achieving energy conservation while minimizing unintended distortions, a tax on energy consumption appears preferable to the present approach of subsidizing particular energy-conserving investments.

Among the alternative energy tax proposals, no single proposal appears to be superior on all counts. A motor-fuels tax scores well in terms of distribution by income class and avoidance of windfall gains and losses, but it tilts conservation incentives toward only one use of oil. A tax on oil consumption has the advantage of focusing on the type of energy consumption for which there is the strongest case for a conservation policy, but it scores badly in terms of distribution by income class and windfall gains and losses. The ad valorem tax on all major energy uses creates fewer windfalls than the oil tax, but does not encourage conversion from oil to other fuels and may create distortions to the extent that the case for conserving oil is stronger than that for conserving other fuels. The broad base and low rate of the ad valorem tax may mean that it is most consistent with Colbert's theory that "The art of taxation consists in so plucking the goose as to obtain the largest amount of feathers with the least possible amount of hissing."

Taxes on Energy Production

The impact of the tax system on energy production results from three basic sets of provisions: (1) the tax treatment of income from energy production, (2) the crude oil windfall profit tax, and (3) tax incentives for various exotic types of energy production.

For many years, there was little doubt that the tax system favored investments in energy production over other types of investments through such controversial provisions as percentage depletion and the immediate writing off (expensing) of intangible drilling costs. Recent changes in the tax law, however, have fundamentally altered the historic favoritism toward energy production: for the first time since the early days of the income tax, many energy investments are now taxed more heavily than other investments. While most foreign countries pursue a conscious policy of taxing energy production more heavily than other activities, it is not clear that the achievement of this outcome in the United States, at least in connection with the income tax, was a conscious policy decision.

Present Tax Laws

Income Taxation of Energy Producers

Before dealing with energy producers specifically, it is useful to review general tax rules.

Overview of general tax rules. As a general rule, individuals and corporations must pay tax on all their gross income but are allowed to deduct the costs properly allocable to earning that income. When costs are expected to help produce income over a period longer than one year, they are generally not deducted in the year the costs are incurred but rather are deducted over some period of time, and the specific method used to spread those deductions may be termed "capital cost recovery."

For ordinary investments in plant and equipment, capital cost recovery now follows the Accelerated Cost Recovery System (ACRS), which was enacted in 1981 and modified in 1982. Under ACRS, the costs of most equipment are written off over a five-year period.[14] Taxpayers receive an investment tax credit equal to 10 percent of the cost of equipment[15] but must reduce the "basis" on which they compute the cost recovery deductions by one-half of that credit (i.e., by 5 percent). Structures are written off over fifteen years and generally receive no investment credit.

A useful way to summarize the incentives provided by a particular system of capital cost recovery is to answer the following question: If the system being analyzed were replaced by a system under which taxpayers could deduct X cents in the year they made an investment for each dollar they invested, how large would X have to be for the taxpayer to be indifferent toward the two alternatives? This "first-year equivalent deduction" (X) will depend on the discount rate at which the taxpayer discounts the tax savings from future deductions: the higher the discount rate, the lower the benefit of a given stream of cost recovery deductions compared with an immediate write-off. Under ACRS, for equipment in the three- and five-year classes, the first-year equivalent deduction is approximately equal to one; that is, the combination of the investment credit and accelerated cost recovery makes the package approximately equivalent to immediate expensing of the costs of investment.[16] For structures, which get no investment credit, and for public utility property, the first-year equivalent deduction is less than one, but because these types of investments tend to be heavily debt-financed and, hence, benefit from the deduction of interest payments, they often receive more favorable tax treatment than do ordinary investments in equipment.

Oil and gas producers. Oil and gas producers are subject to a number of special tax rules, some more favorable and some less favorable than the rules applying to ordinary investments in equipment. Comparing oil and gas wells to equipment is more appropriate than comparing them to structures because their useful life is usually closer to that of equipment and, unlike structures, they tend to be financed with equity, not debt.

To understand the tax treatment of oil and gas producers, it is necessary to distinguish between three types of costs of acquiring or drilling an oil or gas well. First, purchases of equipment used to drill the well, or produce or gather oil from the well, are treated just like purchases of equipment by ordinary taxpayers. Second, costs incurred in acquiring the lease (i.e., the right to produce oil or gas from the property) and geological and geophysical work done prior to drilling are recovered through the depletion deduction. Third, other drilling costs such as labor, fuel, and materials, so-called intangible drilling costs, are recovered under a special set of rules.

The depletion deduction differs significantly between independent producers (i.e., producers with only limited retailing or refining operations) and integrated oil companies. Integrated companies must use cost depletion, which spreads deductions over the life of the oil or gas well. This, of course, is a far less generous system of capital cost recovery than ACRS is for equipment, because ACRS is equivalent to an immediate write-off. Independent producers, however, may use percentage depletion on up to 1,000 barrels per day of oil production or the equivalent amount of gas. After 1983, percentage depletion will be 15 percent of the gross income from the property, and the aggregate deduction is not limited to the amount of cost originally incurred.[17] Thus, for independent producers the depletion deduction amounts to a subsidy to oil and gas production (up to 1,000 barrels per day) equal to 15 percent times the producer's marginal income tax rate (i.e., 7.5 percent for an individual in the 50 percent bracket, and 6.9 percent for a corporation in the 46 percent bracket).

However, while percentage depletion encourages production by independents, it discourages them from making expenditures on lease acquisitions and geological and geophysical work. These expenditures generate no additional deductions for an independent because porcentage depletion is not based on actual expenditures.

The rules for writing off intangible drilling costs also differ between majors and independents. Independent producers deduct intangible drilling costs in the year the costs are incurred. This immediate write-off used to be extremely preferential treatment, but since the enactment of ACRS it is really no more favorable, in present value terms, than the treatment of ordinary investments in equipment. As a result of amendments made in 1982, integrated oil companies may expense only 85 percent of intangible drilling costs, with the rest deductible over thirty-six months. This combination provides a first-year equivalent deduction for the majors of about 98.7 cents, slightly less favorable than the 100.1 cents provided to ordinary investments in equipment.

The tax law, therefore, encourages a division of labor within the oil industry under which the majors specialize in lease acquisitions and geological and geophysical work and the independents specialize in the actual drilling of the wells.

Coal. The coal industry managed to weather the energy crisis without experiencing much adverse tax legislation, and its tax treatment remains significantly more favorable than that of ordinary industries. As with oil, equipment used in coal mining is subject to the ordinary ACRS rules. As a result of amendments made in 1982, furthermore, 15 percent of mining exploration and development costs are treated as investments in ordinary equipment eligible for ACRS and the investment credit. The remaining 85 percent of costs incurred in developing a mine, the coal industry's equivalent to intangible drilling costs, is written off as incurred.[18]

Coal producers are permitted to claim percentage depletion at a rate of 10 percent, subject to the limitation that the depletion deduction for a property not exceed 50 percent of net income from the property. This net income limitation is the binding constraint for most coal mines, so that the depletion deduction, in effect, operates to exempt from tax one-half of the income from profitable coal properties, while ordinary tax rules and the writing off of development costs allow losses to be fully deducted against the other half of the income from profitable properties. A provision enacted in 1982 reduces the allowable percentage depletion deduction by 15 percent, starting in 1984. This will have the effect of reducing the percentage of net income from profitable coal properties that is effectively tax exempt, from 50 percent to 42.5 percent. Nevertheless, the ability of coal producers to deduct all their losses and exclude 42.5 percent of their profits will still amount to a very favorable set of tax incentives relative to those available to other industries.[19]

Electric utilities. As with the oil industry, tax writers cannot seem to make up their minds whether to treat electric utilities more or less favorably than ordinary industries. The ACRS rules for investments by utilities in equipment are relatively unfavorable, with most investments by utilities written off over ten or fifteen years rather than three or five years. However, much utility investment is debt-financed and generates large deductions for interest, so that electric utilities pay relatively little corporate income tax despite the relatively unfavorable ACRS rules. Moreover, shareholders of electric utilities receive a special incentive if they elect to reinvest up to $750 of dividends in the stock of the company: such reinvested dividends are exempt from tax and, in effect, taxed as a long-term capital gain if the stock is held for one year.

Perhaps more important than the amount of actual tax paid by electric utilities are the conditions imposed by the tax law that must be satisfied if utilities are to receive their full panoply of tax benefits. Essentially, the law denies utilities both the investment tax credit and accelerated depreciation if the bodies that regulate utility rates pass through to consumers as lower prices too much of the resulting tax reductions too rapidly. These "normalization" rules amount to an attempt to direct utility regulatory commissions through the tax laws, although the rate-making process is sufficiently complex that it is not clear whether the normalization rules have a significant impact.

Crude Oil Windfall Profit Tax

The windfall profit tax on the production of crude oil in the United States was enacted in 1980 in connection with the phased decontrol of crude oil prices announced the previous year by President Carter. The tax is not directly related to a producer's profits but is rather an excise tax on the production of crude oil based on the extent to which the selling price of the oil exceeds an adjusted base price.

For any barrel of oil, the tax equals the tax rate times the windfall profit, which is defined as the difference between the selling price of the oil and the sum of the adjusted base price and state severance taxes on the windfall profit. Thus, if p is the selling price, b the adjusted base price, s the rate of state severance tax, and t the windfall profit tax rate, the windfall profit tax (WPT) is:

$$WPT = t[p - b - s(p - b)] = t(p - b)(1 - s)$$

The adjusted base price, with some minor adjustments, equals the price at which the oil was controlled in May 1979, adjusted for inflation and, in some cases, an extra 2 percent per year.

Oil is classified into one of four categories, each of which has its own tax rate. Newly-discovered oil bears a 25 percent tax rate in 1983, scheduled to decline to 15 percent by 1986. Heavy oil and oil produced through tertiary recovery techniques has a 30 percent rate. Stripper oil bears a 60 percent rate, and a 70 percent rate applies to all other kinds of oil. Newly-discovered oil produced in a large region of Alaska is exempt from WPT.

The law also provides exemptions or special tax rates to a variety of different types of producers. State and local governments, certain medical or educational charities, and Indian tribes are exempt. Independent producers are exempt on their stripper oil and pay a 50 percent rate in place of the general 70 percent rate on up to 1,000 barrels per day. Royalty owners are exempt on the first 2 barrels per day of their production.

The windfall profit tax is expected to raise $5 billion in fiscal year 1984, net of the reduction in income taxes that arises from the deductibility of the WPT.

Energy Production Tax Incentives

The income tax law also contains an array of tax incentives intended to encourage various kinds of energy production.

These will lead to an estimated revenue loss of $235 million in fiscal year 1984, about 10 percent of direct federal spending on energy supply. The energy production tax incentives fall into three categories: (1) extra investment credits, (2) credits for production of various kinds of synthetic fuels, and (3) tax-exempt financing for certain kinds of energy production facilities.

Most of the energy investment credits expired at the end of 1982. Those that remain through 1985 include an extra 11 percent for low-head hydroelectric facilities, an extra 15 percent for ocean-thermal energy, an extra 15 percent for equipment used to produce solar, wind, or geothermal energy, and an extra 10 percent for equipment used to produce synthetic fuel from biomass. The energy credit for equipment used to make alcohol fuel from biomass is limited to cases where the equipment uses fuels other than oil or gas, a response to concerns that oil- or gas-powered distilleries do not lead to a significant net saving of oil.

The energy production tax credit is an attempt to inject some uniformity into the energy tax incentive program by providing a direct incentive to produce certain exotic types of fuels.[20] The credit equals $3.00 per barrel-of-oil-equivalent (i.e., the amount of energy contained in a barrel of oil), adjusted for inflation since 1979. (The inflation adjustment had increased the credit to about $4 by 1983.) However, the credit phases out as the price of oil rises above $23.50 per barrel in 1979 prices ($31 in 1983) and is eliminated at a price of $29.50 per barrel ($39 in 1983). When this credit was enacted in 1980, Congress anticipated real oil prices considerably above $29.50, and it intended the production credit as a de facto price floor. Because the credit directly increases after-tax income, a $3.00 credit is equivalent to a price increase of $5.56 for a taxpayer in the 46 percent bracket ($5.56 × .54 = $3). Thus, producers of eligible fuels would be guaranteed a real price of approximately $29.00 per barrel as long as real oil prices stayed above $23.50. At the time the credit was drafted, no one involved anticipated that real oil prices would fall below $23.50, but at this writing such a development appears likely.

For alcohol fuels, there is a separate production credit of 50 cents per gallon ($21.00 per barrel) that producers can use in lieu of the gasoline excise tax exemption.

Tax-exempt industrial revenue bond financing is available for low-head hydroelectric generating facilities, facilities to produce alcohol fuel from solid waste, and facilities to generate steam from solid waste.

Summary of Taxes on Energy Production

The tax system embodies a schizophrenic attitude toward energy producers. Coal producers and producers of an eclectic array of alternative fuels receive more favorable treatment than do ordinary taxpayers. Major oil companies now receive less favorable treatment. Independent oil producers receive favorable income tax treatment, counterbalanced by liability for windfall profit tax (except on exempt stripper oil).

Alternative Proposals

Income Tax Reform

The income tax treatment of energy producers, and the resulting incentives for investments in energy production, are only one part of a much broader issue—the appropriate tax treatment of income from capital. The present income tax provides a wide variation in incentives to make different types of investments, and there has been much discussion of tax reforms that would produce greater neutrality. Historically, such tax reform proposals have been viewed skeptically by energy producers because reform would reduce or eliminate their relatively favorable treatment. With the enactment of ACRS and the restrictions on various energy-related tax preferences, however, it is no longer clear that tax reform to achieve greater neutrality would worsen the relative position of the energy industry.

One suggestion is to broaden the base of the income tax so that it approximates the taxpayer's true income, as he might report it to shareholders or to creditors. Such base broadening would permit a reduction in the top tax rate on individuals and corporations to 30 percent without compromising the revenue yield of the individual and corporate income taxes or the progressivity of the individual income tax. Such a broadly based income tax would include capital cost recovery provisions under which capital costs are recovered as an asset declines in value over its useful life, whether that asset is an oil well, a coal mine, or an ordinary piece of equipment. Special tax incentives like the energy credits or percentage depletion would be repealed. These ideas are embodied in a tax reform proposal recently introduced by Senator Bill Bradley and Representative Richard Gephardt (S. 1421 and H.R. 3271).

An alternative suggestion is to convert the individual income tax into a progressive tax on consumer spending and the corporate income tax into a tax on cash flow. In each case, taxpayers would first compute their income and then make two adjustments: (1) they would subtract from the tax base amounts paid to purchase assets and add proceeds from the sale of assets, and (2) they would add proceeds from borrowing and subtract repayments of debt. As a result of these adjustments, the tax base would be income minus saving—that is, consumption. This type of proposal differs from sales and value-added taxes in that the latter are flat-rate taxes collected from businesses whereas a consumption tax collected from individuals could embody progressive tax rates.

Note that certain features of the consumption, or cash flow, tax proposal resemble the present income tax. The deduction for purchase of assets is the equivalent to expensing of intangible drilling or mining development costs. In effect, a consumption tax would put all investments on a par with those that are presently expensed and would restructure the treatment of debt in a manner consistent with expensing.

Windfall Profit Tax Reform

The windfall profit tax has only been in the law for three years (although a version of it was proposed as early as 1973 by President Nixon), so there has been relatively little debate about its structure. Suggested changes include repealing the tax, simplifying it by reducing the categories of oil and of producers who receive different treatment, and converting it into a severance tax.

Even though repealing the windfall profit tax is attractive from the standpoint of increasing incentives for energy production, this option is unlikely to receive serious consideration as long as large budget deficits are projected, since the $5 billion per year raised by the tax is generally perceived by the voters as coming from an industry with ample ability to bear such a burden. A politically more realistic objective might be to exempt newly discovered oil, which would phase the tax out gradually as old oil reservoirs are depleted. Also, the existence of the tax does provide some insurance against the reimposition of price controls during a future shortage, since the high marginal tax rates ensure that the federal government receives about half of any incremental price increases as windfall profit tax receipts.

The tax could be simplified and made more equitable by eliminating the special treatment for certain categories of oil and certain kinds of producers. The tertiary recovery rules, although they provide incentives to produce high-cost oil and encourage experimentation with promising new techniques, are complex and give producers an incentive to use tertiary recovery techniques in preference to other methods. The exemption for royalty owners provides no production incentive, since royalty owners bear no costs of production. Also, there have been suggestions to repeal the special treatment for independent producers.

Another possibility would be to convert the windfall profit tax into a severance tax, in which a lower tax rate would be applied to the entire price of the oil, not just the windfall profit portion of the price. There could either be a uniform nationwide tax rate or different rates for different categories of oil or producers.

A severance tax would be a simplification because it would eliminate the base price and the related inflation adjustment. Further simplification would require reducing the number of categories of oil or producers receiving disparate treatment. The essential difference between a severance tax and a windfall profit tax, however, is the response of tax revenues to incremental changes in oil prices. With a windfall profit tax, the federal government receives a large share of incremental revenues. This has the advantage of providing some "price control insurance" to the industry but the disadvantage of complicating budget planning by making the revenue yield of the tax extremely sensitive to oil prices. (Currently, the tax is yielding $15 billion per year less than was forecast in 1980.)

Oil Import Fee

A third way to modify incentives for domestic energy production would be to impose an oil import fee, which would raise the price of domestically produced oil and exert upward pressure on the price of competing sources of energy. To the extent that price increases would not be recaptured as additional windfall profit tax, the import fee would provide an incentive for increased domestic production. The import fee affects producer incentives in approximately the opposite way as does the windfall profit tax: one raises and the other lowers the net-of-tax price that domestic producers receive for their oil. Thus, it may be unnecessarily complex to have a system in which the two taxes coexist.

Issues

A variety of issues arise in analyzing these and other suggestions for modifying the taxation of energy producers. The major issue is whether national policy ought to be to encourage or discourage investment in various kinds of energy production relative to other investments. Other issues include whether energy producers should be treated alike and whether incentives can be provided in a way that is less complex, more effective, and less distorting.

Should the United States Encourage Energy Production?

The present situation in which investments in oil and gas wells by the major oil companies are treated less favorably under the income tax than are ordinary purchases of equipment deserves re-examination, and consideration should be given to developing a capital cost recovery system under which energy investments are treated no worse than ordinary investments are.

The general case for some sort of positive incentive for investment in energy production relative to other investments is similar to that for a policy to encourage energy conservation. The national security and optimal tariff arguments apply without modification. In addition, some investments in new technologies generate information that becomes generally available, and this knowledge may be worth the costs of a subsidy. However, equity considerations very much weaken the case for subsidizing energy production, since energy production is concentrated in certain states and energy producers are viewed by many voters as having a large capacity to pay taxes. Indeed, the basic reason for the existing crazy-quilt pattern of incentives is our inability to resolve the contradiction between a policy to encourage energy production and a desire to tax energy producers. The best way to resolve the ongoing, divisive debate about the taxation of energy producers would be to accept the principle that they should be taxed no differently than ordinary businesses.

Equal Treatment of Energy Producers

Both the income tax and the windfall profit tax make explicit distinctions between integrated oil producers, independent producers and royalty owners. Independent producers and royalty owners retain percentage depletion; independents retain full expensing of intangible drilling costs; independents receive a special windfall profit tax rate and an exemption for stripper oil; and royalty owners receive an exemption from the windfall profit tax for up to two barrels per day.

That special treatment for independent producers or royalty owners improves the equity of the income tax is problematical. The typical independent producer is likely to be considerably richer than the typical shareholder of a major oil company. Even royalty income appears to be less equally distributed than dividends.[21] Special treatment for royalty owners clearly cannot be justified on the grounds that it provides a production incentive, since royalty owners pay none of the costs of the actual extraction of the oil.

Complexity

There is little doubt that almost any plausible set of incentives for energy production could be accomplished with a system less complex than the present system. The energy production credit and energy investment credit overlap and could be merged into a single credit. Percentage depletion for oil and gas, which is equivalent to a subsidy to oil and gas production equal to a percentage of the price, is offset by the windfall profit tax, a tax essentially equal to a percentage of the price. One way to simplify the law would be to provide windfall profit tax exemptions for independent producers and royalty owners and reduce percentage depletion such that the overall effect was revenue neutral.

Nondistorting Incentives

As with incentives for energy conservation, an important issue is designing incentives for energy production that minimize unintended distortions. Policy toward energy production appears to have at least two separate goals: reducing dependence on oil imports, and encouraging experimentation with new technologies. An oil import fee is the least distorting way to discourage oil imports because it provides energy producers with the maximum leeway to respond to higher oil prices. In contrast, energy investment credits and production credits limited to particular kinds of energy subsidize only certain ways to increase energy production and hence introduce distortions. However, an oil import fee raises all domestic energy prices and thus does not target the incentive toward production of energy using new technologies. For the new technologies, an energy production credit for exotic energy sources is less distorting than an investment credit because it does not provide an incentive to favor a capital-intensive technique of production.

A particular type of distortion that can be introduced with a depletable resource involves distorting the decision about when that resource will be produced—that is, encouraging either excessive conservation or overly rapid depletion of the resource. A temporary windfall profit tax, for example, will give producers an incentive to defer production; a temporary oil import fee will have the opposite effect.

Producers' Expectations

One of the consequences of the frequent changes in energy policy in recent years is that energy producers, when they make their investment plans, make assumptions about the content of future policy changes. Thus, policy changes—or even discussions of policy changes—should take into account their impact on producers' expectations; and the effect of a particular incentive is likely to depend on how it affects expectations.

Examples abound of policies that are less effective than they could be because of their impact on expectations. In 1981, President Reagan proposed repealing the business energy tax credits. Even though Congress failed to act on that recommendation, any business contemplating an investment that is currently eligible for the investment credit or that will produce energy qualifying for the production credit would do well to discount the expected benefit from these credits for the possibility that future legislation will render them unavailable. Similarly, percentage depletion, under attack from tax reformers for decades, may have outlived its usefulness as a production incentive to the extent that producers feel it will be repealed before the end of the useful life of the wells they plan to drill.

Expectational effects may be important in analyzing the windfall profit tax. For example, a low tax rate on newly discovered oil will be less of an incentive to the extent that producers fear that, in the years ahead, Congress will redefine what was once new oil to be heavily taxed old oil. Such paranoia would seem foolish but for the fact that the entire history of oil price controls was characterized by such reclassifications.

When expectations are taken into account, both the oil import fee and the energy production credit, the two incentives that appeared relatively less distorting, become less attractive. An oil import fee is likely to prove unpopular, and many producers may consider it temporary and not fully account for it in their investment planning. (However, an import fee designed to stabilize prices received by domestic producers could reduce uncertainty.) Similarly, a production credit, in effect a promise to pay out a subsidy over the life of a project, may prove less effective than a subsidy paid out when the initial investment is made.

Evaluation of Proposals

There appears to be considerable potential to rationalize and simplify the tax treatment of energy producers. The first step could be to replace the present capital cost recovery provisions for oil, gas, and coal with a system that corresponds to capital cost recovery for ordinary investments in those kinds of equipment whose useful life approximates that of the typical well or mine. Next, some decision should be made about whether there should be a policy to encourage domestic oil production or whether the public perception that the oil industry has a large ability to pay taxes justifies additional taxes on the industry. If the decision is for more production incentive, this can be achieved by reducing or repealing the windfall profit tax, especially the tax on newly discovered oil, and, if that is insufficient, by an oil import fee. If the decision is for extra taxes on oil producers, some attempt should be made to simplify the windfall profit tax by reducing the number of different categories of oil and of producers that receive disparate treatment.

7
COAL AND THE CLEAN AIR ACT

Lester B. Lave

Summary

Coal is America's most abundant energy resource and a key fuel to achieving greater self-sufficiency. It is geographically dispersed and inexpensive to mine and use. Unfortunately, past practices of mining, transporting, and burning coal have caused massive environmental problems. These problems were so great that coal use declined sharply in the post–World War II period as petroleum and natural gas became available in quantity.

The technology exists to mine coal with few accidents, little occupational disease, and minimal environmental damage, to transport it with few accidents, and to burn it with little environmental impact. Even with the cost of control technologies added, coal is still a cheaper fuel than its alternatives.

Federal and some state regulation grew up to cure these problems. While regulation has pushed the invention of technologies that can make coal a safe, environmentally benign fuel, it has created many additional problems. In enacting the statutes to control past abuses, Congress loaded the statutes with unrelated baggage. Compensation for black lung disease was made so easy to obtain that virtually anyone who has worked underground for fifteen years can claim it. Air pollution statutes have attempted to keep plants in the industrial Northeast, rather than allowing them to move to rural areas in the South and West. Construction of new plants has been made unnecessarily cumbersome in the name of environmental protection. Eastern coal interests and environmentalists teamed up to remove the incentive to use low-sulfur western coal, thus preserving some eastern coal-mining jobs at the cost of hundreds of millions of dollars to electricity users and venting additional sulfur dioxide into the environment—a perverse outcome of the 1977 Clean Air Act Amendments.

Coal Opportunities

Coal represents the largest fossil fuel resource in the United States. Coal reserves are estimated at 2.7 trillion tons, (about 80,000 quadrillion Btu). Since the United States currently uses about 17 quads of coal each year, this is a substantial reserve. The United States has the second-largest reserve (behind the USSR), about one-third of world reserves. Indeed, the USSR, the United States, and China together have 89 percent of known world coal resources.[1]

Nor does the good news stop there. Coal is not only abundant but also widely dispersed throughout the nation. This is particularly important since transport costs preclude moving the coal large distances overland to market. Best of all, the coal is easily mined. The price of coal at mine mouth is generally the cheapest of any fuel in the United States and is likely to remain the cheapest fuel for quite some time (see Prologue and Chapter 12).

At one time coal fueled the development of U.S. industry and provided cheap fuel for heating homes. Until well into this century, coal was the dominant fuel that provided abundant, low-cost energy for industrial and domestic needs.[2]

Coal was always a good fuel for boilers, but it was not suited to vehicles or for applications that require an ultraclean fuel such as natural gas. However, proven technology exists for converting coal into gases and liquids to supplement or replace natural gas in home and industrial applications, and gasoline, diesel fuel, and jet fuel in automobiles, trucks, and aircraft (see Chapter 12). Should we so desire, coal offers the United States the wherewithal to achieve energy independence.[3]

The Problems With Coal

Coal is not an unmixed blessing, however. Extracting coal from the ground can leave moonscapes, as when open-pit mining is not followed by restoration.[4] Underground mining can lead to subsidence of land and damage to buildings at the surface. It can occasionally lead to underground fires that are essentially impossible to extinguish and that burn for decades. If the mines are not properly sealed, coal extraction causes acid runoff into streams, as air reacts with the sulfur in the exposed coal and rainwater seeps into the mine, carrying off the resulting acid. Extraction, particularly underground mining, poses risks of accidents and occupational disease.[5] The levels of accident risk are sufficiently high to make underground mining one of the most hazardous occupations.

1. The current risk of underground mining if 0.4 deaths per million man-hours.[6] This means there is a death about every 1,250 worker-years. Over a forty-year working life, 3.2 out of 100 miners would be expected to be killed in underground coal mining. Even this number represents a major improvement from past experience, since in the late 1960s, the fatality rate was nearly three times higher, meaning that almost 10 percent of underground miners would be killed during a forty-year working life. The mortality rates are now comparable to those in underground European coal mines.

2. In addition to the fatal accidents, many other accidents cause disabling injuries.[6] These injuries involve at least one day of work absence, and could indicate permanent disability. The long-term injury rate has averaged 48 disabling (lost-time) injuries per 1 million person-hours, more than a hundred times the rate of fatal injuries. This translates to a disabling injury for every 10.4 worker-years. Over a forty-year working career, an underground coal miner can expect 3.8 lost-time injuries. The injury rate has been rising in recent years.

3. Coal splinters into tiny particles which, when inhaled, lead to coal workers' pneumoconiosis, or CWP (popularly called "black lung"). The prevalence of the disease is approximately proportional to the amount of dust inhaled. One survey found that about 30 percent of underground miners in Appalachia had CWP and that the prevalence increased with time spent underground.[6] Only about 10 percent of those with fewer than ten years underground had CWP compared with 57 percent of those with more than forty years underground.

4. Surface mining of coal is about three times safer than underground mining, per man-hour. Recent statistics show falling rates both for disabling and fatal injuries. Since surface miners are much more productive than underground miners, the fatality and injury rate per ton of coal mined is more than ten times smaller for surface mines than for underground mines.

5. Underground coal miners also risk diseases other than CWP. For example, they have higher risk of lung and stomach cancer, nonmalignant respiratory disease, asthma, and tuberculosis.

6. Transporting coal by train, truck, and barge also leads to many accidents. Coal is no more dangerous to transport than other materials, but large tonnages are involved and so coal makes up a substantial fraction of all cargo carried by rail, truck and barge.
7. Burning coal releases large quantities of air pollutants, principally particles, sulfur oxides, and nitrogen oxides. There is major controversy about the health effects in the general population that might be expected from these emissions.[7] However, it is clear that they are significant.

Table 7-1 summarizes the occupational and public health effects of the coal fuel cycle from mining through combustion. The table assumes that the coal is to be burned in a one gigawatt electricity-generating plant using modern abatement technology.[8] (A one-gigawatt plant is typical of large new plants; the capacity is one-million kilowatts, enough to supply power to 250,000 residences).

Table 7-1
Coal Fuel Cycle Public Health Effects Summary

	Deaths	Nonfatal accidents
	[per GW(e)–yr]	
Transport accidents		
Rail	0.41	0.87
Barge	0.14	0.15
Truck	1.2	20.
Air pollution		
Sulfate Surrogate		
(no threshold)	21 (0–60)	
Sulfate Surrogate		
(10 µg/m³ threshold)	10 (0–30)	
Radionuclides	0–0.3	

Source: Samuel C. Morris, "Health Risks of Coal Energy Technology," in C. Travis and E. Ethier, eds., *Health Risks of Energy Technologies* (Boulder: Westview Press, 1983).

Occupational deaths are expected to be 1.3 each year for this power plant, along with 96 injuries or instances of disease. If the coal were mined underground and transported by truck, 3.09 deaths would be expected; if the coal were surface mined and transported by rail, only 0.324 deaths would be expected.

Current transport patterns would be expected to result in 0.14 deaths (barge) to 1.2 deaths (truck). Air pollution, primarily in the form of sulfur oxides poses the biggest risk, although the effects are subject to the greatest uncertainty and controversy. A no-threshhold model is estimated to indicate that 0 to 60 deaths would result, with a best estimate of 21. Assuming that sulfate levels below 10 micrograms per cubic meter have no adverse health effect, 0 to 30 deaths would be expected, with a best estimate of 10.

Thus, the overall estimate is that 1.3 workers and 12 to 23 members of the general public would die each year from the mining, transporting, and burning of the approximately 3 million tons of coal needed to fuel a one-gigawatt electricity-generating plant. These estimates reflect current experience. If mines are made safer, if prevention of train-car collisions is more effectively handled, and if air pollution control is more stringent, all of these estimates would be smaller. Indeed, if underground mining were proscribed and all coal were transported by train, the estimates would be lower.

Large-scale use of coal in the last century gave it the reputation of being a dangerous, dirty fuel. Despite its abundance and low cost, the availability of natural gas and petroleum products in this century led to a shift away from coal. Coal continued to be used only in industries where its chemical properties were required, such as in making steel, or in large industrial and electricity-generating boilers where cost was important and where the large scale permitted some of the inherent problems to be overcome. Domestic use of coal fell from 454 million tons in 1950 to 374 million in 1961. It then increased slowly (3 to 4 percent per year) to 515 million tons in 1970 and 729 million tons in 1981.

Solving the Problems

Obviously, some way must be found to reconcile the problems and opportunities offered by coal. The primary solution has been to regulate the mining, transport, and burning of coal. Congress has passed various statutes to protect miners against accidents and black lung disease. The statutes and actions of the Mine Safety and Health Administration (MSHA) require procedures to lessen the dangers of mine collapse and explosions. They also set limits on the concentration of coal dust in the air and thus decrease the incidence of coal workers' pneumoconiosis (CWP); they provide for compensating miners who are disabled. The result has been a decline in the number of mining fatalities, as shown in Table 7-2. Presumably, the

Table 7-2
Fatalities Decrease Substantially After the Passage of the NSHA Act of 1969

Year	Underground mines[a]			Strip mines[b]		
	Number of fatalities	Fatalities per million man-hours	Fatalities per million tons	Number of fatalities	Fatalities per million man-hours	Fatalities per million tons
1961	256	1.6	0.9	17	0.4	0.13
1963	245	1.5	0.8	22	0.5	0.14
1965	223	1.4	0.7	18	0.4	0.10
1967	174	1.1	0.5	22	0.5	0.12
1968	268	1.8	0.8	24	0.6	0.13
1969	149	1.0	0.4	28	0.6	0.14
1970	206	1.3	0.6	29	0.6	0.12
1971	141	0.9	0.5	23	0.4	0.09
1972	122	0.7	0.4	20	0.4	0.08
1973	99	0.6	0.4	16	0.3	0.06
1974	90	0.5	0.3	24	0.3	0.08
1975	99	0.5	0.4	32	0.3	0.10
1976	104	0.5	0.4	23	0.2	0.06
1977	91	0.4	0.4	27	0.2	0.07
1978	67	0.4	0.3	17	0.1	0.04

[a]Includes underground workers only.
[b]Includes only strip mines.
Source: U.S. Department of Labor, Mine Safety and Health Administration, "Injury Experience in Coal Mining," Tables 2 (1978), 7 (1972–77), and 5 (1961–71).

incidence of CWP will decline in future years as miners who began work after conditions improved dominate employment.

Transport safety is regulated by the Interstate Commerce Commission (ICC). The regulations attempt to decrease the number of collisons, derailments, and other accidents as well as to mitigate the damage should a crash occur.

Federal laws and regulations also attempt to ensure that land is restored after mining. Provisions govern strip mining and relate to acid runoff from both strip and underground mining. Regulations also govern emissions of pollutants into the air and water.

Nonregulatory Alternatives

Much of the role of MSHA in preventing accidents could be assumed by a nonregulatory approach.[9] Perhaps the simplest approach would be to levy a fine related to the severity of each injury. The combination of the workers' compensation premium and this fine would provide each mining company with a substantial incentive to enhance safety. This approach would be superior to merely increasing the workers' compensation payments and requiring each company to insure itself. As payments to workers increase, there is the problem of moral hazard, of workers either taking less care or even of deliberately setting out to be injured. A fine could be justified as the social investment in keeping each miner healthy, over and above the interest each individual has in keeping himself healthy. After all, society must stand ready to support the miner's family and would lose tax revenue in the event the miner becomes disabled.

A number of small mining companies are unlikely to be susceptible to these economic incentives. Having little or no assets, these companies would go bankrupt rather than pay a substantial fine. Furthermore, such small companies do not have the safety expertise to be aware of the dangers. Thus, the system of fines could be substituted for regulation only in financially solvent companies above some minimum size. For insolvent companies or those of small size, MSHA regulation is presumably necessary.

CWP presents more difficulties for a nonregulatory solution than do accidents. The long time between initial exposure and the development of the disease means that the "present discounted value" of future disease is essentially zero for usual market interest rates. Nonetheless, society still desires to lower the incidence of disabling disease. One alternative might be a system of fines related to dust concentrations in mines, rather than to the ultimate incidence of CWP. There are at least rough dose–response relationships of the expected incidence of CWP for each level of coal dust in the mine. One could then work out the appropriate fine for each dust level averaged over one year (or one day). Given a random inspection process with known frequency, the fines could be calculated to provide the proper economic incentive. If inspection were moderately frequent, even marginal and small mining companies could be governed by the system.

Solutions as Problems

Unfortunately, solutions have a way of turning into new problems. For example, the regulation of underground mining leads to an increase in its cost. This prompted the expansion of strip mining before there were adequate regulations concerning restoration. A second example is the regulations restricting air and water pollution; these increase the amount of waste that must be disposed of on land. The current state-of-the-art control technology for sulfur dioxide emissions is a lime-limestone scrubber. A mixture of lime and limestone reacts with the sulfur dioxide and other combustion gases to form a gypsum sludge. A one-gigawatt (one million kilowatt) coal-fired electricity-generating plant produces several hundred thousand tons of this sludge each year, which will not solidify for decades, or even centuries. At the Mansfield plant in western Pennsylvania, this sludge is being pumped into a river valley and is expected to fill it in ten to twenty years.

Abating the emission of sulfur dioxide, nitrogen dioxide, and other combustion products into the air is desirable. However, doing so will create residuals that will have to be disposed of in other ways. The best intended solutions tend to push the problems off into other areas instead of solving them.

Why this is so can be seen from the notion of materials or mass balance.[10] Matter is not destroyed in our economic processes. Rather, its location and chemical composition are changed. Thus, when coal is burned, it does not disappear. Rather, we are left with the ash (generally about 15 percent, in good quality coal), carbon dioxide, and other combustion gases, including sulfur dioxide. The ash and sulfur in the coal cannot be wished away. Generally, the cheapest way for the boiler operator to get rid of the unwanted residuals is to dump them in the air as fly ash and sulfur dioxide. However, the resulting damage to health, plants and animals, property, and aesthetics means that venting the residuals into the air is not the cheapest disposal method from a social viewpoint.[4] Technologies exist to trap almost all of the fly ash, but they then create a new disposal problem. The trapped ash could be dumped into rivers, disposed of on land, or used in some cases in road building.

A 1,000 megawatt power plant uses about three million tons of coal each year. If that coal contains 15 percent ash and 3 percent sulfur, 450,000 tons of ash and 90,000 tons of sulfur are constituents of the coal. They must be put someplace in some chemical form. Clearly, some chemical forms and places are much less desirable than others, but rarely are there desirable places and chemical forms for this material. Ash and even sulfur are in excess supply.

Unfortunately, our environmental legislation focuses on one problem at a time, giving rise to regulations that restrict the dumping of a particular chemical compound in a particular place. But we cannot put corks in smokestacks or drainpipes. A materials-balance approach is needed to decide where and in what form to dispose of unwanted products.

Coal and the Clean Air Act

The above difficulty is inherent in environmental regulation and conditions all attempts to solve the problems. In practice, a host of immediate difficulties have more influence on environmental regulation. These revolve around the notions of due process and how Congress and the regulatory agencies approach the problems. A brief history of the Clean Air Act and how it treats stationary sources of air pollution will reveal the difficulties.[7]

Redundant Multiple Regulations Cause Difficulties

Three aspects of the current structure are particularly important. The first is that the 1970 Clean Air Act Amendments mandate redundant regulation. The Environmental Protection Agency (EPA) was required to define *primary air quality standards* to protect public health (of even the most sensitive group in the population) and more stringent *secondary standards* to protect plants and animals, trees and buildings, aesthetics and visibility. Each state with areas that did not attain these National Ambient Air Quality Standards (NAAQS) was required to draw up and obtain EPA approval for a State Implementation Plan (SIP); attainment of the NAAQS was required by a stated date. The deadline for attainment of the NAAQS has been delayed several times, but many areas still fail even to attain the primary air quality standards.

Apparently distrusting this process, Congress also required EPA to set emissions standards for new sources. For example, sulfur emissions from new power plants were limited to 1.2 pounds per million Btu. These new source performance standards (NSPS) were to reflect best available control technology (BACT). As plants were replaced or modernized, these emissions standards would act by themselves to attain vast improvement of air quality, even neglecting the SIPs. These emissions standards obtained regardless of whether the area had already attained the NAAQS.

The second aspect aims at protecting areas whose air quality was much *better* than the requirements of the NAAQS. An early suit resulted in EPA being ordered to protect the air quality of pristine areas and not allow it to be degraded, even though air quality was far better than the NAAQS.[7]

The prevention of significant deterioration (PSD) was formalized by Congress in the 1977 amendments. Areas that had air quality better than the NAAQS were divided into three groups: class 1 areas are national parks and other areas in which no deterioration of air quality was allowed; class 2 areas consist of farm and undeveloped areas in which a small amount of deterioration in air quality is permitted; class 3 areas are those designated for industrial and urban development, where air quality might be allowed to deteriorate, possibly down to the NAAQS. States could reclassify class 2 into class 3 areas.

EPA then determined allowable increments of emissions that they believed were consistent with the goals for that area. A company wanting to locate a plant in an area had to (1) show that its plant would meet the new source performance standards, (2) do modeling of the diffusion and transformation of emitted pollutants to show that ambient concentrations of pollutants didn't increase appreciably, and (3) show that its emissions were within the increments allowed by EPA for that area.

The third aspect also comes from the 1977 amendments. Congress debated, and was at the point of enacting, a requirement that all large new boilers fired by coal be subject to a uniform requirement that 90 percent or more of the sulfur in the stack gas be removed. This additional technology requirement was imposed because EPA claimed to have demonstrated that lime-limestone scrubbers could remove 90 percent of the sulfur in the stack gases. As a technology requirement, the removal was not tied to the amount of sulfur in the gas. For example, a plant burning 5 or even 8 percent sulfur coal would be subject to the same 90 percent removal requirement as a plant burning 0.5 percent sulfur coal.

These three aspects of the current structure have served to complicate air pollution abatement and to increase cost. Congress clearly did not trust individual states to exert the proper amount of effort in meeting deadlines. However, EPA review of the SIPs has proved a time-consuming, cumbersome problem. Any time a state negotiates a compliance decree with a polluter, a modification in the SIP is required, and EPA must approve. In recent years the delays in approval have been so great that some annual SIPs never get approved.

It makes little sense for EPA to have responsibility for the process by which a state attains the NAAQS, save for long-distance transport of pollutants, discussed below. Instead, EPA should ensure that states meet air quality goals and that penalties are levied if they do not.

In practice, the emissions standards for new sources (defined by EPA) are much more stringent than the emissions standards for existing sources (specified under the SIPs). In some cases the penalties for constructing a new plant are so great that firms are motivated to continue using an old plant, despite the economic incentives to scrap an obsolete plant and build a new one (in the absence of such a pollution-control penalty). One analysis for automobiles shows that the new source penalties for automobiles succeeded in delaying for years improvements in air quality that would have been obtained by having slightly less stringent new source standards.[11]

These procedures have been costly and cumbersome. More important, they have not succeeded in improving air quality dramatically. Indeed, the data suggest that air quality has improved only modestly.[12]

Regional Politics Complicates the Issues

Just as politics, in the form of federal government distrust of the motives of state government, was behind the cumbersome SIP review, so regional politics is responsible for the old source–new source dichotomy. Certainly, it costs more to retrofit an old plant than to install pollution control in a new plant. For this reason, existing sources should be subject to less stringent standards. However, making the new source standards excessively stringent undermines incentives to build new plants. Since new plants are predominantly built in the South and West, stringent new source standards have the effect of keeping existing plants in the Northeast and north central regions operating.

The prevention-of-significant-deterioration legislation has many of the same motives. By making it more difficult to build a new plant in a rural area, Congress encourages firms to keep their old plants operating or to build new plants on the old sites. The suspicion that this motivation was present is boosted by two bits of evidence.[13] The first is an analysis of the home states of congressmen supporting or opposing these amendments. The analysis shows that the strongest supporters were those from old industrial areas in danger of losing their industry. Since these areas are located far from the pristine areas to be protected, one might infer motivations other than protecting the wilderness. The other piece of evidence is the extraordinarily cumbersome procedure established for the firm wanting to build a plant in a rural area. The procedure is expensive and time-consuming, but it rarely has an outcome other than requiring the new plant to meet new source performance standards, which it is required to meet in any case.[7]

Although there is legitimate argument over the extent to which rural areas should be protected, there is no justification for the current cumbersome approval process.

The Environmentalist–Coal Operator–United Mine Workers Alliance

Ackerman and Hassler document the political motivations behind the 1977 Clean Air Act Amendments requiring that 70 to 90 percent of the sulfur be scrubbed from coal-fired boilers.[14] Environmentalists who wanted to reduce sulfur emissions found themselves allied with eastern coal operators who were afraid that low-sulfur western coal would take over their markets. When the new source performance standards allowed no more than 1.2 pounds of sulfur to be emitted for every million Btu of energy from burning coal, there were several ways of achieving the objective: (1) one could burn low-sulfur coal ("compliance" coal) and take no further steps to abate sulfur emissions; (2) alternatively, some moderate-sulfur coals can have enough sulfur removed in a washing process to meet the standards (coal washing is relatively inexpensive); (3) the third alternative is lime-limestone "scrubbing" of the flue gases. However, even 90 percent scrubbing will not remove enough sulfur to have some high-sulfur coals meet the standard. In addition to problems with expense, there are problems with reliability, since scrubbing equipment requires maintenance and can break down. If the plant cannot be operated when the scrubber is not working, the plant will be considerably less useful, since it will be available less often.

Using low-sulfur coal, or medium-sulfur coal with washing, can have considerable economic advantages over installing a scrubber. Eastern coal interests feared that they would lose markets since low-sulfur western coal did not have to be scrubbed in order to meet the new source performance standards. The United Mineworkers Union shared this concern.

Environmentalists' goals might have been met by tightening the NSPS and prohibiting plants upwind of class 1 areas. However, they wanted best available control technologies applied to plants in the West, even those using low-sulfur coal, in order to further reduce sulfur emissions.

All three groups found themselves supporting a requirement for mandatory scrubbers on all new plants. Apparently in the name of political expediency, the environmentalists did not go on to demand that limits be placed on sulfur emissions, in addition to the scrubbing requirements. If so, owners of the high-sulfur coals would have opposed the package.

Studies sponsored by the Department of Energy and by utilities showed that requiring scrubbers and not tightening the ceiling on sulfur emissions would be more expensive and would lead to greater sulfur emissions than would a moderate tightening of the ceiling on sulfur emissions with no scrubber requirement.[15] Sufficient pressure was placed on Congress to get it to modify the proposed 1977 amendments. Congress gave EPA the power to require a less expensive scrubber that would take 70 percent of the sulfur out of the stack gas if the total emissions of sulfur fell below a stringent ceiling. It is this compromise package that is now in force. Unfortunately, for plants that cannot get their emissions below the ceiling by using low-sulfur coal or washed medium-sulfur coal, there is no reason not to use the cheapest, extremely high sulfur coal, since a scrubbber is required and there are no emissions limits.

Another element of the legislative history is that the members of Congress were not thinking of all large boilers when they passed the 1977 amendments. Rather, they were thinking of public utility boilers. They presumed these utilities could pass along the higher costs to their customers and were not in danger of going out of business; the additional costs did not seem so onerous. In fact, public utilities have found it difficult to raise the capital to build scrubbers or to pass along the costs to users.[16]

Local Autonomy versus Regional Problems

President Reagan has proposed taking responsibility away from the federal government in many areas and returning it to the states. This philosophy is particularly appropriate for environmental quality. Some people desire a pristine environment and are willing to pay the price of achieving it. Others prefer higher incomes and an expansion in local employment to superclean air. I have no stake in how clean the air is in Los Angeles, Denver, and Houston. If the residents of these cities want to pay the costs of vast improvements in air quality, that is fine with me; it is also fine with me if they choose lower costs and less clean air, at least as long as the air does not get so dirty as to violate the approved minimum standards. Indeed, it seems likely that residents of different cities will choose different levels of air quality.

A more important aspect of local determination stems from the fact that communities start with very different levels of air quality. Industrial areas have problems with suspended particles and sulfur oxides; urban areas have problems with photochemical oxidants related to automobile emissions. Unfortunately, new source emissions standards apply equally to all areas (except where they may be more stringent for class 1 areas). In cases such as automobile emission standards in rural areas, regulation is expensive and produces little or no benefit.[17]

As long as there are no important spillover effects from local decisions, there is little reason to reach a national decision and impose it on local jurisdictions. However, there are areas of important spillovers. Tall stacks project emissions so high into the atmosphere that almost none of the emissions reach ground level locally. Instead, they affect people living hundreds or even thousands of miles away. Thus, the decision to abate local pollution problems by building tall stacks is a decision to shift the costs to someone else. Tall stacks are not an acceptable method for handling local pollution problems.

In some locations, pollution is almost entirely contained within the local area, as with Los Angeles. In other places, local emissions drift hundreds or thousands of miles, as with emissions from coal combustion in the Ohio River Valley. When emissions do not remain in the local area, local residents cannot be given the responsibility for balancing local abatement (and its associated control costs) against the local benefit of abatement.

A good example is a local decision to allow emissions into a local lake. Since the community must bear the costs of both emissions abatement and of the polluted water, giving it responsibility for these decisions will not give rise to important spillover effects. In contrast, local residents should not be given responsibility for decisions concerning emissions into a river, since the pollution is quickly carried downstream to cause discomfort for distant people.

An important case involving long-distance transport of air pollution is that of acid rain.[18] Emissions of sulfur and nitrogen oxides are oxidized in the atmosphere into acid aerosols. These then find their way to the ground through dry or wet deposition. The acids can damage vegetation and acidify streams and lakes, even to the point of killing all fish in a lake. Since oxidation takes many hours, it tends to take place as pollutants are transported considerable distances. Thus acid rain is a regional or even international problem rather than a local problem. Local decisions about emissions would not be expected to give rise to a socially acceptable solution.

Decreasing Air Pollution Efficiently

Efficiency was not a prime criterion when Congress considered environmental legislation. Congress was determined to improve air quality and to do so quickly. Practicality intruded principally in allowing existing pollution sources to face a more lenient test. Congress debated and rejected instructing EPA to make some trade-offs between the benefit from abatement and the social costs. Noncompliance was to be enforced by the courts, with substantial fines, plant closings, and even criminal penalties in some cases.

Predictably, the Clean Air Act has not produced air pollution abatement at low cost. It has not even produced rapid abatement, since the courts have been loath to close plants or fine companies that claimed they already had too few resources to spend on abatement.[7] Unfortunately, the threat to close a plant down was too severe to be credible.

Economists have argued the benefits of more efficient abatement mechanisms, especially the use of effluent fees or marketable discharge licenses.[19] Both mechanisms give flexibility to decrease emissions from sources that are inexpensive to control and to ignore sources that are extremely expensive to control. The result is a vast cost saving in achieving a given level of control, compared to an abatement scheme that calls for uniform cutbacks among all emission sources.

Effluent fees work by charging the emitter a fee for each pound of pollution dumped into the air. The fee presumably differs by type of pollutant and is meant to reflect the cost to society of having slightly more polluted air. The justification is that dumping pollution into the air creates an externality by imposing costs on others. The effluent fee acts to charge the polluters for the social cost of dumping the pollution and thereby gives them the proper incentive to decide how much pollution to control and how rapidly to do so.

Marketable discharge licenses are based on a determination of how much pollution should be allowed to be emitted into a particular "airshed." Each license is a permit to emit a stated amount of a pollutant; the total emissions add up to that determined to be socially desirable. Under this system, someone opening a new factory would be required to purchase the required discharge licenses from current polluters, along with purchasing land for the factory and acquiring the other means of production.

Effluent fees control the price, letting the quantity of emissions be determined by the marketplace. Marketable discharge licenses control the quantity of emissions and allow the price for the licenses to be controlled by the marketplace. Both methods use economic incentives rather than civil or criminal penalties to decrease the emissions of pollutants. Both can be shown to achieve a given level of emissions reduction more cheaply than the current system. However, effluent fees have been characterized as a "license to pollute," allowing rich companies to continue fouling the air, should they so desire. While it is certainly true that rich companies or individuals can now purchase expensive products or property and destroy them, we rarely see this occur. It is at least as unlikely that companies would decide to waste their wealth by emitting pollutants when it would be less expensive for them to decrease pollution.

Both of these economic implementation systems depend on public and judicial acceptance of their fairness. The current system of civil fines and criminal penalties has not worked because it was often found to be inequitable in particular cases. If the courts refused to enforce effluent fees because companies complained that they could not afford them, or if the courts permitted some companies to emit more than their licenses allowed in order to preserve the jobs of some workers, these economic incentives could not work either.

Recommendations

Coal can and should assume a larger role in the nation's energy future. However, many of the ad hoc responses to past problems must be reviewed and discarded.

In order to realize the potential contribution of coal to the energy supply of the United States the following steps are necessary:

1. More complete recognition of the environmental, occupational, and public health hazards associated with coal mining, transport, and use.
2. Agreement to take steps to cope with these problems.
3. A greater emphasis on efficiency than has prevailed in the past in handling these problems.
4. Eschewing "politics as usual" in confronting the issues. The programs must not be held hostage to preserving jobs or company profits in one region.
5. Reform of the Clean Air Act to place greater emphasis on achieving air pollution abatement at lower cost, and on deciding what levels of air quality are desired in view of social benefits and costs. Many of the cumbersome procedures must be scrapped or at least simplified.

8

FEDERAL LEASING POLICY

Walter J. Mead
and
Gregory G. Pickett

Summary

From the beginning of federal oil and gas leasing on the Outer Continental Shelf in 1954, the government has used cash-bonus bidding as a means of determining who gets the lease and what price will be paid. The federal government has collected approximately $48 billion in revenue from these sales and subsequent production.

All but a handful of the sales were conducted under auctions based on cash-bonus bidding with a fixed one-sixth royalty. Critics have charged that big oil companies have obtained leases at less than fair-market value and these critics persuaded Congress in 1978 to mandate use of other bidding systems, including royalty, profit-share, and work commitment bidding, paired with various fixed payments.

Economic analysis shows that the premise for the 1978 act is faulty. There is no evidence to support the contention that the government has failed to collect fair-market value for its leases or that large companies have obtained leases at bargain prices.

Our analysis indicates that pure cash-bonus bidding serves the national interest better than any alternative system considered. If a fixed payment is required, it should be a royalty, to be reduced and then eliminated as a reservoir approaches the end of its economic life.

Introduction

Outer Continental Shelf (OCS) oil and gas resources are controlled by the United States on behalf of all U.S. citizens.

The federal government's jurisdiction over the OCS extends from the limits of state boundaries seaward to a water depth of 200 meters or, according to the 1958 Geneva Convention, beyond that point "to where the depth of the superadjacent waters admits of the exploitation of the natural resources."[1]

Periodically the right to explore for and produce any oil and gas is sold to the highest bidder. The OCS Lands Act of 1953 authorizes the secretary of the interior (1) to grant mineral leases on the OCS, and (2) to prescribe regulations for their administration. Under this legislation, the vast majority of oil and gas leases have been auctioned under a cash-bonus bidding system with a fixed $16 \frac{2}{3}$ percent royalty. Experimental use was made of royalty bidding procedures.

Congress Prescribes Alternate Bidding Systems

In the later 1970s, there was widespread criticism that bonus procedures failed to collect fair-market value for the leases and that large companies where "ripping off the government" by obtaining leases at less than fair market value. As a consequence of this criticism, Congress passed the OCS Lands Act Amendments of 1978.[2]

Whereas the 1953 act authorized only cash-bonus or royalty bidding, the 1978 amendments specified four alternative bidding systems and then authorized any combination of these bidding arrangements plus any other system that might occur to the secretary of the interior. The specified bidding systems included (1) cash-bonus, (2) royalty, (3) net profit-share, and (4) work commitment bidding. Each of these bid variables must be paired with one or two required payments, which are fixed at the time that bidding takes place.

Most bidding forms are options. They confer on the high bidder the right to produce any oil or gas that might be found on the lease. The bonus bidding system compels the bidder to pay for that right. In contrast, profit-share bidding and royalty bidding require that the government grant the option free of charge. Neither the profit-share nor the royalty system requires any up-front payment for the option.

Congress specified that bidding systems other than the conventional cash-bonus bid with a fixed royalty must be used for not less than 20 percent and not more than 60 percent of the total area offered for leasing each year during the first five years following enactment of the legislation. This constraint ended on December 31, 1983. Congress may again prescribe leasing systems, which must be used; but if no legislation is forthcoming, then the interior secretary is free to use either bonus or royalty bidding as authorized by the 1953 legislation.

Plan of Chapter

The purpose of this chapter is to evaluate the strengths and weaknesses of alternative bidding systems. The analysis is intended to be useful (1) to Congress in debating alternative leasing systems, (2) to administrators in evaluating the leasing record, and (3) to the public in its consideration of the alternatives.

The analysis will begin with a discussion of the background of oil and gas leasing. In this section we will specify the standard by which leasing policy alternatives should be evaluated from an economics perspective. Alternative standards that might be important to Congress and the public will also be discussed. The heart of the paper is in the second section, which provides an analysis of the alternative bid variables. The dominant issue is the efficient allocation of resources. If an *efficient* leasing system is used, then an *optimum* amount of oil and gas will be produced at the *lowest possible cost*, and the federal government will collect the *greatest possible economic rent*. In order to evaluate the policy options, we will examine:

1. Whether cash-bonus bidding in the past has produced effectively competitive results
2. Whether the leasing system maximizes and collects the full economic rent
3. How administration costs may differ under alternative leasing regimes
4. Whether some leasing systems lead to abandonment of valuable oil and gas resources
5. Whether the bidding system identifies the most efficient firms
6. Whether the bidding system yields a correspondence between the amount paid for leases and the value of oil and gas produced
7. The use of fixed payments in the lease agreement

In the final section, conclusions will be drawn and policy recommendations offered.

General Considerations

It is probably true that most of the oil and gas resources remaining under U.S. jurisdiction will be found in the marine environment rather than on shore. It is probably also true that most of the hydrocarbons will be found in the Outer Continental Shelf area, which is totally controlled by the federal government. From 1953 through 1981, more than 5.7 billion barrels of oil and 53.5 trillion cubic feet of gas have been produced from the OCS, with a total market value of $81 billion. The federal government has collected approximately $48 billion in revenue from sales of OCS oil and gas leases.

Leasing Procedures

Selling oil and gas leases on the OCS involves several steps. First, the government issues a call for tract nominations and invites comments from industry and the public. On the basis of these responses, geographic areas identified for leasing are defined by the Interior Department. Environmental impact studies are then initiated. When draft environmental impact statements become available, public hearings are held in which industry, environmental groups, state and local governments, and other interest groups are invited to comment. On the basis of witness testimony, final environmental impact statements are then prepared and submitted to the President's Council on Environmental Quality. The secretary of the interior makes the final decision as to whether or not to hold a sale, what tracts to offer, and what environmental restrictions to include in the lease-sale terms. The secretary's decision, of course, is subject to court challenges and delays. If a lease-sale is authorized, interested parties are invited to submit bids using the bidding system specified by the government. A lease is then issued to the highest bidder, subject, however, to rejection by the government of any and all bids.

It should be made clear that when a firm purchases a prospective oil or gas lease it buys not one but two assets: (1) it buys the obvious right to explore for and produce any hydrocarbons found on the lease; (2) in addition, it buys information not only for the tract acquired but for adjacent tracts as well. Prior to bidding, firms will normally conduct or otherwise obtain geological and geophysical data that might identify prospective hydrocarbon-bearing structures covering specific leases. Figure 8-1 shows two such potential structures centered on tract E. The firm that obtained tract E in a lease sale may have

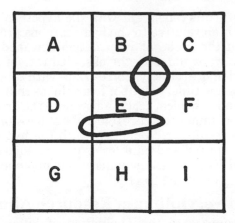

Figure 8-1. Diagrammatic illustration of potential hydrocarbon-bearing structures crossing several lease tracts.

Identified two potential structures either before or after winning the tract. Drilling at the site might have established that the smaller structure was dry and as a consequence the lessee has obtained valuable information indicating that unleased tracts B and C are of little or no value. Hence, in any subsequent sale, the lessee of tract E would use his information to avoid overbidding on tracts B and C. The larger of the two potential structures may, on the other hand, have shown promising prospects on the basis of post-sale drilling. In this event the lessee of tract E could use his superior information to bid more wisely (more aggressively) on any subsequent lease-sale offering of tracts D and F.

Evidence in support of the above points is available from rate-of-return analyses of Gulf of Mexico OCS lease-sale returns. Table 8-1 shows that, for wildcat leases, the observed

Table 8-1
After-Tax Rate of Return by Lease Class

Lease class	After-tax rate of return[a] (percent)
All 1,223 leases	10.74
1,109 wildcat leases	10.04
63 drainage leases won by neighbor lessees	16.20

[a]The first year of OCS lease sales, 1954, is treated as year zero. Cash flows are aggregated by calendar year and discounted to 1954 regardless of the year of the lease sale.

low rate of return is consistent with the hypothesis that high bids were too high relative to the hydrocarbons produced from such leases; if the value of information obtained from such wildcat leases were added to the observed rate of return, the results would more closely correspond with a competitive norm. At the same time, neighbor firms (firms that owned part or all of productive wildcat tracts adjacent to the drainage tract and sharing the same structure) earned a super-normal rate of return. These results are consistent with the hypothesis that information acquired and paid for as part of adjacent wildcat leases yielded compensating gains when neighbor firms acquired adjacent drainage tracts.

The Objective: Optimum Resource Allocation

As a prerequisite to the evaluation of different bidding alternatives for leases, policy goals must be identified as clearly as possible. There is wide agreement among natural resource economists that a single objective of "optimum resource allocation" should guide resource management policy. A statement of that objective has been carefully and concisely stated by S. L. McDonald as follows:

> The Department of the Interior, as custodian of the affected federal lands, should lease lands for minerals production on such terms and conditions, and at such a rate, as will tend to maximize the present value of the pure economic rent derivable from them.

McDonald then defines economic rent as

> any surplus in the income of a factor of production—land, labor, or capital—where *surplus* is the excess of the factor's income over the minimum amount necessary to call forth its productive services. . . . Clearly, such a surplus could arise only where the supply of the factor of production is less than perfectly elastic; for if the supply were perfectly elastic, the quantity supplied would expand in response to the least surplus and immediately wipe it out.[3]

McDonald presents his statement of objective in terms of government leasing policy, but the statement applies equally well to privately owned resources. Firms that maximize profit over time also maximize the present value of their resources, and society's interest will be served subject to the conditions that (1) no monopoly power is exercised, and (2) either no externalities exist or they have been effectively internalized. These same conditions also apply to government management of the nation's resources. However, to the extent that the government exercises its considerable monopoly power as the sole owner of OCS reserves, it produces the same resource misallocation losses that would occur when a private firm exercises monopoly power.

The concept of economic rent is easily illustrated in Figure 8-2. The height of the bar represents the total revenue obtain-

TOTAL REVENUE
Economic Rent
Necessary costs

Figure 8-2. Model of economic rent estimation. All values are discounted to obtain net present values. Note that economic rent is not a return on invested capital; the latter is treated as a necessary cost.

able, subject to a cost constraint, from any given oil or gas lease. This gross revenue is reduced by the necessary costs of exploration and production shown in the lower segment of the bar. Under conditions of effective competition, the operator is forced by the market to avoid unnecessary costs over which he exercises control. Government regulations that impose social costs in excess of social benefits on lease operators should be avoided. Such constraints either reduce production and revenue (lower the height of the bar), or impose unnecessary costs on the operator. In either case, the residual economic rent is reduced, and society suffers a loss.

The future revenue stream, economic rent, and all necessary costs are represented in Figure 8-2 to be in present value terms. Given the above assumptions regarding competition and externalities, the interest rate used to compute the present values corresponds with the discount rate a firm would use in computing its optimal bid value. This discount rate represents a normal competitive risk-adjusted rate of return on the capital employed.

Economic rent is paid in several forms under bonus bidding as practiced by the federal government. First, specified annual "delay" rent payments are required. These payments cease when production begins. Second, royalty payments, normally amounting to one-sixth of the wellhead value of any oil and gas produced, are required. Third, the discounted present value of the *remaining economic rent is theoretically collected in the form of the cash-bonus bid.* The leasing method used affects the amount of economic rent available for collection because the leasing method may affect the future revenue stream and the amount of the exploration and development costs imposed on the lessee.

Optimum Speed of Development: How to Allocate Production Over Time

Oil and gas resources are nonrenewable natural resources. Consequently, they must be allocated over time in a manner that allows their price to perform a conservation function so that as exhaustion is approached, substitute and higher-cost energy sources such as solar, oil shale, and possibly nuclear fusion become available. The operating rule for optimum resource allocation requires that the in situ net value of oil and gas increase at a compound interest rate corresponding with the opportunity cost of capital.

The record of federal stewardship with respect to optimum resource allocation over time is not good. A congressional bias favors "speedy development" of oil and gas reserves owned by the federal government. But no study within the government has addressed the question of the optimum allocation of the nation's resources between present and future uses. Nevertheless, the long-standing five-year rule requires that lessees begin producing oil or gas from federal leases within a five-year term or forfeit their lease back to the government. Thus any lessee is forbidden to delay production even if he believes that it would be profitable to do so. The five-year rule forces current production independent of future supply and demand prospects and probable price behavior. Thus, "speedy development" should be rejected as an operating objective for federal lease management because it violates rational resource economics and thereby the needs of a future generation.

Aiding Small Business

Similarly, one finds frequent congressional reference to aiding small business as an objective of federal leasing policy. From an economics perspective, leases should be awarded to the most efficient firms. If legislation biases lease allocations in favor of small firms that are less efficient, then the nation's scarce resources are not used effectively and the result will be a slower economic growth rate, an inefficient allocation of scarce resources (the opposite of resource conservation), and a lower standard of living than would otherwise exist. It should be noted that no evidence indicates that the large multinational oil companies are more efficient operators in the OCS environment than are smaller firms. On the contrary, the record shows that the big eight international oil companies not only paid more for their leases, but more of their acquired leases were dry, and they earned a lower rate of return on their lease investments than the smallest firms did. This record is shown in Table 8-2.

Table 8-2
Results of Lease Operation for 1,223 OCS Oil and Gas Leases, Gulf of Mexico, Issued from 1954 through 1969

	Average bonus per lease	Aggregate internal rate of return after taxes (percent)	Percent of leases
Big 8 firms	$2,310,499	10.37	61.70
Big 9–20 firms	2,354,070	11.26	63.71
All other lessees	1,740,377	11.15	57.90
Average all firms	2,228,332	10.74	61.57

Source: W. J. Mead, A. Moseidjord, and P. E. Sorensen, "Toward Efficient National Policies for Leasing Oil and Gas Resources," *The Energy Journal*, Vol. 4 (1983).

Stimulating Competition

Another objective frequently stated by Congress is that leasing policy should stimulate competition. Competition is essential if lease-sale markets are to perform effectively. Steps might be taken to increase the number of competitors able to participate in the lease auction market and thereby increase competition for leases. These steps involve primarily reducing barriers to entry in the form of front-end payments to the government. Whether these steps are appropriate depends on whether competition is effective under the present leasing system.

From 1954 through 1981, on 19,074 tracts offered for lease in the Gulf of Mexico, the average number of bids received per tract was 3.2. Only 47 percent of the tracts offered for lease actually received bids. Thus 53 percent of the tracts offered received no bids. The government reserves the right to reject any and all bids and exercises this right when it believes that the high bid received is inadequate, or for any other reason. The record shows that from 1954 through 1981, the government rejected high bids on 12 percent of the tracts receiving bids.

From the above information one might erroneously conclude that competition is inadequate, inasmuch as 53 percent of the tracts offered received no bids and 12 percent of the bids offered were rejected. However, the record of 1,223 tracts actually leased through 1969 shows that 61.9 percent of those tracts were dry. Another 16.3 percent did not produce enough revenue to cover their costs. Thus the tracts receiving no bids were probably of no net value *in the aggregate*. It is also likely that at least 80 percent of the tracts where bids were rejected by the government were of negative value (the in situ value of any reserves was negative). Supporting these generalizations (that bidders in the aggregate are able to distinguish productive from unproductive tracts) is the finding from a multiple regression analysis that there is a positive and significant correlation between high bids, as a dependent variable, and the value of oil and gas actually produced from tracts leased.[4]

Additional evidence on the basic rationality of the bidding process is shown in Table 8-3. Tracts receiving very low bids have a very high percentage of dry leases and a very low value of production. Correspondingly, tracts receiving very high bids have a low percentage of dry leases and a high value of production.

Evidence of effective competition for leases actually issued was given in Table 8-2, showing that the after-tax rate of return on such leases was 10.74 percent. This is clearly not in excess of the average 11.8 percent earnings after taxes for all

Table 8-3
Relationship of Bid Prices to Production Results

Bonus bid class	Number of leases issued	Percent dry leases	Average bonus per lease	Undiscounted avg. gross value of production per lease (actual through 1979)
$250,000 or less	354	81.36	126,450	4,996,521
$250,001–$1,000,000	367	64.58	524,998	9,660,978
$1,000,000–$3,250,000	285	48.77	1,874,621	20,807,215
More than $3,250,000	217	41.01	9,002,511	42,295,839

Source: W. J. Mead and G. G. Pickett, "An Economic Analysis of Oil-and Gas-Leasing Experience under Profit-Share and Bonus Bidding with a Fixed Royalty," in M. Neiman and B. Burt, eds., *The Social Constraints on Energy-Policy Implementation* (Lexington, Mass.: Lexington Books, D. C. Heath), pp. 39–61.

U.S. manufacturing companies. If one uses a 12.5 percent discount rate, the present value per tract leased after taxes was negative $192,128.[5]

In sum, the evidence indicates that lease-sale markets are effectively competitive. Additional steps that might be taken to increase competition for leases need to be evaluated relative to the costs of such policies. Congressional debate concerning leasing policies that tend to lower barriers to entry and thereby increase competition often are expressed in terms that indicate a second interest—aiding small business. Given effective competition, the issue of subsidizing small business should be debated separately. The latter issue raises a question of whether the public interest would be served by the entry of smaller firms into OCS oil and gas exploration, drilling, and production. The record of bidding for OCS leases shows that from 1954 through 1973, a twenty-year period, 128 separate firms won OCS leases either as solo bidders or as part of a bidding combino.[6] Through 1969, the big twenty firms won 84 percent of the leases issued. The remaining 16 percent of the leases were won by firms that range widely in size below the big twenty group. Oil and gas drilling and production in the marine environment require a high degree of technical competence. Concern for the environment raises questions about the wisdom of subsidizing small firm entry. Furthermore, in the event of an oil spill, society's interest in prompt action to halt the spill and to effectively clean up the recoverable oil requires long-term financial responsibility on the part of lessees. Very small firms may not qualify in terms of technical competence or in financial strength, or both.

Other Concerns

One frequently encounters congressional and political statements of leasing objectives that are virtually meaningless. For example, a recent U.S. House of Representatives report specified that the bidding regime should "strike a proper balance between securing a fair market return to the Federal Government for the leasing of its lands, increasing competition in the exploitation of resources, and providing the incentive of a fair profit to the oil companies, which must risk their investment capital."[7] This statement of objectives is meaningless inasmuch as "proper balance," "fair market return," and "fair profit" are undefined.

It is important to understand that trade-offs occur between (1) amounts bid for the bid variable, and (2) required payments to the government, which are fixed. For example, lease-sale contracts commonly require payment of an annual rent per acre that is fixed by the government in advance of bidding and has varied between $3 and $8 per acre. Where the bid variable is a cash bonus, there will be a dollar-for-dollar trade-off in terms of present values such that the higher the required rent payment the lower will be the bonus bid.[8]

However, in another trade-off setting one should not expect a dollar-for-dollar relationship. For example, as we will see later, royalty payments lead to reduced production of oil and gas from leases. Consequently, where the government doubles its ordinary one-sixth royalty to a one-third royalty, total lease revenue as shown in Figure 8-2 will decline. On a present-value basis, bidders will reduce their bonus payments by more than a dollar-for-dollar increase in royalty payments.

It should also be understood that any additional costs which the government imposes on lessees are paid dollar-for-dollar in terms of present value by the government. That is, in calculating the net present value of tracts in order to submit bids, bidders subtract from expected revenues all costs of exploration, development, and production.

Finally, any additional uncertainty concerning the terms of a lease-sale, such as rules and regulations which may be imposed in the future by the government, leads bidders to lower the expected present value of the lease and reduce their bids. Uncertainty will be reflected either in higher discount rates or in assumed additional costs. Either avenue leads to lower bids.

Analysis of the Bid-Variable Alternatives

The primary purpose of this section is to analyze the four major alternative bid variables: (1) cash bonus, (2) royalty, (3) net profit share, and (4) a work program to be performed by the lessee. We will then discuss the fixed payment alternative. The analysis will be in terms of four important performance criteria: (1) whether the bidding system maximizes and collects the economic rent; (2) the effect of the bid variable on administration and other costs; (3) whether the bid variable provides an unambiguous test of the high bidder; and (4) whether the bid variable equates payments to the government with the net value of production.

Criterion No. 1: The Optimal Bidding System Alternative Maximizes and Then Collects the Economic Rent

We have previously explained the economic rent concept and pointed out that it is a residual after all necessary costs are subtracted from revenue. Both revenue and costs are estimated in advance of bidding by all lessees. Expected dry tracts, or tracts expected to have costs in excess of revenue (including option value and the value of information about adjacent tracts as implicit revenue), receive no bids. Under effective competition, bidders in the long run will be forced to bid away all of the economic rent. However, the bidding system, together with any fixed payment and regulations, has an important effect on both production (and therefore revenue) and costs, and hence on the amount of economic rent available and collected by the government.

Pure Cash-bonus Bidding

A major advantage of a pure cash-bonus bid is that it transfers economic rent to the government at the time of the auction. Upon payment of the bonus, the economic rent becomes a "sunk cost" to the lessee and thereafter has no effect on his investment or output decisions. This result is optimal because economic rent is not a social cost. Rather it is a transfer payment to the government and as such it should not be allowed to affect investment or output decisions.

Pure Net Profit-sharing Bidding

Under pure profit-share bidding, the rent transfer function may or may not affect investment or output decisions, depending on how profit is defined. At least four choices are available. First, profit may be defined in the conventional IRS sense—that is, in the same manner in which taxes are calculated. In this event, investment decisions will be distorted because the profit-share payment is, in effect, an addition to income taxes beyond the approximately neutral effect of the latter. Investment decisions are made on an after-tax basis. Some otherwise supermarginal investments will fall into the submarginal category and will be rejected. Similarly, some leases that promise after-tax profits above the cutoff point will fall below that point and be abandoned due to the higher "tax" rate. One might argue that the oil or gas resource will still be there for later production. However, at a later point, the tract must be redrilled and bear new marginal cost, whereas on the first occasion the drilling cost had become a sunk cost and would not affect the initial decision to produce or not produce.

Alternatively, profit subject to sharing may be defined in accordance with the second choice, the "fixed-capital recovery plan"; the third option, the "investment account system"; or the fourth choice, the "annuity capital-recovery system." In all of these last three systems, the capital outlay, with or without an imputed interest charge, is subtracted from revenue before computing the profit-share to be paid to the government. In this event, new investments that might enhance recovery would not be incorrectly restricted. Instead, some otherwise uneconomic investments would be made. This will occur because these systems allow full recovery of the investment, plus a return on the investment, to come out of the government's share of the profit. Because the lessee is allowed full recovery of investments that fail to yield revenues in excess of the investment outlay, the lessee is likely to become less cautious than he would be under bonus bidding where he bears the full cost (before taxes) of his mistakes. Whether economic investments are discouraged or uneconomic investments are encouraged, the effect on economic rent is the same: the available economic rent is reduced.

A potential advantage of profit-share bidding is that over-zealous and uneconomic regulation might be restrained for the reason that such regulation reduces profit, and the government shares in that loss. There is a subtle point here that escapes laymen and legislators. There appears to be a misconception that regulations, particularly environmental regulations, can be imposed without regard to social costs and benefits. This misconception appears to be based on the incorrect conclusion that oil companies bear the cost. Some of the private cost, in the short run, may be borne by lessees. In the long run, it is highly probable that consumers will bear this private cost. Any social cost in excess of the social benefit is a net loss to society as a whole. The imposition of regulations having social costs greater than benefits should be avoided, regardless of the leasing regime.

Pure Royalty Bidding

In the case of royalty bidding, the loss of output and economic rent can be extremely large, relative to profit-share bidding. The loss follows from the fact that a royalty payment is a percentage of *gross* income, not *net* income. Three negative consequences follow for the lessor's collection of the economic rent.

First, because royalty payments are levied on each unit of production, they are added to other incremental costs of production and lead to early abandonment of production. Because royalty payments are transfer payments and not real social costs, *premature* abandonment occurs. This point is illustrated in Figures 8-3a and 8-3b. The production history of a typical oil

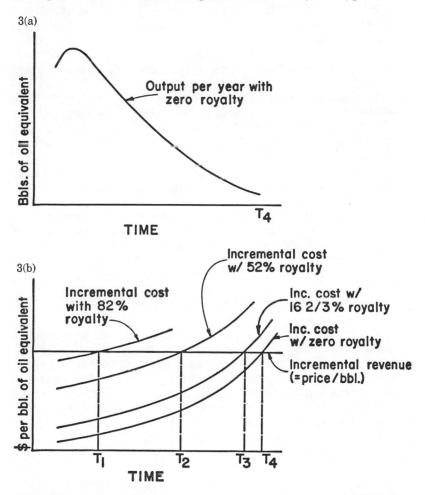

Figure 8-3. Diagrammatic illustration of oil well productivity over time, and incremental cost and revenue flows over time.

or gas well is shown in Figure 8-3a. Production begins at a high level and proceeds along some exponential decline rate as reservoir pressure diminishes. As production falls, incremental costs of production per barrel increase. Figure 8-3b shows that, without a royalty payment, production will cease at point T_4 when incremental costs rise to equality with the price of oil. This point is a societal and a private optimum—with no obvious externalities and a competitive domestic oil market.

However, with the conventional OCS 16²/₃ percent fixed-royalty requirement, production would be abandoned at point T_3 when incremental costs, including the royalty payments, rise to equality with the price of oil. Even with this relatively low royalty, a well will be prematurely abandoned, leaving socially valuable oil or gas in the ground.

In lease-sale No. 36 of October 16, 1974, ten tracts were offered under royalty bidding with a fixed bonus of $25 an acre. The royalty bids ranged from 51.8 percent up to 82.2 percent. We have plotted royalty payments of 52 and 82 percent in Figure 8-3b to show the approximate magnitude of the premature abandonment problem under high bonus requirements. We find that when incremental costs amounting to 52 percent of gross wellhead value are added to the incremental lifting costs, abandonment occurs at T_2. With an 82 percent royalty requirement, abandonment would occur at T_1, leaving a large amount of valuable oil in the ground. One might argue that the oil will be there for subsequent development and production, but again, once a field is abandoned, reopening it at a later date requires new incremental costs in the form of drilling and development that would have been unnecessary if the initial production had continued.

Congress gave the secretary of the interior authority to reduce the fixed royalty rate down to zero in order to avoid the premature abandonment problem. However, no interior secretary has ever exercised this authority. The primary consideration here appears to be a political hazard—in other words, the public might not understand the economic reasoning indicating that society would benefit by such action and might instead see it as a "giveaway" to oil companies.

A second negative consequence follows from the fact that royalty payments add to the operating costs resulting from any investments designed to *enhance* oil or gas recovery from a reservoir. Many such investments will become uneconomic as a consequence. Thus, royalty payments lead to suboptimal levels of investment in reservoir development.

Finally, the higher the royalty requirement, the greater will be the number of wells that will be plugged and abandoned rather than developed. One can see from Figure 8-3b that incremental profits per barrel are very small for high-royalty levels. The incremental costs of production do not include the cost of field development. Hence only giant fields will be produced under the type of royalty bidding used in the October 1974 experimental sale.

In all three instances, the collection of economic rents will be reduced under royalty bidding. This is a major negative factor relative to royalty bidding. As a variation on the royalty bidding system as outlined above, the state of California and the federal government have both authorized and experimented with a sliding-scale royalty where bidding is on some factor equal to or greater than 1.0 times the specified sliding scale.

A sliding scale would be an improvement over an inflexible royalty rate if the low-end rates are below the conventional 16⅔ percent fixed rate, but it creates a new set of problems. Lower rates must be "triggered into effect" by some objective measure. This measure is commonly the output per well, over some time period. First, the operator is in control of output and can reduce his payment to the government by avoiding high output levels, thereby stretching output over a longer period of time. But a social optimum requires that the output profile be such as to maximize the present value of the resource. This problem is a minor one because some other producers who are free to set their annual output levels so that the value of oil in the ground increases at the opportunity cost of capital (the condition that maximizes the present value of oil reserves) will set their annual output levels in an offsetting manner.

Second, production-enhancing investments are discouraged by the fact that increased production may escalate royalty rates into higher brackets on the sliding scale and some of the incremental benefits will be taken by government.

In sum, royalty bidding, even if modified and improved by a sliding scale, reduces the recoverable oil and gas from productive fields. In terms of Figure 8-2, total revenue is reduced, without corresponding reductions in costs, with the result that the nation loses economic rent.

Work Program Bidding

Work program bidding is a system wherein the bidder is invited to state exactly what he would do if awarded the lease. The government may suggest areas in which the bidder is required to specify his proposed work program. For example, proposals may be requested concerning the number of wells that would be drilled; whether new onshore investments would be made in separation plants, refineries, petrochemical plants, and so forth; whether subsea completions would be used; and the like. In Alaska, bidding on a proposed work program regime required a statement of the dollar amount for the program. Presumably this meant that the bidder would cost out his entire exploration, development and production program. A winning lessee would then be required to post a bond to ensure that he would faithfully discharge the commitment.

The disadvantages of this system for economic rent maximization are overwhelming. The optimum level of investments in well drilling, onshore facilities, field development, and other areas cannot be specified in advance of knowledge about drilling results. If initial drilling is unsuccessful, subsequent drilling should normally be scaled down, relative to highly successful initial drilling. Similarly, optimal field development investment plans can be determined only when the extent, quality, and value of reserves are known.

Work program bidding is based on the assumption that the leasing bureaucracy is in a position to know the socially optimal line of development.[9] However, this group is generally neither trained nor experienced in making investment decisions, either from a private or a social perspective (there may be no difference). Yet it must decide which program is to be awarded a lease. Experience with this system in the North Sea fields has shown that bidders are forced to determine what kind of work plan is favored by the decision-making bureaucracy and then bid generously in that (or those) favored area(s).[10]

To the extent that work program bidding results in wasteful investments or operating costs, dollar-for-dollar reductions (in present value terms) occur in cash-bonus payments or other cash payments to the government. Thus, economic rents are sacrificed in the work program bidding system.

Summary

From the perspective of maximizing and collecting the economic rents, a cash bonus appears to be the superior bid variable, and profit-share bidding a distant second best. Royalty bidding and work program bidding should be rejected without further experimentation on the argument that these systems substantially reduce the economic rents available to society.

Criterion No. 2: The Bid Variable Should Minimize Administrative Costs (Relative to Their Benefits)

Cash-bonus Bidding

From the viewpoint of keeping administrative costs at a minimum, cash-bonus bidding is preferred. With this system, the winning bidder pays for his lease with a lump-sum payment. Under a pure cash-bonus bidding system, no further payments are required. There is nothing for the government to monitor. The lessee is entitled to produce any oil or gas found on the lease. An optimizing lessee will produce from the lease, provided that the present value of production exceeds the present value of the additional costs. He will cease production when incremental recovery costs reach the incremental value of the product. In the absence of any externalities, this private judgment corresponds with the social optimum.

Royalty Bidding

Second best, from the viewpoint of minimizing administration costs, would be a royalty bidding system. The only cost of administration is the requirement that production be measured, plus the need to place a wellhead value on that production. Problems occur in the latter function where the oil or gas produced is not sold at the wellhead in an "arm's-length" transaction. This situation frequently occurs when the producer processes the oil or gas in his own facilities, or when products are traded rather than sold. If a sliding scale is used in royalty bidding, administrative costs are likely to increase substantially. The lessor may determine that the lessee is constraining output in order to reduce his royalty obligation, and the lessor may want to introduce a production-monitoring program to enforce production at a level that maximizes current output, subject to physical recovery constraints.

Profit-share Bidding

Profit-share bidding entails relatively heavy administrative costs. However, no reliable estimates are available on the administrative costs borne by either the lessor or the lessee. In its proposed net profit-share leasing regulations under the "fixed capital recovery rule," the Department of Energy (DOE) speculated that a lessee firm would incur accounting system modification costs estimated at between $50,000 and $150,000. In addition, annual administrative costs were estimated at between $25,000 and $30,000 per lease. It is not clear whether costs of litigation were considered by DOE. Also, DOE provided no estimates of increased administrative costs for the government. Firms responding to the proposed regulation countered that administrative costs were underestimated, but no estimates were given by the responding firms.

The source of the heavy administrative costs is the agency problem. Where there is a divergence between the interest of the lessor and the lessee as an agent, one must expect that the agent will behave in a manner that serves his own interest first. The agency problem leads to monitoring costs by the lessor in an effort to prevent the lessee from pursuing such opportunistic behavior. The divergence of interest arises out of the fact that the costs of some projects are deductible from revenues and, where a lease is profitable, the costs are shared according to the profit-share percentage. However, the benefits of the project may accrue primarily or entirely to the lessee firm.

The agency problem is compounded by income taxes and any severance taxes, the latter being like income taxes in that they are levied against wellhead values (delivered prices less all costs from wellhead to delivery point). Let us take a simple example involving only corporate income taxes at a 46 percent rate and a 50 percent profit-share. The incentive to operate efficiently is a reward of 54 cents per dollar of profit after income taxes. Paired with a 50 percent profit-share requirement, one dollar of profit reward generated by an efficiency effort is reduced to a reward of 27 cents after taxes and profit-share. The other half of the coin would state that government bears 73 percent of the cost of wasteful activity. This is a special example of the externality problem. Where benefits accrue primarily to the lessee, but the costs are shared between government and the lessee, some investments and expenditures will be made because the benefits to the lessee exceed the cost-share born by the lessee. The nation learned this lesson during the Korean War when an excess profits tax was levied on corporate income with the result that the highest tax bracket became 82 percent. Under such conditions, leasing of a corporate jet, business conferences in Hawaii, and similar "perks" that benefited management, were commonplace for the reason that they cost only 18 cents on the dollar after taxes.

We will provide illustrations of these potentially wasteful expenditures; however, it should be understood that the list is not exhaustive:

1. When a company wishes to do some research and development (concerned with oil exploration and production) where offsetting benefits are assigned a low probability of success, it is more likely to do so on leases involving profit-share payments—that is, where the losses are shared with the government. If the experimental work is successful, it can then be adopted on other properties controlled by the lessee (thereby producing benefits for the lessee).
2. Where a company has a mixture of high-quality and low-quality drill rigs or drill ships, it is likely to use the poor equipment on the profit-share lease and reserve the best equipment for other company operations.
3. Where a company needs to train crews in drilling and reservoir development, it is likely to do its training on profit-share leases.
4. "Gold-plating" is likely to occur on profit-share leases where the share paid to government is high and the retained share is low. Evidence of this practice may be found in the Long Beach (Wilmington) field where profit-shares paid to the government are extremely high. In this instance, attractive palm tree–studded "islands" have been created to shield the production equipment, in lieu of the usual and frequently unsightly production equipment, towers, and piping. Substantial public relations benefits flow to the operating firms, but the expense is borne almost entirely by the city of Long Beach.
5. In the event of oil equipment supply shortages such as occurred in 1973 and 1974, profit-share lessees are likely to allocate available supplies to their non-profit-share leases first. The opportunities to shift expenditures to profit-share leases are limited only by the imagination of the lessee.

Some of the shifts suggested above are merely the equivalent of income transfers and therefore involve no social costs to society. However, any shift that leads to inefficient investments or operations creates social costs. In any case, the economic rent available to the lessor is reduced by shifting of the type illustrated above. Such shifting forces the lessor to protect his interest against agency behavior by monitoring investment and operating decisions. This leads to monitoring costs Administrative interference in the operation of a profit-

share agreement becomes a necessity. The Long Beach–
Wilmington contract provides as follows:

> The City Manager . . . shall exercise supervision and control of all
> day-to-day unit operations . . . and . . . shall make determinations
> and grant approvals in writing as he may deem appropriate for the
> supervision and direction of day-to-day operations of the Field
> Contractor, and the Field Contractor shall be bound by and shall
> perform in accordance with such determination.[11]

The federal government has recognized that the divergent
interest problem leads to increased monitoring. In the Energy
Department's "Final Rulemaking" regulations, the point was
made that

> in a very real sense, the USGS [U.S. Geological Survey] represents
> the interests of the public in the exploration and development of its
> resources and already has an active role in OCS decisions. As
> future sharer in net profits, USGS must ensure that any sharing
> arrangement make provisions for monitoring expenses incurred
> during exploration, development, and production, in order to en-
> sure an equitable division of net profits.[12]

The regulations require approval of all expenditures subject
to fixed capital recovery. The federal system allows full recovery of these capital expenditures, plus a specified rate of return
for the operator, provided that offsetting revenues are subsequently generated from the lease. In the case of subsequent
operating expenditures that are fully deducted against income,
but without an additional payment for the use of capital, the
regulations state that "The lessee retains absolute discretion to
incur such additional costs as appear warranted." However, all
expenditures are subject to audit within the three years following the expenditure. In instances where the government attempts to deny prior outlays, legal challenge must be expected.
Such litigation will impose heavy costs on both the lessee and
the government, and serve to reduce economic rents.

In the case of wages and salaries charged to the profit-share
lease, the regulations state vaguely that "employees need be
engaged in [profit-share lease] operations continually only for
a specific period of time, such as a month, week, or pay period.
Their wages and salaries as well as other enumerated personnel costs are then includable."[13] The lessee will have a clear
incentive to charge wages and salaries to the profit-share
lease, to the limit of credibility. One must expect costly litigation to follow.

Oil service companies perform many of the essential functions of post-lease exploration, drilling, development, well reworking, and the like. Some of their costs may not be specific to particular leases (transportation for example). This leads to the problem of allocating costs between leases on which they operate. The usual incentives arising out of the agency problem will lead to litigation.

The lessee normally owns equipment that will be used on profit-share leases. In the absence of an arm's-length rental transaction, problems will arise as the lessee attempts to maximize his rental charge on the profit-share lease. The regulations allow equipment rents to be based on "actual costs of acquisition, construction, and operation . . . subject to the ceiling of average commercial rates for similar equipment and facilities prevailing in the vicinity.[14] The ceiling will be difficult to apply due to differences in location, environmental hazards, state of the market, type of equipment, etc. The divergent interests of the parties will lead to litigation.

Finally, there will be litigation over what litigation costs may be charged against the profit-share lease. Where the lessee uses in-house legal counsel, allocation of such joint costs becomes highly subjective. Regulations must fall back on prelitigation wording: joint costs must be "reasonable and equitable," but these words will have to be interpreted by the courts.

In sum, the agency problem leads to detailed regulations. The lessor must specify operating behavior as well as the accounting of all costs in order to protect his interest. This necessity leads to litigation as well as administrative costs to be borne by both the lessee and the lessor. In the first event, economic rents, as illustrated in Figure 8-2, are reduced. In the second, economic rents are dissipated by the government in ways that do not benefit the public. There are few or no comparable problems of this sort in the cases of bonus or royalty payments.

Simulation models have been constructed to estimate the economic rent that would be collected under profit-share bidding relative to the conventional bonus bidding with a fixed royalty. D. K. Reece concluded that the government would capture substantially greater rent under profit-share bidding.[15] However, he assumed that "the total rent recovered by society would be the same under alternative leasing systems." In effect, Reece assumed that the lessee's exploration and development program was *independent,* whether royalty, profit-share, or bonus bidding was used. Furthermore, Reece ignored the problems associated with *defining* a competitive equilibrium under the pure royalty and pure profit-share bidding systems. Finally, Reece abstracted from the *sequential* attributes of exploration by assuming that the lessee would incur only a one-time *fixed cost* in (1) evaluating the lease after the auction, (2) exploring the lease, and (3) developing the lease. However, because of the no-cost option attributes of pure royalty and pure profit-share bidding, the lessee may exercise his right to abandon the lease without penalty at any time. Clearly, if firms differ in their efficiency attributes, the exploration and production profile of the lease will depend on *who* is initially assigned the right to explore and develop the lease.[16] These assumptions are sufficient to tilt Reece's conclusions in favor of profit-share bidding. The critical issue in such models is the validity of the assumptions. Our evidence indicates that administrative costs for profit-share operations are both high and wasteful. A discussion of time-preference rates appropriate for industry and government is not within the scope of this chapter.

Work Program Bidding

Work program bidding, like profit-share bidding, imposes heavy administrative requirements on both lessor and lessee. The program proposed by each bidder must be carefully defined in the bid statement. It must then be carefully evaluated by the lessor in order to identify the "high bidder." Following designation of the winning bid, the lessor must monitor the lessee to ensure that he performs in a manner that is not less than his bid statement guarantees. Performance of this duty requires that the work program be specified in great detail in order to minimize the disputes as to the promised work program. The same agency problems exist here as identified above relative to profit-share bidding. After winning the lease, the lessee has an interest in minimizing his costs for projects that may be of importance to the bureaucracy but are of little importance to the lessee. One must expect that monitoring the work program will lead to heavy administrative costs, including expensive litigation.

Summary

From the viewpoint of minimizing administrative costs, bonus bidding is preferable, with royalty bidding a close second. Profit-share and work program bidding involve high and wasteful administrative costs for both lessor and lessee and reduce or dissipate economic rent. No reliable estimates are available indicating the amount of such costs.

Criterion No. 3: The Bid Variable Should Unambiguously Identify the High Bidder and Indicate the Most Efficient Operator

The high bidder is clearly identified under either pure bonus, royalty, or profit-share bidding.[17] Work program bidding may not indicate the high bidder where several work program projects are identified. For example, Firm A might guarantee to drill five wells, with subsea completions of any productive wells, while Firm B might guarantee to drill six wells and build a petrochemical plant in a high unemployment zone onshore. Obviously, there is no objective means of determining which firm is the high bidder.

An interesting variation in bidding occurs in the state of Louisiana where bidding is permitted on two or more variables simultaneously. Firm A bids $1 million in cash-bonus payments paired with a 14 percent royalty, and Firm B bids $1.1 million in bonus, but with a 13 percent royalty. Again, no objective test of the high bidder is possible.

Cash-bonus Bidding

While the high bidder is clearly identified in bonus, pure royalty, and profit-share bidding, only in the pure bonus-bidding case is there a theoretical basis for asserting that, in the long run, high bidders are the most efficient operators. In this instance, the high bidder indicates his expectation that costs are low relative to the expected income flow. This expectation in the long run must be correct, or the firm will face bankruptcy in its lease development operations. The bonus is paid in advance, and the government knows exactly what its revenue will be when it accepts the high bid.

Royalty Bidding

In the case of royalty bidding, the payment to the government depends on how much oil is found and produced. An inefficient firm may find less, or may produce a lower percentage of the oil (or gas) in place. Consequently, the government may receive less income flow from the inefficient firm that has submitted the higher royalty bid.

Profit-share Bidding

This same problem exists in the case of profit-share bidding. In the example given below, Firm A, bidding a 50 percent profit-share, will be the winner over Firm B, bidding a 40 percent profit-share. However, Firm A is less efficient (produces less revenue at higher cost) and generates less profit. Thus the payment to the government by the apparent high bidder is less than the second highest bidder would have paid.

	Profit-share bid (percent)	Efficiency test Rev. − Cost = Profit	Payment to government
Firm A	50	100 − 60 = 40	20
Firm B	40	115 − 55 = 60	24

If the government is to have confidence that by accepting the apparent highest bid it will receive the largest payment, it must conduct a separate analysis in which it evaluates the efficiency attributes of the bidding firms. As a practical matter, this is impossible. Consequently, the high-bidding firm in royalty and profit-share systems will get the lease but will not necessarily produce the highest payment to the government.

Criterion No. 4: The Bid Variable Should Approximately Equate the Payment to the Government With the Net Value of the Lease

The primary merit of royalty and profit-share bidding is that if no oil or gas is found, no payment is made to the government. When commercial quantities of hydrocarbons are found, then payments to the government are correspondingly large. However, bonus bidding is superior to royalty bidding in this respect because in bonus bidding, payments are based on flows of revenues minus costs, whereas in the royalty case, payments are based on gross income only, leading to premature abandonment, abandonment of productive reservoirs, and rejection of socially desirable investments.

Bonus bidding fails in this test, on a tract-by-tract basis. For example, Exxon, Mobil, and Champlin bid and won the Destin Anticline sale offshore west of Florida in 1973 with a high bid amounting to $212 million, and never found commercial deposits. In contrast, the Prudhoe Bay field, containing the largest oil reserve ever found on the North American continent, was obtained at a bonus cost of only about $950 million for its estimated 9.6 billion barrels of oil and 26 trillion cubic feet of natural gas. Again, in contrast, this event was followed by enthusiastic bidding for North Slope leases in 1969 where bidders paid over $900 million for leases without even one commercial discovery being announced.

While bonus bidding clearly fails to produce cash-bonus bids approximating real values on a tract-by-tract basis, there is a close correspondence in the aggregate, as we have demonstrated. Thus diversification available to the federal government (i.e., selling many leases to many buyers) yields the same benefits as if tract-by-tract correspondence existed. Similarly, diversification produced approximately normal returns to the firms. On the basis of the 1,223 leases issued from 1954 through 1969 in the Gulf of Mexico, all of the big twenty firms had positive rates of return on their leases. The returns ranged from 3.37 percent to 20.88 percent, and the number of leases per company ranged from 14 to 165.[18] From a national economic perspective, the lack of correspondence on a tract-by-tract basis is of negligible importance. However, politicians appear to be embarrassed by the occasional Prudhoe Bay situation where the lessee obtains a bargain, even though the larger record shows that approximately 62 percent of the tracts leased were dry and worthless while another 16 percent were productive but not profitable.

In the case of work program bidding, little can be said about the correspondence issue. The terms of payment depend on the criteria of the decision-making bureaucracy and the bidders.

Fixed Payments Should Cause No Reduction in the Economic Rent

In the foregoing, we have discussed the four bid variables. Leasing regimes in use by the federal government as well as in oil states have commonly paired a bid variable with one or more fixed payments. Most commonly, a cash-bonus bid variable has also required a fixed rental payment that must be paid annually until production begins. It is then replaced by a royalty payment, normally 16⅔ percent on the OCS.

The first purpose of the bid variable is to determine the winning bidder. Its second purpose is to specify part of the payment flow to the government. But the latter is determined jointly with any fixed payment requirements. The terms of the fixed payments are determined in advance of bidding by the government, whereas the bid variable is determined in the auction by the high bidder.

The economic analysis given above applies equally to the selection of fixed payments. The important difference is that the amount of the fixed share, being controlled by the government, can be set in a manner that avoids the problems of very high royalty or profit-share percentages. The use of fixed royalty or profit-share payments leads to reductions in economic rents and should be avoided unless there is evidence of compensating benefits.

Governments, both federal and state, commonly specify work program requirements in the lease contract. Unless these requirements produce social benefits in excess of social costs, there will be a reduction in the economic rents available to society. In this event, the requirements waste resources and eventually cause reductions in the living standards of U.S. citizens.

In formulating public policy relative to bidding regimes, it should be understood that trade-offs exist between the specified bid variable and the fixed payments. If the traditional 16⅔ percent royalty is eliminated, competition will force firms to bid higher bonuses. Because economic inefficiency occurs as a result of royalty payments, its elimination would cause the present value of bonus bids to increase more than the loss in present value of royalty payments. Similarly, if a zero profit-share is replaced by some positive profit-share payment requirement, the present value of the bonus bid will be reduced by more than the present value of the profit-share payment.

Conclusions

1. On the basis of resource allocative efficiency, the optimum leasing system would be a *pure* cash-bonus bidding regime (i.e., without royalty or other fixed payments). It would maximize the economic rent and minimize the government's administrative costs, thereby minimizing the dissipation of its economic rent. To the extent that a fixed payment must be added, an annual rental payment per acre would involve only minor additional social costs. Fixed royalty payments should be deleted from the lease sale regime because such payments reduce oil and gas production that would otherwise produce gains for society in excess of social costs. Nor should a fixed profit-share be added. Its primary social cost occurs as a result of heavy administrative costs for both lessee and lessor. The main negative feature of a pure bonus bidding system is that, on a tract-by-tract basis, payments to the government do not correspond with benefits received by the lessee. However, due to the fact that the government sells many leases, returns to the government correspond closely with the amount of economic rent that is embodied in the resource.

2. As a second-best system, we recommend the customary federal system of bonus bidding with a fixed royalty, provided the secretary of interior exercises his legal power to reduce the fixed royalty as exhaustion of the well, field, or tract approaches. This action would eliminate the premature abandonment problem. It would continue, however, the social costs of discouraging otherwise optimal investments in enhanced production, and of plugging and abandoning slightly supermarginal newly discovered reservoirs. The latter problem could be mitigated or avoided entirely if the secretary would, upon discovery of the problem, reduce or completely eliminate the royalty requirement on such leases.

3. The third-best system would be bonus bidding with a fixed profit-share requirement. The principal social cost, causing a reduction in economic rents and their dissipation, is the high and wasteful administrative cost of collecting the profit-share.

4. As a poor fourth-best system, profit-share bidding would have the merit of a close correspondence between payments to the government and the value of the lease to the lessee. However, high profit-share payments involve heavy administrative costs; and the higher the percentage, the more litigation is likely to occur due to disputes between lessor and lessee. Further, profit-share bids must be accompanied by work program commitments in order to attain a competitive equilibrium objective. This point applies equally to royalty bidding.

5. Systems with little or no merit that should be avoided entirely include royalty bidding and work program bidding. Both impose heavy social costs on the nation.

9

NUCLEAR POWER ECONOMICS AND PROSPECTS

Bernard L. Cohen

Summary

The costs of nuclear power in the United States have escalated far more rapidly than inflation in recent years. As a result, no new nuclear power plants have been started since 1975, and in the 1990s electricity may be twice as expensive in the United States as in Europe. The cause of this price escalation is traced largely to constant tightening of safety requirements in spite of new scientific and technical developments favorable to more relaxed safety requirements. For example, the LOFT program has shown that the emergency core cooling system performs much better than expected in preventing a meltdown; and chemistry studies have found that iodine and some other elements would be released almost totally in nonvolatile form in any catastrophic accident, which reduces the expected number of deaths manyfold.

Introduction

Nuclear power was economic before 1973 when the price of oil was only $3 a barrel. It has never caused a single death among the public, observed or calculated, and it has produced negligible pollution. What has changed?

Nuclear power is still safe and nonpolluting. But the cost of plants has escalated dramatically, by a factor of five more than general inflation. Why? Because of increased public concern about health and safety—a concern that has been greatly exaggerated by zealots and by the media.

The real issue is cost—and this subject is addressed in what follows. But because I cannot ignore the public concern, I have added a factual appendix that deals with the issues of health and safety.

Cost Increases for Nuclear Power Plant Construction

Data on Cost Increases

The cost per million kilowatts* of capacity of American nuclear power plants[1] is plotted vs. their date of completion in Figure 9-1. The most striking feature of these data is the very rapid cost increase over the past decade, far exceeding the effects of inflation. Our first task is to understand this feature of the data.

A large part of the problem can be addressed with reference to a related but somewhat simpler subject: the average cost of plants as estimated at the time the project was initiated. The Philadelphia office of United Engineers periodically develops such estimates under the assumption that construction will proceed in an orderly and efficient manner based on a design that satisfies all regulatory requirements in effect on the initiation date. Their estimates[2] have increased from $350 million in 1971 to $4,400 million in 1982, during which time period the consumer price index increased by a factor of only 2.3. We begin by trying to understand this dramatic cost increase.

The cost of a power plant is defined as the total amount of money spent up to the time it goes into operation. It is the product of three factors:

1. *The cost of all materials and labor if purchased at the time the project is started.* This is called the energy economics data base (EEDB), and is colloquially referred to as the "overnight" cost, as if assuming the plant was built overnight. Its estimate by United Engineers represents a large accounting effort, involving a listing of the thousands of individual undertakings required to construct a plant, determining the quantities of engineering, materials, equipment, and labor needed for each of them, estimating the costs of these on a national average, summing these costs, and adding appropriate contingency expenses derived from experience in comparing this type of esti-

* 1 million kilowatts = 10^9 watts = 1000 megawatts = 1 gigawatt.

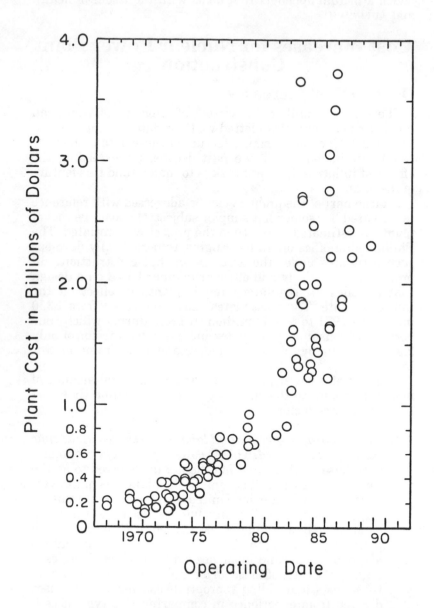

Figure 9-1. American Nuclear Power Plant Cost (per million kilowatts of capacity) Versus Dates of Completion.

mate with actual costs. This EEDB increased from $220 million to $1,300 million between 1971 and 1982.

2. *A cost escalation factor, which takes into account the inflation of costs with time.* For each item, this depends on how long after the project initiation date it is purchased. For example, the basic plant design will be done shortly after the project is begun, with a cost escalation factor close to 1.0, whereas the plant start-up testing costs will be incurred many years later, yielding a larger cost escalation factor.

3. *A factor covering the interest charges, incurred on a given expenditure from the time it is spent until the date the plant goes into operation.* This is often expressed as an additive cost and called "allowance for funds used during construction." In the above example, the interest would be large for the basic design work, but would be little more than 1.0 for the start-up testing. Since interest rates are normally a few percentage points higher than inflation rates, delaying purchases until they are needed reduces their interest costs more than it increases cost escalation factors and is therefore economically advantageous.

The contribution of both escalation and interest to the total cost of the plant are determined by two factors, the duration of the construction project and the inflation rate, assuming that interest rates on borrowed capital are closely related to the latter. There is, therefore, no need to treat escalation and interest separately, so we consider only their product (escalation times interest). It was only 1.17 in 1967 when project times were 5.5 years and the consumer price index (CPI) was increasing by only 4 percent per year. Escalation times interest increased to 1.45 in 1973 when the duration of projects stretched to 8 years but inflation rates were still only 4 percent per year. It went up to 2.1 in 1975–1978 when estimated time requirements reached 10 years and the CPI was increasing by 7 percent per year, and jumped to 3.2 in 1980 when project durations extended to 12 years and the inflation rate soared to 12 percent per year.

It is important to understand that an escalation-times-interest product of 3.2 in 1980 means that the final cost of a plant started at that time was expected to be more than triple the costs for labor and materials at that time; that is, 69 percent of the total estimated final cost of the plant was due to inflation and interest on funds used during construction.

We now turn to a discussion of the other factor that determines the cost of a plant, the EEDB. The increase with time of this "overnight" cost is, of course, partly due to price inflation between the various project initiation dates. This inflation rate has been about 2 percent above the rate of increase of the CPI because the cost of labor and materials used in construction has been rising more rapidly than other prices.[3] For example, the average annual price increase between 1973 and 1981 was 11.5 percent for concrete, 10.2 percent for turbines, and 13.7 percent for pipe,[4] as compared with only 9.5 percent for the CPI. This 2 percent added inflation rate for labor and materials used in power plant construction was also included in the cost escalation factor discussed previously.

After the EEDB is corrected for inflation, it still doubled between 1971 and 1982. This means that the actual quantity of labor and materials used in constructing a plant doubled during that time period. This conclusion can, of course, be understood on a more direct basis. Between the early and late 1970s, according to one study,[5] the quantity of steel required to construct a plant increased by 41 percent, the amount of concrete by 27 percent, the footage of piping by 50 percent, the length of electrical cable by 36 percent, and the number of man-hours of labor increased by 50 percent. Commonwealth Edison, the utility that services the Chicago area, completed its Zion plants in 1973–74 with 5.3 man-hours of labor per kilowatt of plant output; its Byron plants, scheduled for completion in 1984–85, will use an estimated 12.8 man-hours of labor per kilowatt.[6] For plants started in 1982, United Engineers estimates 23 man-hours per kilowatt.

Regulatory Ratcheting

What has caused this large increase in the labor and materials required to build a nuclear power plant? It is widely referred to as "regulatory ratcheting." In order to be licensed, a plant must have a long list of safety features required by the Nuclear Regulatory Commission (NRC), and every piece of equipment must satisfy specific standards. Every new feature NRC adds to its requirements and every new upgrading of standards leads to an increase in the amounts of labor and materials. Under normal circumstances, a regulatory agency can be expected to change its licensing requirements in response to operating experiences and new research developments, sometimes increasing and sometimes reducing the requirements. In the case of NRC, however, the changes have all been in the direction of increased requirements. Like a ratchet wrench that is moved back and forth but always tightens and never loosens a bolt, NRC has been constantly tightening requirements; hence the term "regulatory ratcheting."

Regulatory agencies normally consider costs before imposing new requirements to decide whether they are cost-effective. NRC has given nominal recognition to such procedures, but NRC personnel privately concede that their cost estimates have been very crude and generally on the low side. In fact, NRC has recently set up a Cost Estimation Group in an effort to ameliorate this problem. But even with the best of intentions, it is difficult to estimate costs of things that have never been done, and there seems to be a strong tendency even by architect-engineers and constructors, to underestimate them. United Engineers has frequently been forced to increase their cost estimates as field experience has accumulated.

In addition to increasing the quantity of labor and materials required, regulatory ratcheting has increased the length of time required for construction. According to the United Engineers estimates, the time from project initiation to ground breaking increased from 16 months in 1967 to 32 months in 1972, to 54 months in 1980. These are the times needed to do the basic engineering and design, to prepare environmental-impact and reactor-safety analyses, to have these reviewed by NRC, to satisfy NRC with responses to its criticisms and reservations, to do the same with the Advisory Committee on Reactor Safeguards, to go through public hearings (under the supervision of an Atomic Safety and Licensing Board), to respond satisfactorily to objections raised in that process, and finally to receive a construction permit. The time from ground breaking to operation testing, according to the United Engineers estimates, went from 42 months in 1967, to 54 months in 1972, to 70 months in 1980. This is the period when all actual construction work is carried out. Extending the time required for the project, of course, increases the cost of the plant by increasing the interest and escalation factors.

We have now identified the three basic factors that have increased the costs as estimated at the project initiation date: (1) increased labor and materials requirements; (2) increased time requirements; and, (3) inflation. The first two of these result from regulatory ratcheting, while the third is due to external matters. The total cost increased by a factor of 12.4 between 1971 and 1983, while the labor and materials requirements doubled. This leaves a factor of 6.2 to be divided between lengthened project times and inflation. If project times had stayed as estimated by United Engineers in 1971, project midpoints would have been 1974 and 1985. Between these dates, inflation, experienced and projected, was a factor of 2.5. This leaves a factor of $(6.2 \div 2.5 =)$ 2.5 as the cost increase due to lengthened project times. In summary, the two factors that are due to regulatory ratcheting have increased the estimated costs by a factor of $(2 \times 2.5 =)$ five. This means that the plant started in 1982 costing $4.3 billion when completed in 1994, could have been completed in 1989 as at a cost of $860 million, were it not for regulatory ratcheting.

Regulatory Turbulence

All of our discussion thus far has been based on estimates made at the time of project initiation, considering only regulations in force at that time, and assuming reasonably efficient construction procedures. It does not include the effects of changing regulations during the construction period. This is one aspect of what we call "regulatory turbulence."

As new regulations are promulgated, plants under construction must be modified to comply with them. This can often be a very expensive process. For example, if the concrete walls of a building have already been poured as new regulations are issued mandating additional equipment inside, compliance requires either that the walls be knocked down to increase the available space, or that the space within be reallocated. The former alternative is very expensive and would only be adopted as a last resort, but the latter can be exceedingly difficult if the original design made efficient use of the space. At the very least, a great deal of extra engineering is required, and in many situations great expense is incurred in relocating pipes and electrical cables.

One solution to this problem, of course, is to design with lots of extra space, but this also adds to the expense, since the space will not be needed in most cases. Another approach is to try to anticipate new regulations by including equipment and using the standards that they may require. This can also be an expensive game since those requirements may never develop.

Beyond the promulgation of new regulations, regulatory ratcheting has been especially devastating in the area of inspection and enforcement. Plant construction has traditionally involved a considerable amount of decision-making in the field based on engineering judgment. Regulatory ratcheting has reduced this practice to the point where it has now all but disappeared.[7] Every detail of drawings and procedures prepared by people far away and unfamiliar with the local field problems must be rigidly followed, to a degree far beyond what was intended by the preparers. Workers in England have demonstrated that they can bring their railroad system to a halt by rigidly following every rule in every detail, and analogous consequences have resulted from rigid regulatory enforcement in U.S. nuclear plant construction. The morale of construction personnel has suffered greatly from being forced to do things that they know are wrong or wasteful, and for not being allowed to exercise their judgment. They have become obsessed with keeping careful records so as not to be held personally responsible for some of the things that go on.[8] As a result of rigid enforcement, trivial noncompliance issues arise frequently, and precious time is wasted in their adjudication, often leaving workers idle as they await decisions. As management is forced to focus attention on trivia, its efficiency in organizing worker assignments suffers, resulting in further decreased labor productivity.

For a $2-billion plant built with money borrowed at 15 percent interest, each day of delay in the date of start-up costs close to a million dollars. Needless to say, with delay so expensive, all sorts of unorthodox and otherwise uneconomic measures are often used to reduce it. Strikes must be avoided at almost any cost. Suppliers cannot be held accountable if doing so causes delay.

A major source of delay in plants has been legal maneuvering by antinuclear activists.[9] The Midland plant in Michigan was stopped for three years at one point and later delayed for several shorter periods through suits that were eventually thrown out by the U.S. Supreme Court with a ruling chastising a lower court for "obstructionism." These delays doubled the cost of the plant. The Shoreham plant on Long Island was also delayed for three years by overt and shameless legal obstruction. The Seabrook plant in New Hampshire was delayed two years by regulatory indecision and legal maneuvering. Since these activities are part of the regulatory process, the delays are basically due to regulatory turbulence.

Frequent design changes in the course of construction can lead to confusion and, in some cases, to errors in construction. These can be expensive to correct, and even more important, can lead to costly delays if they are not discovered until the plant is nearly ready to start up. This is what happened at the

Diablo Canyon plant near San Luis Obispo, California, where it caused over a year of delay, costing hundreds of millions of dollars.

Antinuclear activists delight in pointing out such errors and ascribe them to the incompetence of nuclear power plant constructors. However, these are the same constructors who were building nuclear power plants so rapidly, efficiently and cheaply ten to fifteen years ago. It is therefore difficult to explain why they have suddenly lost their competence. Construction in a climate of constant design changes is a harrowing ordeal; it is hardly surprising that the frequency of mistakes has increased.

An important source of cost increases has been construction stoppage by utilities due to cash-flow problems. As a result of regulatory turbulence, costs of many plants often exceed what was originally planned. With no available source for the extra money, the utility has no alternative but to halt construction temporarily. In the long run, this delay greatly increases the cost of the plant, but for the short term, it solves the cash-flow problem.

As a result of such problems and general cost escalations, the credit ratings of utilities have fallen substantially over the past decade. This means that they must pay higher interest rates on the money they borrow. Since interest charges are one of the important contributors to plant costs, this is another source of cost escalation due to regulatory turbulence.

The foregoing discussion explains the rapid increase in costs of nuclear plants with time. It also allows us to understand the wide variations in costs among plants completed at the same time, as evident in Figure 9-1. Delays caused by legal maneuvers, major construction errors, and utility cash-flow problems have driven up the costs of some plants, but not of others. Some constructors have been more successful than others in coping with regulatory ratcheting. Those who have kept in close touch with NRC have fared better than those who have waited for new regulations to be promulgated. Labor rates vary considerably from one part of the country to another. Some plants have been delayed by strikes while others have not. There are major differences in design among plants, and some designs have been more seriously affected by regulatory ratcheting than others. For example, those with a larger containment building found it easier to add newly required equipment.

A few examples will illustrate the escalations in cost and time. In 1971, Consumers Power Company of Michigan[9] completed its Palisades plant four years after project initiation at a total cost of $186 million, but its Midland plant, with only 15 percent larger electrical output, is expected to be completed in 1984, fifteen years after project initiation at a cost of $1,500 million. In 1969, Niagara Mohawk completed its Nine Mile Point–1 plant in five years at a cost of $163 million; it hopes to complete its Nine Mile Point–2 plant, 1.7 times larger in electrical output, in 1986, fifteen years after the project initiation date, at a cost of $4,200 million.[10] (Its original cost estimate was $425 million.) The Tennessee Valley Authority completed plants of about the same size in 1974–77 for $295 million, in 1981–82 for $865 million, and in 1984–85 (projected) for $1,570 million.[1] (The four TVA nuclear power plants completed in 1981 were originally scheduled for completion in 1973–76.)

These and dozens of similar horror stories are largely the consequences of regulatory ratcheting and regulatory turbulence. Another consequence is that no new nuclear power plants have been ordered by U.S. utilities since 1978, and none of those ordered since 1975 are actively under construction. Several dozen plants previously ordered and under construction have been canceled, representing an enormous waste of money. In short, regulatory turbulence for which it is partly responsible (other factors introducing regulatory turbulence have been activities by antinuclear groups, NRC procedures, and failure by utilities to adapt to these) have been an unmitigated disaster for the nuclear industry. They have stopped the growth of nuclear power dead in its tracks, and in so doing, they have caused coal-burning plants to be constructed instead of nuclear plants. Each year that this practice continues, it causes many thousands of eventual unnecessary deaths.

What Has Caused Regulatory Ratcheting?

What has been the driving force behind this economically disastrous process of regulatory ratcheting? It has all been done in the name of improved safety. But the nuclear regulators of 1970 were confident that their licensing regulations were adequate to protect the public. What new scientific or technical information has come to light to indicate that all of this tightening of regulations was necessary?

New Scientific and Technical Information Contrary to Regulatory Ratcheting

Of course, there was the Three Mile Island accident. It pointed out certain deficiencies in instrumentation and control-room design, but their correction has added only a few percent to the costs of power plants. (It also indicated that improvements were needed in operator training, but that does not affect construction costs.) The Brown's Ferry fire and a few minor incidents in other power plants suggested adding some new equipment, but these added less than 1 percent to the price of a plant.

The most important nuclear safety issue in the early 1970s was whether the emergency core cooling system (ECCS) would avert a meltdown in case of a large loss of coolant accident (LOCA). The controversy over this problem spawned the principal antinuclear activist organization, the Union of Concerned Scientists (UCS). That issue got the lion's share of media publicity from 1972 to 1975 when the antinuclear movement came into prominence and developed its political clout. Settling the ECCS issue cost the U.S. government hundreds of millions of dollars, but it *was* settled. The LOFT reactor constructed in Idaho to study the problem found[11] that the cooling system performed far better than had been expected, which in turn was far better than the pessimistic estimates used for safety analyses. The cooling system issue was thus laid to rest. This should have *reduced* regulatory requirements.

Probably the next most important nuclear safety topic of this period was the development of probabilistic risk assessment to quantify the risks of nuclear accidents. The Rasmussen Study[12] came out in 1975, the UCS critique[13] was published in 1976, an NRC Review[14] took place in 1978, and there have been several new and improved studies in the past few years. The conclusion of the Rasmussen Study was that if all U.S. electricity were nuclear, there would be an average of five deaths per year from reactor accidents, and they would reduce the life expectancy of the average American by 0.02 days (30 minutes). These numbers are roughly consistent with the results of other studies carried out more recently. The only exception is the antinuclear activist UCS analysis, which gives an average of 600 deaths per year from an all-nuclear power industry, reducing life expectancy by 2 days. Since the only viable alternative to nuclear energy is coal-burning, which kills an estimated 10,000 Americans each year[15] (some analyses[16] give numbers five times higher) with its air pollution, the results from nuclear risk analyses would surely be viewed objectively as very favorable. Even if we accept the UCS position, building a nuclear plant instead of a coal-fired plant *saves* 25 lives per year, or close to a thousand lives over the life of the plant. Surely this does not indicate a need to tighten nuclear safety regulations, especially if doing so has the effect of causing utilities to build coal-fired plants instead of nuclear plants— clearly counterproductive if saving lives is the object.

Probably the most important but least publicized new information over the past dozen years is that there have been about 1,500 reactor-years of commercial light-water reactor (LWR) operation plus nearly an equal number of reactor-years of naval LWR operation without a meltdown and with only single large loss of coolant accident (Three Mile Island). This experience essentially eliminates the need to depend on the probabilistic risk assessments: even if we assume that we have been very lucky not to have had more such incidents, it confirms that nuclear power is at least much safer than burning coal. Again, this should give reason to relax rather than to tighten the safety requirements in force in 1970 when very little evidence on nuclear safety was available.

Another important issue relevant to regulatory ratcheting since 1970 has been the health effects of radiation. If new evidence had indicated that radiation is more dangerous than previously believed, that would be reason to require more equipment in nuclear plants to reduce emissions. But no such evidence has developed. Based strictly on scientific data and analyses, the National Academy of Sciences Committee on Biological Effects of Ionizing Radiation (BEIR) gave lower risk estimates in its 1980 Report[17] than in its 1972 Report.[18] The United Nations Scientific Committee on Effects of Atomic Radiation (UNSCEAR) has not increased its 1972 estimates,[19] and its 1982 Report expresses confidence in them.[20] The International Commission on Radiological Protection came out with its first estimate in 1977, and it was slightly lower than those of BEIR and UNSCEAR.[21] There were surely no new developments here to justify regulatory ratcheting; if anything, the increased confidence should have allowed some relaxation of regulatory requirements.

The most important new information bearing on consequences of possible reactor accidents was the discovery that iodine, cesium, and a few other elements would not be released in their volatile forms.[22] The situation is somewhat analogous to chlorine in the ocean. In its elemental form, chlorine was used as a poison gas during World War I. If the chlorine in the ocean were gaseous, it would be impossible to live near the coast; there could be no New York, Boston, New Orleans, or Los Angeles. However, as we know, the chlorine in the ocean is in the form of sodium chloride, common table salt, which is not volatile. In the early 1970s it was believed that iodine and cesium released in a nuclear accident would be largely in their highly volatile elemental forms. It has been found, instead, that they form cesium iodide, which behaves like table salt. Thus, the estimated consequences of a nuclear accident are greatly reduced. This finding gives good reason to relax regulatory requirements rather than tighten them.

A few unexpected problems with nuclear plants have come to the surface, such as pressurized thermal shock[23] and steam generator tube failures.[24] But the measures required to control these problems do not affect the construction costs of nuclear plants. In fact, they are not even safety issues, but rather problems that increase the operating and maintenance costs for nuclear plants. They provide no justification for the regulatory ratcheting we have been discussing.

In summary, new scientific or technical developments do not suggest that regulatory requirements should be more strict now than they were in 1970. On the contrary, with a few exceptions they suggest that regulations should have been relaxed.

Not only is the need for tightened regulations impossible to justify on technical grounds, but there are serious questions as to whether they actually do improve safety.[7] For example, NRC became obsessed in the late 1970s with protecting plants against highly improbable earthquakes. One aspect of this program was to tie down pipes very accurately and rigidly at many points to prevent dynamically amplified forces in an earthquake. The specifications, derived by a complex computer analysis of "amplified response spectra," are so tight that they require use of machine-shop practice in the field, which is exceedingly expensive; in fact, these specifications often exceed the tolerances in the pipe manufacture, and therefore require machining of the pipe. Moreover, piping supports in the field are often found to interfere with other components of the system; this requires extensive reanalysis (and consequently delays, as these analyses cannot ordinarily be done in the field) and relocation of piping. In one study[7] of 336 nuclear grade piping systems, 539 reanalyses were required, consuming an average of about 250 man-hours each. Over 60 percent of the piping supports require redesign, consuming an average of 50 man-hours per support. All of these changes require documentation reports; two plants at which this was studied generated over 15,000 such reports apiece. Largely ignored in this obsession with seismic safety is the fact that piping systems must be allowed to expand and contract with temperature changes and that the majority of piping failures are associated with water hammer, corrosion, fatigue, and fabrication or metallurgical deficiencies, whereas extensive literature searches do not reveal a pattern of piping failures during earthquakes. It is widely believed that these extremely expensive procedures on pipe support have *reduced* overall reactor safety.

Media Distortion and Public Concern

If scientific and technical factors are not responsible, what has caused the regulatory ratcheting that has stopped the growth of nuclear power dead in its tracks? NRC, as a government agency, must be responsive to public concern, and public concern about the perceived dangers of radiation and of nuclear power has been growing by leaps and bounds. It is this mounting public concern that has been the driving force behind regulatory ratcheting.

But what has caused the increase in public concern? The primary sources of information on nuclear power for the public are the news media, both electronic and print. How have they reported on developments?

Let us go over the items discussed above:

- In 1971–73, the issue of whether the emergency core cooling system could protect a reactor was raised, and hearings were held on it over a period of more than a year; there was tremendous media coverage. But I have never seen a media report on the many years of research and the very expensive LOFT tests that resolved the issue and showed that cooling system performance was not a problem. All that the public heard about the issue would indicate that nuclear energy is dangerous.
- The results of the probabilistic risk analyses are always quoted in the media in terms of the number of people who will die. Very rarely do they discuss the averages; they nearly always talk about the *worst* accident, expected once in 100,000 meltdowns. They rarely discuss its probability, and most of the public consequently believes that *all* melt-

downs are like this worst one. They never compare the risks with those of coal burning. One poll[25] shows that 80 percent of the public believes that coal burning is safer than nuclear electricity, although every scientific study[26] has indicated the contrary. The media never mention these studies.

- The public has been given every reason to believe that new evidence indicates that radiation is more dangerous than previously estimated. The few reports[27] tending to favor that position have been given wide publicity. Each of these reports has been devastatingly criticized and rejected by the scientific community,[28] but this fact has received no media coverage. Even the official reports by BEIR, UN-SCEAR, and ICRP have received little coverage; and the coverage given was usually so distorted as to give false impressions.
- I have never seen media coverage of the important good news about the nonvolatile nature of iodine and cesium in an accident.
- Pressurized thermal shock has been treated by the media as an important safety issue, whereas it is really more of an operations cost problem.
- The steam generator tube rupture in a Rochester, New York, power plant was the lead news item on national news for over twenty-four hours. It was treated as a nuclear accident, broadly hinting that there would be widespread health consequences. It was never mentioned that *no one* received as much radiation exposure from it as he receives every day from natural sources.

An increasingly common tactic is to publicize individual failures or substandard aspects of nuclear power plants. But the basic principle of nuclear safety design is "defense in depth." If one system fails, there is another backup system to protect the reactor; if the backup system fails, there is another backup system behind it; and so forth. For example, the most widely discussed potential accident is for the system to burst open, allowing the cooling water to escape; this causes a meltdown of the reactor fuel, releasing airborne radioactive dust that could harm public health. The lines of defense to protect against that sequence are:

1. Elaborate quality control on materials and equipment whose potential failure would result in opening up the system.

2. Elaborate inspection programs involving X-ray, magnetic particle, and ultrasonic testing, plus extensive visual inspection.
3. Leak-detection systems to detect small cracks; large cracks ordinarily start out as small cracks, so detecting and repairing the latter averts large cracks that could cause an accident.
4. Emergency core cooling system (ECCS) to rapidly restore water to cover the fuel if the original water should be lost.
5. Containment, the very powerful structure housing the reactor system, which would normally prevent the escape of radioactivity even if there were a fuel meltdown.

Each of these lines of defense would have to fail, one after the other, before there could be harm to the public. Moreover, each of them is, in itself, a defense in depth. For example, the ECCS consists typically of six separate pumping systems, any *one* of which would be sufficient to prevent meltdown. Occasionally, there is a failure in one component of this defense in depth in some power plant: an operator closes a valve that should be left open, an X-ray inspection is found to be inadequate, a pump is discovered to be inoperable. This is reported by the media as though a disaster had been narrowly averted, completely ignoring all of the other lines of defense that protect against release of radioactivity.

I could go on endlessly discussing how the media have distorted nuclear energy issues, always enhancing the impression of danger. For example, they have never told the public that the Three Mile Island accident was *not* a near miss—that even if there had been a meltdown, there very probably would have been no public health consequences. That is the conclusion of all the detailed studies of the accident.[29] Three Mile Island also provides an excellent example of the operation of defense in depth. In that worst accident in the history of nuclear power plants, where so many things went wrong causing one line of defense after another to fail, there were still lines of defense left standing to protect the public.

Our conclusion is that the regulatory ratcheting that has crippled the nuclear industry has been driven not by scientific or technical findings, but by ever-growing public concern about the safety of nuclear power. And this growth in public concern has been caused by media distortion.

The Situation in Other Countries

While regulatory ratcheting driven by mounting public concern about nuclear safety in the United States has had its counterpart in other countries, it has generally been less devastating. The most successful nuclear power program is in France where government decision-making on technical issues is less closely tied to public opinion. In 1981, for example, 13 percent of U.S. electricity was nuclear, whereas in France it was 38 percent, projected[30] to reach 60 percent by 1985. Twelve new plants went into operation in France during 1981 after an average construction period of six years, whereas in the United States a total of six plants became operational in 1980 and 1981 combined, after an average construction period of 10.5 years. These French plants were started in 1974–75 and cost an average of $500 million[31] (all costs in this discussion are per million of kw plant capacity). None of the American plants started in that time period is completed, but their costs are estimated to average $1,500 million ($1,188 million in 1981 dollars). The French started construction of seven plants in 1979, one in 1980, and six in 1981, planning for completion in six years.[32] According to official government estimates, French plants started in 1982 will be finished in 1988 at a cost (in 1981 dollars) of $700 million,[33] whereas U.S. plants, if any had been started at that time, would have been completed in 1995 at a cost of $1,700 million.[34] French plants, incidentally, are built according to American designs under license from Westinghouse. I have never heard it claimed that French plants are less safe than U.S. plants.

The situation in other countries has been less favorable than in France but generally more favorable than in the United States. Japan got a late start in nuclear power, but is accelerating rapidly. The six plants completed there since 1979 averaged less than seven years in construction.[35] In accordance with a program set up in 1975, 17 percent of Japanese electricity was to be nuclear by 1985, but that goal was reached in 1982. Not a single reactor has been canceled (as opposed to nearly a hundred cancellations in the United States), and political and legal delays rarely occur. The West German nuclear program has been especially hard hit by antinuclear activism, but three new plants were started there in 1982. The Atommash plant in the USSR has been commissioned to produce eight new reactors per year for the foreseeable future.[36]

In Austria and Sweden, antinuclear activists have stopped

the growth of nuclear power by direct political action, and such action was narrowly defeated in Switzerland. But in the major nations of Europe, nuclear power is faring better than in the United States.

Many smaller or less developed countries have been under-taking nuclear power programs by purchasing power plants from suppliers in the major nations. American manufacturers had 70 percent of this business in the early 1970s, but in recent years they have had *none* as regulatory turbulence in our government's export policies have given American suppliers a reputation for being unreliable (see Chapter 10). In 1981, the French sold four reactors to foreign countries.[36] Germany and Canada have also shared in displacing the United States in the export business, and Japan is on the threshold of joining the competition.

Costs of Nuclear-generated Electricity to the Consumer

So far discussion has centered on the cost of building a nuclear power plant. But the important issue from the na-tional viewpoint is the cost of electricity to the consumer, expressed in cents per kilowatt-hour. We next discuss that question and how this cost compares with the cost of electricity generated by other technologies.

The charge for electricity to the consumer is calculated as the sum of three elements:

1. Fixed charges, which are not related to the amount of electricity produced. This includes amortizing the cost of construction, paying interest on bonds and dividends on stock (the two principal ways of raising capital), paying for insurance and taxes, and providing for the ultimate decommissioning of the plant.
2. Operation and maintenance, which includes salaries and wages, materials, and supplies needed to keep the plant operating.
3. Fuel charges, including the cost of fresh fuel and manage-ment of the waste.

The sum of these three charges for one year is divided by the total number of kilowatt-hours (kwh) the plant is expected to produce in that year to obtain the cost per kwh. According to the 1982 analysis by the U.S. Energy Information Agency

(EIA),[34] a nuclear plant started in 1982 would go into operation in 1995 producing electricity with the following charges in 1980 dollars:

	Nuclear	Coal-fired
Fixed charges	2.9¢	1.7¢
Operation and maintenance	0.5¢	0.4¢
Fuel	0.9¢*	2.2¢
Total (per kwh)	4.3¢	4.3¢

The EIA analysis for a coal-fired plant completed in 1995 is listed for comparison. The total cost, 4.3¢ per kwh, is essentially double the 1980 costs, which were 2.3¢ per kwh for nuclear and 2.5¢ per kwh for coal. Note that coal burning has also had its problems with cost escalation, which have been only slightly less disastrous than those for nuclear power.

Recently completed French nuclear plants are producing electricity at 1.7¢ per kwh[37] (still in 1980 dollars). It is estimated[38] that European plants started in 1982 and completed in 1990 (five years earlier than U.S. plants started at that time) will produce electricity for 2.3¢ per kwh in France, 2.2¢ in Italy, 2.8¢ in Belgium, and 3.1¢ in Germany.

Historically, electricity and other forms of energy have been much cheaper in the United States than in Europe. This fact has often been given a large part of the credit for our economic success. We see that situation reversing; by the mid-1990s, electricity may be nearly twice as expensive here. This drastic change is bound to have consequences for our economy, probably wiping out whole industries and accentuating unemployment and balance-of-payment problems.

Since most of the cost of nuclear power is due to fixed charges closely related to the construction cost of the plant, the latter has a major role in determining the cost per kwh. We have already shown that higher construction costs are due to regulatory ratcheting. But regulatory ratcheting has also escalated the price of electricity in other ways: (1) it has required frequent shutdowns for inspections and plant modifications, which reduce the number of kwh generated per year and thereby raise the cost per kwh; (2) new regulations have substantially increased the work force required both for operating and for maintenance; (3) regulatory ratcheting has also

* For comparison, the OPEC price for oil that may be burned to produce electricity is 4.7¢ per kwh.

had its effects on the cost of mining and milling uranium, and of converting it into fuel; (4) it has greatly escalated the cost of waste management, which also reflects in the fuel costs. We can therefore assign a large part of the high price per kwh of future U.S. electricity to regulatory ratcheting.

The Utilities' Viewpoint

The projected costs of nuclear and coal-fired electricity are roughly equal; in fact, most analyses still show some cost advantage for nuclear power.[4] One might wonder, therefore, why a few U.S. utilities are continuing to start construction of coal-fired plants rather than nuclear plants. Part of the reason is the uncertainty in the regulatory ratcheting process. The rate of regulatory ratcheting would be greatly accelerated by a nuclear accident; it might even be accelerated by a new popular book or a shift in political philosophy that gains public support. These events can stir up public concern that drives regulatory ratcheting. Another reason is the long time required to plan and construct a nuclear plant, now estimated at twelve to fifteen years. It is very difficult for a utility to predict the amount of electric power it will need that far in advance. And if plant completion is delayed long beyond what is estimated when the project was initiated—which has been the rule rather than the exception in recent years—planning becomes even more difficult and uncertain.

Another important deterrent to a nuclear project is the large amount of capital that must be raised. A nuclear plant costs much more than a coal-fired plant; the reason the price per kwh is similar for the two is that the fuel costs are much higher for coal. But the fuel costs do not have to be paid until the electricity is generated; fuel costs can then be added to consumer bills. Effectively, fuel costs are collected directly from the customers. Plant construction costs, on the other hand, must be paid by the utility. This money is not collected until many years later. Of course, an appropriate interest charge is included in the money collected, reflecting the fact that the utility has to pay interest through the intervening years. In the long run, no money is lost in the process. But the utility's ability to raise capital is heavily strained during the construction period. It is much easier to build the cheaper coal-burning plant, and thereby avoid the financial risk.

This problem could be eliminated if utilities could include costs for construction work in progress in calculating the rates they charge for electricity. That way, the customers would pay for the capital costs as they are incurred. However, public utility commissions generally have not allowed this. With the existing policy, present-day consumers enjoy low rates based on power plants built many years ago when costs were much lower. But when new plants come on line and their cost plus interest is figured into the rate base, bills will rise precipitiously. The guiding principle seems to be to keep rates as low as possible for as long as possible, and not worry about the burden this imposes on future consumers—or about the problems of utilities in raising capital.

The easiest path for a utility is not to build any new power plants, and that is the path they have largely been following. Reserve capacity is ample now, although it is not clear to what extent this is due to our depressed economy. In the long run, however, this hiatus in construction cannot continue. If prosperity returns, our use of electricity is bound to increase substantially. Our population is increasing by 2 million per year, adding that number of new consumers. Our work force is expanding, with more than a million new jobs each year—and more jobs usually mean more electricity. Old plants are wearing out and will have to be replaced. Oil and gas provide about 25 percent of our electricity; they are becoming scarce and, in the long run, will become quite expensive. Oil-fired electricity is already more than twice as expensive as coal or nuclear.

A serious problem here is that when the need for new power plants becomes apparent, it may be too late. It takes many years to plan and construct a plant, and these will be years of increasing shortage of electricity causing depression of our economy and general inconvenience. Ironically, the utilities will probably be blamed for these ills just as they are now being blamed for rising prices of electricity.

What Can Be Done to Improve the Situation?

The situation we have outlined is a very bleak one. Our nuclear option is being thrown away, at a devastating economic cost and at a very substantial cost in the areas of health and the environment. Recall that every time a coal-fired plant is built instead of a nuclear plant, several hundred people are condemned to an early death even according to the reckoning of the antinuclear activists. What can be done to improve the situation?

Regulatory Reform

The obvious solution is to reverse the regulatory problems that have ruined our nuclear industry. Regulatory reform has been advocated in high places for many years. The Carter administration moved in that direction, but without success. A Nuclear Licensing and Regulatory Reform Act developed by the Department of Energy and a Nuclear Power Plant Licensing and Reform Act developed by NRC were introduced in Congress in 1983, with some hope for passage of a law including many of their features in 1984. They do not directly address the regulatory ratcheting problem as that is the internal business of NRC. Rather they attempt to reduce some of the regulatory turbulence that has amplified the problem, and to streamline some of the licensing procedures that have extended the time required for a construction project: (1) they propose stringent review procedures to be followed by NRC before it sets up a new requirement for plants already licensed to proceed with construction or with operation; it is hoped that this would reduce the number of these new requirements; (2) they would speed up the licensing process by combining to some extent the procedures for obtaining construction permits and operating permits; (3) they would substantially limit the use that can be made of public hearings by intervenors to delay projects; (4) they would allow facilities or major facility subsystems to undergo NRC review, and once approved, to be used in other plants without further review except on matters where there are relevant differences; (5) they would simplify the procedures needed for a utility to make design changes; (6) they would allow approval of plant sites before designs for the plants are offered, or even begun. This would be an obvious time-saving measure.

In many ways, these measures represent attempts to give the U.S. program some of the advantages of the more efficient programs in other countries. An analysis by a Los Alamos group[39] indicates that the Department of Energy proposal would reduce average nuclear construction time from 12.7 to 7.5 years, and would reduce costs in 1981 dollars from $1,342 per kw to $964. The NRC proposal is less far-reaching and therefore offers somewhat fewer benefits. It takes a great deal of optimism to hope that a new law will get through Congress in 1984, and that it will not be so crippled by amendments as to reduce the benefits substantially. There is a large contingent in Congress that is politically committed to oppose anything beneficial to the nuclear industry. However, it seems reasonable to hope that there may be some improvement in the situation.

On the other hand, I find it difficult to believe that this will be sufficient to spur new starts in construction of nuclear power plants. Undertaking such a project entails a long-term commitment spanning several presidential and congressional elections. Given the hysterical and continually escalating fear of nuclear power and of radiation in the body politic, it is only too easy to foresee possible adverse impacts on such an undertaking. Moreover, the regulatory ratcheting has not been stopped, let alone reversed, and no one seems to believe that it will not continue. The driving forces behind it have not abated.

Media Distortion

In most situations, the way to solve a problem is to understand its basic cause and remedy it. We have previously identified the basic problem with nuclear power as regulatory ratcheting driven by mounting public concern caused by media distortion. But what has brought about this media distortion?

Part of it is due to the paradox that although the media are the principal source of education for the public, they are *not* in the education business. Their business is *entertainment;* they live or die by the effectiveness with which they entertain their audiences. I have been told that for the network evening news, one point in the Nielsen ratings is worth $7 million a year in advertising revenue. In such an atmosphere, how long would a producer last if he gave proper education a higher priority than grabbing and holding an audience? Scaring people about radiation and nuclear accidents seems to serve their entertainment function well, and they play it to the hilt. Moreover, even if a segment of the media wanted to educate the public about scientific and technical issues, it has no resources for doing so. Science education in universities, even for nonscience students, is taught by scientists with Ph.D. degrees who have contact with the scientific literature. There is typically one such scientist for every ten to one hundred students. But I doubt if the television networks employ even a single Ph.D. scientist for their hundreds of millions of "students." Even in high schools, where it is widely recognized that science education has been a disaster area, the subject is taught by professional educators who have had several college courses in science and generally read journals with articles written by scientists. In the media, however, science is usually "taught" by scientific illiterates.

But I am convinced that the source of media distortion runs much deeper than priority for entertainment and scientific illiteracy. It is my impression that most media people actually believe, deep down in their hearts, that nuclear power is very dangerous and that the government and the scientific establishment are trying to cover up that fact. How that situation came about is a story worth telling. It is the story of a battle fought in this country in the 1970s between the nuclear scientists on the one hand and the antinuclear activists on the other.

The Battle Lost

The two sides consist of entirely different types of people. For a typical antinuclear activist, political battling is a primary interest in life, whereas the vast majority of nuclear scientists have no interest in or inclination toward that activity—and even if they did, they have little native ability or educational preparation in it. While the typical antinuclear activist was taking college courses in writing, debate, and social psychology, the typical nuclear scientist was taking courses in calculus, radiation physics, and molecular biology. After graduation, the activist gained worldly experience by participation in political campaigns, environmental activism, and anti-Vietnam War protests, while the scientist was working out the mathematical complexities of neutron transport or radiation carcinogenesis, or devising solutions to some of the multitudinous technical problems in power plant design. While the activist was making political contacts and developing know-how in securing media cooperation, the scientist was absorbed in laboratory or field problems with no thought of politics or media involvement. At this juncture the activist went out looking for a new battle to fight and decided to attack the nuclear scientists. It was like a lion attacking a lamb.

Nuclear scientists had longed agonized over safety measures needed in power plants and over health impacts of radioactivity releases. All the arguments were published for anyone to see. It took little effort for the antinuclear activists to collect, organize selectively, and distort this information into ammunition for their political battle. Anyone experienced in debate is well prepared to do that. When the activists charged into battle wildly firing this ammunition, the nuclear scientists first laughed at the naiveté of the charges, but they did not laugh for long. They could easily explain the invalidity of the attacks by scientific and technical arguments, but no one would listen to them. The phony charges of the attackers, dressed up with their considerable skills in presentation, sounded better to the media and others with no scientific knowledge or experience. When people wanted to hear from scientists, the attackers supplied their own: there are always a few available to present any point of view. Who was to know that they represented only a tiny minority of the scientific community. The battle was *not* billed as a bunch of scientifically illiterate political activists attacking the community of nuclear scientists, which was the true situation. Instead, it was represented as "environmentalists"—what a good, sweet, and pure connotation that name carries—attacking "big business" interests (the nuclear industry), which was trying to make money at the expense of the public's health and safety. Jane Fonda refused to debate nuclear scientists; her antagonists, she said, were corporate executives.

The rout was rapid and complete. In fact, the nuclear scientists scarcely were able to get onto the battlefield. The battlefield here was the media. When an issue arose, the media often sought information from both sides, but they seldom called a nuclear scientist. Instead, they chose a utility executive, a nuclear industry public relations specialist, or a government bureaucrat. Even if these people were well informed, which they often were not, they had low credibility with the public.

On the rare occasions when a nuclear scientist got a hearing, the obstacles in his path were formidable. His opponent was usually one of about five anti-nuclear scientists in the United States who have largely abandoned their scientific careers to participate in antinuclear activities. But that point does not come across in the introductions; the debate is billed as one scientist vs. another, with no indication that one represents the vast majority of the scientific community while the other represents only himself. When a scientific point must be explained in thirty seconds in terms understandable to the public, the antinuclear scientist can make his views seem just as convincing as the nuclear scientist can. Since the activist has far more media experience, he often does better.

Media people get most of their science education from other media productions. A few of them may read books, but nearly all popular books about nuclear energy are written by antinuclear activists or by reporters heavily influenced by them. Active scientists generally have no experience in writing popular books, and they rarely have time or talent for such activities. As a result of this unbalanced exposure, media people have been slowly but surely won over by the antinuclear activists. The nuclear scientists lost the battle. The public was driven insane over fear of radiation; its understanding of radiation dangers has completely lost contact with reality. The public has come to believe that potential reactor meltdown accidents, which are statistically projected to kill an average of five people each year, are a bigger problem than automobile accidents, which kill fifty thousand a year. They have come to believe that the technically trivial operation of burying radioactive waste is more of a problem than air pollution, one of the wastes from coal burning, which is killing about ten thousand Americans per year, many thousands of times the number who might ever die as a result of exposure to nuclear waste.

Recommendations

What can be done about the situation? Nuclear scientists must continue the battle by writing magazine articles and popular books. We must try to contact the media and show them how they have gone wrong. Our professional societies must help, and we must seek help wherever it is available: the media are not easy to contact.

On the political side, I believe the best hope is to involve the government in providing information for the public, and to correct the misleading information spread by the media. There were attempts in this direction during the Carter administration, but they were blocked by the political clout of the antinuclear activists. I have often wondered why there are no such efforts by the Reagan administration. Such programs have been very productive in Japan where the public information job was especially difficult because of the atomic bombings of Hiroshima and Nagasaki. As a result of the successful program, the Japanese public is now embracing nuclear power.

We should work on the younger generation by contacting teachers, writing articles in the journals they read, and speaking at meetings they attend. I have found that high school science teachers are generally sympathetic and can understand our arguments.

We face a long uphill battle, and it is easy to become discouraged. But I am convinced that until the battle is won and the public comes to realize that widespread use of nuclear power is by far the safest, most healthful, and environmentally most acceptable energy strategy, nuclear power has a dismal future in our country. Perhaps Shakespeare was right when he said, "Truth will come to light," and perhaps Lincoln was correct in saying, "You can't fool all of the people all the time." If so, we may yet prevail.

Appendix: Health and Safety Impacts of Nuclear Power

How Dangerous is Radiation?

Nuclear radiation consists of subatomic particles that travel at speeds approaching the velocity of light—186,000 miles per second. They readily penetrate deep inside the human body where they damage or destroy some of the molecules of which it is composed. A typical particle of radiation damages 100,000 of these molecules. Under certain circumstances, the damage to one of these molecules can result in cancer or in genetic defects in our progeny.

In view of these facts, one might think that being struck by a particle of radiation is a terrible tragedy. However, this cannot be the case, because every person who has ever lived has been struck by about a million particles of radiation every minute of his or her life. This radiation comes from outer space (cosmic rays), and from naturally radioactive materials in the ground, in our bodies, in the air we breathe, and in all materials we come close to. All materials in nature contain radioactivity. Why, then, are we not all dead from cancer or afflicted with genetic disease? The reason is *not* that it takes a certain minimum number of particles of radiation to do us harm. As nuclear fearmongers like to say, no level of radiation is perfectly safe; even a single particle of radiation striking us can cause cancer or a genetic disease in our progeny. Fortunately, however, the probability for this is very small—only about one chance in 40 quadrillion (40 million billion). Since a human life span is about 40 million minutes, our risk of developing cancer from the million particles per minute of natural radiation that strike us is 40 million million divided by 40 million billion: one chance in a thousand. As each of us normally has one chance in five of dying from cancer, this means that one out of two hundred cancer deaths is probably due to natural radiation.

People living close to nuclear power plants receive exposures from those plants about 1 percent larger than their exposure from natural radiation. Thus, spending a lifetime residing near a nuclear plant gives a person one chance in a hundred thousand of dying from its radiation. By comparison, we have one chance in sixty-five of being killed in a motor vehicle accident, making that latter 1,500 times more of a risk. This means that if moving away from a nuclear plant increases a person's travel mileage by more than one part in 1,500, about 30 yards per day, it is safer to continue to live near the nuclear plant. This does not include the risk from accidents, which increases the equivalent risk to something like 100 yards of extra driving per day.

The important thing to understand about radiation hazards is that they must be treated *quantitatively*. Qualitative treatments can be entirely misleading. Unfortunately, news media coverage, which is the public's principal source of information on radiation hazards, is invariably qualitative, and has been grossly misleading.[40] As a result, the public's concept of the dangers of radiation has essentially lost contact with reality.

Any quantitative treatment of radiation health effects must start with the *dose* received. The most common measuring dose is the *millirem* (mrem). Receiving one mrem corresponds to being struck by about 5 billion particles of radiation, but this number varies somewhat, depending on the type of radiation, its energy, and the size of the person. An exposure of 1 mrem to the whole body gives about one chance in 8 million of developing cancer, and a similar chance of causing a case of genetic disease in later generations. These are the estimates given by the National Academy of Sciences Committee on Biological Effects of Ionizing Radiation (BEIR),[17] and they are very similar to the estimates of the United Nations Scientific Committee on Effects of Atomic Radiation (UNSCEAR)[41] and of the International Commission on Radiological Protection (ICRP).[21]

How Do We Know the Dangers of Radiation?

Since radiation-induced cancers may take as long as fifty years or more to develop, how do we know the health consequences of the Three Miles Island accident? The answer is that we have measured the radiation doses in millirems received in that accident, and we know the cancer risk from a millirem of radiation. How do we know this risk? Basically, from forty years of extensive scientific research costing $2 billion in the United States alone. However, we attempt here to give a brief summary of the results.

A great deal is known about the effects of high levels (i.e., large doses) of radiation from studies of exposed groups. The best-known example is found in the survivors of the A-bomb attacks on Japan, but there are many others from medical exposure to X-rays, radium, and other radiation diagnoses and therapeutic procedures, from workers inadvertently exposed in industrial activities, and so forth. Several dozen such situations have been studied and analyzed to give quantitative estimates of the effects of high-level radiation, above 100,000 mrem. The problem is to extend these estimates down to lower levels.

The simplest procedure is to assume that effects are proportional to dose, the "linear" hypothesis. The basic justification for this is that our understanding of radiation carcinogenesis indicates that a single particle of radiation striking a single molecule in our bodies can cause cancer. If all radiation-induced cancer were due to this type of event, the risk should be simply proportional to the number of particles of radiation, and hence to the dose. However, there are also processes known in which two particles of radiation combine to cause cancer, whereas neither particle alone would have had that effect. Since these latter processes would contribute importantly only to effects at high levels, the linear hypothesis based on our knowledge of high-level radiation tends to overestimate effects at low levels. An alternative way of stating this is to say that the curve of cancer risk versus dose is concave upward.[42]

There is abundant supporting evidence for this concave upward curve of risk versus dose. This behavior is exhibited in studies of chromosome breaks[43] (a type of damage observable under a microscope); in transformation to malignancy for mouse cells exposed to radiation in vitro[44]; and in observed cancer rates in animals exposed to radiation.[45] The data are less clear but highly suggestive of a concave upward curve for leukemia among the Japanese A-bomb survivors,[17] and for industrial workers who ingested small amounts of radium in the 1920s.[46] For exposure to naturally occurring radioactive radon gas, a linear risk versus dose dependence, based on high dose data for uranium miners, substantially overpredicts the incidence of lung cancer among non–cigarette smokers in the general population due to our environmental exposure to radon—that is, the linear hypothesis based on high-level radiation overestimates effects of low-level radiation.[47]

All of this evidence has convinced the great majority of involved scientists and the prestigious committees that represent them that the linear hypothesis certainly does not underestimate effects of low-level radiation and probably gives a moderate overestimate of these effects.

A few reports have offered evidence that the linear hypothesis substantially underestimates effects of low-level radiation. The best-known of these—by Mancuso, Stewart, and Kneale[27]—has received extensive media coverage. However, it has attracted over twenty critiques[48] and has been rejected by all prestigious committees and all official agencies charged with responsibilities in radiation protection.

How Dangerous Are Reactor Meltdown Accidents?

Probably the most widely feared aspect of nuclear power is the possibility of a reactor meltdown accident, often referred to by the media as "the ultimate disaster." In order to put this problem on a quantitative basis, we give two estimates of its probability and consequences, one by a large study group sponsored by the U.S. Nuclear Regulatory Commission widely known as the Rasmussen Report,[12] and the other by the principal antinuclear activist organization in the United States, the Union of Concerned Scientists (UCS).[13]

For the frequency of reactor meltdowns, NRC estimates one per 20,000 plant-years, whereas the UCS estimate is one per 2,000 plant-years. After 1,000 plant-years of commercial operation around the world and over 2,000 equivalent plant-years of naval reactor operation, all without a meltdown, the UCS estimate implies that we are very lucky.

There is a widespread misunderstanding of the consequences of a meltdown. We often hear that it would kill tens of thousands of people and contaminate a whole state, but such statements are grossly misleading. As protection for the public in the event of a meltdown, reactors are enclosed in a powerfully built "containment" building, which would ordinarily contain the radioactive dust long enough (about one day) to clean it out of the air. For example, investigators of the Three Mile Island accident all agree that even if there had been a meltdown, there would have been little harm to the public, because there is no reason to believe that the integrity of the containment would have been compromised. In most meltdowns, no fatalities are expected.

Certain events could break open the containment, releasing radioactive dust into the environment. If this happens, the consequences depend on the timing and on weather conditions. In the most unfavorable conditions, with a large containment break early in the accident, NRC estimates 48,000 fatalities, but this unusual combination is expected only once in 100,000 meltdowns.

According to NRC, the average number of fatalities in a reactor meltdown is 400; according to UCS, it is 5,000. A median estimate of the fatality rate due to air pollution from coal-burning power plants is about 5,000 each year. Thus, for reactor meltdowns to be as harmful as coal-burning power plants, we would need a meltdown every month (5,000 ÷ 400 ≅ 12/year), according to NRC, or once each year, according to UCS. We have ample reason to believe that meltdowns will not occur that frequently.

When the frequency and consequence estimates are combined, NRC concludes that we may expect an average of 0.02 fatalities per plant-year; UCS predicts 2.4. Note that even the UCS figure, given by the leading antinuclear activist organization, is still far less than the 20 fatalities per plant-year due to air pollution from coal-burning electricity generation.

Of course, these fatalities from air pollution are not detectable in the U.S. population in which 2 million people die every year. But the same is true of 98 percent of the fatalities from reactor meltdown accidents. For example, in the worst such accident considered by NRC, there would be 45,000 extra cancer deaths in a population of 10 million over fifty years. For each of these 10 million, the risk of dying from cancer would be increased from the normal risk of 20.5 percent to 21.0 percent. The present risk in different states varies between 17 percent and 24 percent, so the cancer risk in moving from one state to another is often many times larger than the risk from being involved in NRC's worst-case nuclear accident.

Detectable fatalities occurring shortly after the accident and clearly attributable to it would be rather rare. According to NRC, 98 percent of all meltdowns would cause no detectable fatalities, the average number for all meltdowns is 10, and the worst meltdown (a 1 in 100,000 occurrence) would cause 3,500. The largest coal-related incident to date was an air pollution episode in London in 1952 that caused 3,500 fatalities within a few days. Thus, as far as detectable fatalities are concerned, the worst nuclear accident (expected only once in 100,000 meltdowns) has already been equaled by coal burning.

The extent of land contamination in a reactor meltdown accident depends on one's definition of "contamination." The whole earth can be said to be contaminated, because there is naturally occurring radioactivity everywhere. Many areas, like Colorado, can be considered contaminated because they have larger than normal radiation levels.

But if we use the internationally accepted definition of the level of contamination that calls for remedial action, the worst meltdown considered in the Rasmussen Report (one in 10,000) would contaminate an area equal to a circle of 30-mile radius. About 90 percent of this could be easily decontaminated using fire hoses and plowing open fields, so the area where relocation of people is necessary would be equal to that of a 10-mile radius circle.

Forced relocation of people is not an unusual circumstance. It occurs in building dams where large areas are permanently flooded, in highway construction, in urban redevelopment, and so forth. In such situations, the major consideration is the cost of relocating the people. Therefore, it seems reasonable to consider land contamination by a nuclear accident on the basis of its monetary cost.

According to the Rasmussen Report, the cost in the worst 0.01 percent of accidents can exceed $15 billion, but the average cost for all meltdowns is $100 million. Air pollution from coal burning also does property damage by soiling clothing, disintegrating building materials, inhibiting plant growth, and so on. Estimates of the annual costs of this damage are in the range of $10 billion per year. Thus, at an average of $100 million per meltdown, we would need a reactor meltdown every four days to be as costly as the property damage from coal burning.

It is frequently asked why damage due to nuclear accidents is excluded from homeowners' insurance policies. This is because such damage is covered by separate liability insurance, provided partly by private insurance companies (about $200 million), and partly by a fund of $5 million each assessed against owners of nuclear plants, which brings the total to $550 million now, and close to $1 billion when plants now under construction are completed. Insurance companies are limited by law in the amount they can risk in a single accident (they also have about $1 billion insurance against damage to the plant itself), but beyond that, limiting risks from single events is the basic principle of the insurance business. It should be pointed out that few if any accidents are insured to the same extent as possible nuclear accidents; there is, for example, no such insurance against collapse of a bridge.

There has been extensive publicity about evacuation planning for areas around nuclear plants, a recent requirement imposed by NRC regulatory ratcheting. Such plans add something to safety, but far more important is the fear of nuclear power they engender in the public. Note that there are no evacuation plans around chemical plants or around railroad tracks traversed by tank cars carrying toxic chemicals, although such accidents have frequently required evacuation in the past. For example, shortly after Three Mile Island, the most serious nuclear accident to date—which, incidentally, did *not* require general evacuation—a chemical tank car accident near Toronto required evacuation of over 100,000 people.

High-Level Radioactive Waste

Several types of radioactive waste are produced by the nuclear power industry, but we limit our discussion here to the one that has received by far the most publicity—the "high-level" radioactive waste. It is the material derived from spent fuel after it has been removed from the reactor; it contains the bulk of the radioactivity produced by the industry. The plan is to convert it into a rocklike material (glass or ceramic) and bury it deep underground, *the natural habitat of rocks.* Under such circumstances, it seems reasonable to expect it to behave like other rock.

All analyses seem to agree that the principal concern is that the buried waste-converted-to-rock may someday be contacted by ground water, dissolved and carried with the groundwater back to the surface, and thereby get into food and water supplies. That is, the principal hazard lies in the possibility that some of this material may get into human stomachs.

We know a great deal about how rocks behave. For example, it may readily be shown that an atom of average rock now submerged in groundwater has about one chance in a trillion per year of entering a human stomach via the above described route.[49] If we assign this probability to the buried waste, the total eventual effects of the waste produced by one large power plant in one year is to cause 0.018 cancer deaths over the next several million years.[50] One perspective on this result is to compare it with the health effects of air pollution from coal burning. A median estimate for those health effects is about twenty-five deaths per year[15] per large plant. This is over a thousand times larger than the 0.018 deaths from a nuclear plant generating the same quantity of electricity.

If one goes further and recognizes some of the long-term health effects of chemicals released in burning coal, the toll of the latter technology mounts. Some of these materials—cadmium, arsenic, beryllium, chromium, and nickel—are carcinogens. They will eventually cause about eighty deaths,[51] several thousand times the number expected from the nuclear waste.

Another of these materials is uranium, which serves as a generator of the radioactive gas radon, which will eventually cause an additional thirty deaths, again over a thousand times the number caused by nuclear waste. Nuclear power, on the other hand, removes uranium from the ground and thereby reduces radon exposure to the public. It has been shown[52] that it thereby *saves* about five hundred lives each year of plant operation, 30,000 times the number lost due to the waste.

Questions have been raised about the danger of transporting radioactivity (in the form of used fuel elements or reprocessed material from such units). The casks in which this material is shipped cost several million dollars; one can surmise that a great deal of protection against traffic accidents can be obtained at that price. In endurance tests these casks have been hurled into solid walls at 80 miles per hour, then immersed in a gasoline fire for thirty minutes, and then submerged in water for eight hours, all without release of radioactivity. Safety against sabotage was tested by attacking a cask with a military bomb; results indicate that if this were done in midtown Manhattan an average of only one death would result. Elaborate probabilistic risk analyses find that if all U.S. power were nuclear, we might expect an average of one death every thousand years from release of high-level waste in transport. Of course, there would be many times that number of deaths from the normal traffic aspects of this transport; for example, transport of coal is now causing several hundred traffic deaths each year.

10
INTERNATIONAL NUCLEAR POLICY

Petr Beckmann

Summary

Past policies of trying to prevent nuclear proliferation have been hampered by technical and political misconceptions.

The chief technical misconceptions are (1) that the main danger of proliferation is associated with plutonium, and (2) that the electric power fuel cycle is a reasonable option for the manufacture of nuclear explosives, or at least for its concealment.

In the political field, it is not widely realized that President Carter's international nuclear policies were a gross and unilateral violation of the 1968 Non-Proliferation Treaty; so was the U.S. Non-Proliferation Act of 1978. These policies led to a worldwide mushrooming of reprocessing and enrichment plants, and tarnished the reputation of the United States as a reliable trading partner.

It is suggested that a policy based on a strong and credible military defense will minimize the dangers, or at least the effects, of nuclear proliferation, without undue government interference in foreign trade.

Introduction

Two closely intertwined issues affect international nuclear policy: proliferation and foreign trade.

The aspect of nuclear proliferation to be discussed is the diversion of fuels from the nuclear electric power cycle to the manufacture of nuclear explosives. The problem is essentially one of international policy, since only a government, or possibly a terrorist group (such as the PLO) with the support of some government, could plausibly acquire a nuclear weapon in this way. Manufacture of a bomb by a clandestine group working in a garage without government or military support is not, as yet, a credible danger; it is, at any rate, a negligible danger compared with that of diverting a ready-made military weapon—itself a very unlikely event.

The problem of proliferation is widely misunderstood, mainly in its technological aspects. The major misconceptions are:

1. that the main danger is associated with plutonium;
2. that the generation of electric power is a reasonable option for the manufacture of bomb-grade plutonium, or at least for the concealment of its production.

These technical misconceptions are compounded by the idea that nonproliferation is a technical rather than a political problem, and by most Americans' unawareness that during the Carter presidency, the United States was violating the 1968 Non-Proliferation Treaty.

Overview of World Nuclear Power

At the end of 1982, the world had 276 nuclear power plants operating in 23 nations with a total capacity of 176,000 MW; roughly one-third of them (77 reactors with 59,000 MW) were in the United States. Although 18 reactors were canceled in the U.S. in 1982, the world total increased by 8 percent during that year, and both world and U.S. capacities are expected to double by the end of the decade.[1]

In the Western world, the leader is France, more than 40 percent of whose electric power generation is nuclear; this is to increase to 70 percent by 1990. Both France and Britain practice full-scale, commercial uranium enrichment, breeding, and reprocessing, and France does so with strong public support. Japan has a strong nuclear program, including export of nuclear reactors, and South Korea is among the world's leaders in the speed of introducing nuclear power.

In Germany and Britain, nuclear power generation has also gained public acceptance, though breeders, reprocessing, and waste disposal are still controversial; in the smaller European countries, as in the United States, public acceptance has not yet been secured.

The USSR and the rest of the communist world are rapidly introducing nuclear power, considering it the principal source of electric power in the future. In the Soviet Union public acceptance is irrelevant, and proliferation is not a problem: the USSR is the sole supplier of nuclear services, such as enrichment or reprocessing, and sole owner of fuel rods, which it rents out in a strict one-for-one issue/return system that no genuinely sovereign state would be likely to accept.[2]

Technical Brief

With the presently unimportant exception of thorium, the raw material for all nuclear energy is uranium, which is found virtually everywhere on the globe, though only a relatively few places have it in sufficiently high concentration to make mining *commercially* profitable. However, if dollar cost is no obstacle, then every country has enough uranium for weapons.

The Isotopes of Uranium

Natural uranium is a mixture of uranium with atomic weights 235 and 238; the two isotopes have different physical properties, but they cannot be chemically distinguished or separated (as two different elements can). Only U-235 is fissionable, and it amounts only to 0.7 percent of the natural mixture. U-238, which accounts for the remaining 99.3 percent, is not itself fissile, but can be bred into plutonium, another fissile element. Unlike uranium or thorium, plutonium is not found in the natural environment.

Natural uranium must be enriched—in other words, its originally tiny U-235 content must be increased to make it useful in light-water-cooled reactors—up to 3.5 percent for fuel (not enough for an explosive chain reaction), or to more than 90 percent for a weapon of reasonable size.(Lesser enrichment for weapons is possible, but requires much larger masses.) Enrichment always involves separation of the two isotopes, and since only physical processes, not chemical separation, can be employed, this step now requires much energy and large industrial facilities. As we shall see, this is likely to change soon.

It is also possible to use unenriched natural uranium in power reactors (as do the Canadian CANDU reactors), but this requires the use of heavy water (or highly purified graphite) produced in industrial complexes no simpler than enrichment plants.

Plutonium Theft

When about 75 percent of the U-235 content in the fuel rods has been burned up, they are "poisoned" by fission products and must be reprocessed to utilize the remaining energy in the fuel. This means removing unwanted fission products, recycling the unused uranium-235, and also extracting the valuable plutonium, a fuel that is formed in the rods during burnup. The uranium and plutonium can then be combined into a nonexplosive mixed-oxide fuel, conserving a significant part of the original energy.

Plutonium is also produced in breeder reactors, which breed it from U-238; however, unlike production reactors (producing plutonium for weapons), fuel breeders and conventional electric power reactors operate under different conditions and produce plutonium of an isotopic mix that is not suitable for explosives, as will be explained shortly.

The sources of the plutonium—whether the reactor core, breeder blanket (where U-238 is bred into plutonium), or spent fuel rods—are all thermally and radioactively so "hot" that they are accessible only by remote control. And since a breeder need not produce more plutonium than its client conventional reactors burn, there need never be much plutonium in storage for operational reactors. That leaves only the short time during which the reprocessed fuel is in transit as an allegedly vulnerable point; this perceived danger receives considerable attention in Willrich and Taylor's well-known book.[3]

Much has been written on possible safeguards in transportation, which are, in fact, easy to implement: collocation of nuclear facilities (e.g., breeder and fuel processing in the same building, as is done with the EBR-II breeder in the Idaho National Engineering Lab), trucks that can be made immobile in case of attack, and so forth.

But these largely artificial issues have always been irrelevant, and in the last few years they have probably become obsolete as well.

The Power/Bomb Confusion

It is indeed *possible* to gain some material for the manufacture of bombs from the nuclear power fuel cycle, just as it is possible to manufacture TNT from the ingredients used in the manufacture of chocolate. The reason that chocolate factories do not represent a credible danger of illicit munitions is that this would be an utterly inept way of making explosives when so many easier ways are available. The same is true of nuclear explosives that might be made in the electric power fuel cycle.

Disregarding the alternative of acquiring a ready-made nuclear weapon by stealth or force (which would probably be the least difficult way), there are several ways of making nuclear explosives, as shown in Table 10-1. The one that is by far the

Table 10-1
Eight Known Ways of Producing Nuclear Explosives[4]

Method	Cost	Required technology	Required industry
Research reactor	small	small	small
Military plutonium production reactor	medium	medium	medium
Civilian electric power reactor	large	large	large
Diffusion cascade*	large	large	large
Centrifuge cascade*	medium	medium	medium
Aerodynamic jet cascade*	medium	large	large
Electromagnetic separation*	medium	large	large
Accelerator*	medium	medium	medium

*Enrichment methods.

most protracted and most expensive is associated with the generation of electric power.

There is a good reason for that: electric power production is merely an additional and unnecessary obstacle for the production of explosives; for example, reactor-grade plutonium would have to be chemically and physically purified to be as effective as the plutonium bred in a production reactor. No country has ever made explosives using a power reactor; in particular, India, which is often incorrectly alleged to have done so, bred plutonium in a research reactor, using natural uranium and heavy water.[2]

The quite unjustified nuclear power/bomb confusion is simply a part of the antinuclear campaign, whose main tactic is intimidation by fear of the unknown. The campaign thrives on misinformation. Talk of the "nuclear police state" is a typical element of that tactic. The existence, storage, and transportation of large numbers of actual nuclear weapons (not just some possible raw material for them) has not led to police states in the United States, Britain, or France.[5]

The Plutonium Obsession

Fear of the unknown and exploitation of ignorance seem to have paid off well for the antinuclear campaigners in the case of plutonium, which they have made into a scare word.

Actually, as the fuel that concentrates more energy in a unit of volume than any other, it represents one of humanity's greatest achievements. As a poison it is less impressive. Describing it as "the most toxic substance known to man" is ludicrously inaccurate.[6] Plutonium also appears to be a dismal failure as an explosive when extracted from the spent fuel rods of a power reactor. Indeed, the plutonium scare in connection with weapons proliferation is triply misleading: politically, because the danger of proliferation is not a technological, but a political problem; technologically, because reactor-grade plutonium is unsuited for an explosive; and logically, because the greater threat arises from uranium-235.

Consider the last two assertions.

The Plutonium Route

Suppose some country with a fuel-reprocessing industry chemically extracted the plutonium from the acid bath in which the fuel rods must be dissolved. What it would have on its hands is a mixture of at least five plutonium isotopes, of which only two (239 and 241) are fissile, and only one (239) is free from a phenomenon that Taylor and Willrich cavalierly dismiss with just about two sentences, but that in effect "poisons" or "denatures" reactor-grade plutonium for use as an explosive.

The phenomenon is that of "spontaneous fission neutrons," a type of radiation that is emitted by the other isotopes at a rate about 100,000 times greater than by plutonium-239.[7] This neutron radiation constitutes a radiological hazard (especially in the presence of lighter elements such as oxygen, in which it will induce a far higher level of neutron radiation); at the same time, it also precludes an effective explosion—that is, an explosion of sufficient force at a predictable time. The radiation level is not immediately lethal and might not deter a group with suicidal determination and a willingness to lose their hands or their lives after completing the job; however, the high neutron flux would also lead to spontaneous explosive chain reactions. Such predetonations are dangerous above all to their immediate neighborhood, for they "fizzle" by extinguishing themselves: they blow the critical mass apart again before it has a chance to undergo a full explosion.

It is true that a nuclear "device" made of reactor-grade plutonium was detonated beneath the Nevada desert in the 1970s (the quotation marks are meant to express doubt as to whether it had the small size and transportability of a bomb), but the force of the explosion has never been made public, nor is it known whether the device exploded at the intended time. It must, in any case, have been an exacting task; no one with lesser resources than the U.S. government has ever attempted it.

In the weapons industry, these difficulties are evaded by

using highly purified plutonium-239 (with minimal spontaneous neutrons), and by sophisticated implosion techniques
achieving uniform and significant compression (implosion) of
several subcritical masses, which must be held together
against the force of a nuclear explosion for several microsec-
onds until the explosive is consumed. But the crucial point is
this: the production of pure plutonium-239 requires special
techniques incompatible with the conditions that apply in a
power reactor. Figure 10-1 shows why: within hours of the first
plutonium-239 appearing as the U-235 burns up, it is contaminated with other isotopes, particularly plutonium-240.

But could the plutonium-239 not be cleansed of the "spiking"
isotopes after it has been extracted from the reactor? Or could a
power plant not be clandestinely adapted for the production of
weapons-grade plutonium (though it could not produce electricity at the same time)? The answers to these questions are
no more known to this writer than whether bread can be made
edible again after being soaked in detergent, or whether a
washing machine can clandestinely be turned into an oven.
The more relevant question, of course, is why not bake uncon-
taminated bread in a regular oven?

The Uranium Route

The bread, in this case, is highly enriched uranium, and the
bakery any one of several enrichment processes. Uranium was
the explosive used in the first nuclear explosion ever (Los
Alamos, 1945), and the enrichment then was performed by
means of large mass spectrometers (Calutrons) at Oak Ridge
National Laboratory. This is an inefficient and costly method,
but a simple one, open to any nation or organization. Once the
"what if" and "worst case" approach is adopted, why assume
that Colonel Qaddafi is a stickler for energy conservation or
cost effectiveness?

There are, however, more ominous enrichment methods in
existence, or on the horizon.

In all cases, enrichment involves the separation of two
uranium isotopes, the wanted U-235 and the abundant U-238.
The two atoms (usually bound in the molecules of the gaseous
uranium hexafluoride) are chemically identical and differ only
in the masses of their nuclei, so that they must be separated by
a physical process. In the Manhattan Project, the most obvious
method, separation by centrifuge, was rejected for a less efficient diffusion process (used to this day in the United States),
because at the time there were no materials that could with-

stand the stress on the centrifuge at the required velocities. That objection no longer applies, and ultracentrifuges are now used for enrichment by the European Community (Euratom).

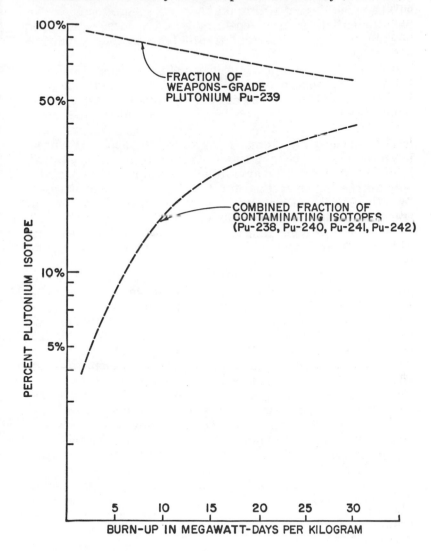

Figure 10-1. Contamination in a power reactor of (weapons-grade) plutonium-239 by other plutonium isotopes, as a function of the degree of "burn-up" (measured in megawatt-days from the first energy conversion of the fissile uranium-235).

As the efficiency and the strength of materials used in centrifugal enrichment grow, the ultracentrifuges will decrease in size, and the required quantity (working in a battery) will be reduced. That will make them ever more conducive to concealment. Even now, costs and intelligence agents are probably a greater deterrent to proliferation than technical difficulties.

It is also possible to use a cruder method developed in West Germany, and independently in South Africa, where the centrifugal principle is realized not by a rotating container, but by a gas ejected from a jet and forced into a multicurved path. Brazil bought an enrichment plant that works on this principle from West Germany. Neighboring Argentina apparently does not have one, but bought 100 kg of 20 percent enriched uranium and an unspecified amount of heavy water—the ingredients for the Indian method for creating nuclear explosives— from the Soviet Union in 1982. On June 8, 1982, Argentina, a nonsignatory of the Non-Proliferation Treaty, informed the International Atomic Energy Agency that it "reserves the right to use nuclear energy for non-proscribed military purposes."

There are other methods of uranium enrichment, but one promises to be particularly compact and efficient: laser enrichment. When a gas containing uranium is irradiated with laser light of a certain precise wavelength, one of the two uranium isotopes can (after some further treatment) be made to acquire an electric charge, while the other remains neutral. The two can then be separated by an electromagnetic principle used in any TV picture tube. This method promises to be compact, cheap, and low in energy consumption. It was realized in the laboratory some years ago, and its large-scale application may possibly be in operation behind closed doors already.[8]

No matter which of these methods a country or terrorist group chooses for the clandestine manufacture of nuclear bombs in the future, it will have two effective helpers: the carefully nurtured and widespread bomb/power confusion and the plutonium obsession.

The Policy of Self-Mutilation

The misunderstandings about proliferation in the political field rival those in technology; they hinge on the 1968 Non-Proliferation Treaty, the 1978 U.S. Non-Proliferation Act, and the effectiveness of international safeguards. To understand the present sorry state of affairs, one must briefly go over some recent history.

The 1968 Non-Proliferation Treaty

The 1968 Non-Proliferation Treaty is one of the most unequal treaties of all times. It divides the world into "nuclear-weapon states" and "non–nuclear weapon states," with the latter promising not to do what the former are doing; this aspect takes up most of the treaty's eleven articles. In return for their forbearance, the non-weapon states received nothing but a promise expressed in Article IV:

> 1. Nothing in this treaty shall be interpreted as affecting the inalienable right of all the Parties to the Treaty to develop research, production and use of nuclear energy for peaceful purposes without discrimination. . . ..
> 2. All the Parties to the Treaty undertake to facilitate, and have the right to participate in, the fullest possible exchange of equipment, materials, and scientific and technological information for the peaceful uses of nuclear energy. Parties to the Treaty in a position to do so shall also co-operate in contributing . . . to the further development of the application of nuclear energy for peaceful purposes, especially in the territories of non-nuclear States party to the Treaty, with due consideration for the needs of the developing areas of the world.

More than forty countries—including France, India, and Israel—refused to become a party to this one-sided treaty; others signed it, presumably in the hope that Article IV would guarantee them technical assistance, fuel supplies, and access to enrichment and reprocessing services.

That hope proved wrong. In 1977, Article IV was, in effect, unilaterally and precipitously abrogated by one of the "Nuclear-Weapon State Parties"—the United States of America.

In 1976, the Ford Foundation's report "Nuclear Issues and Choices," reflecting the views of the anti-growth constituency, claimed that there was no need for fuel reprocessing or breeders and that both should be delayed while awaiting technical and political solutions of the proliferation problem, which the authors, once again, believed to be limited to plutonium, and to reactor-grade plutonium at that.

The following year, President Carter adopted this policy, prohibiting the completion of an almost finished reprocessing plant in South Carolina, banning reprocessing in the civilian sector (the ban has since been lifted by the Reagan administration), and attempting to block the Clinch River Breeder Reactor program. At the same time Carter announced that current contracts for fuel and enrichment with other countries would be "renegotiated" to bring them in line with this new policy.

On the face of it, this was merely an inept policy resulting from the plutonium obsession. (A year earlier, Carter had not objected to new uranium enrichment capacity based on the more proliferation-prone and less easily detected ultracentrifuge method.) In reality, much of the world took it as the abrogation of a treaty with an arrogance and hypocrisy hitherto seen only in Soviet treaty abrogations.

The arrogance of the United States was exemplified by events in Japan, whose $200 million reprocessing facility was nearing completion. Japanese power plants at the time were running exclusively on fuel enriched in the United States. As became customary after the 1974 Indian explosion and the subsequent safeguard guidelines drawn up by the industrial countries, the United States required prior consent for any reprocessing of this fuel. However, it had been well understood by both sides that the United States required the Japanese reprocessing plant to be built so as to facilitate safeguards; now this same agreement of prior consent was unilaterally reversed from a requirement to a prohibition.[2] The Japanese public was so outraged that the U.S. State Department had to grant a two-year extension of the authorization to reprocess; but Japan had learned its lesson and did not again count on the United States as a sole supplier of nuclear fuel or other services.

The hypocrisy was even more immediately apparent in the Carter policy; presented in lavish moral terms, it prohibited the reprocessing of all but military fuel and banned all plutonium except that intended for bombs.

If the policy was highly damaging to the U.S. nuclear industry, it was disastrous for U.S. prestige abroad. With the effective abrogation of Article IV, America's detractors could at last make charges of U.S. imperialism stick, and the standing of the United States as a reliable international business partner remains tarnished to this day.

The 1978 U.S. Non-Proliferation Act

But worse was to come. President Carter had invited all interested countries to participate in an International Fuel Cycle Evaluation as relevant to nonproliferation. Several hundred experts from fifty countries participated after first extracting a promise that current international policies should be pursued during a two-year "truce" while research was being carried out for the evaluation without prejudging it. Within one year, this "truce" was broken by the U.S. Congress, which in April 1978 passed legislation misnamed the Non-Proliferation Act, and Carter broke his pledge of awaiting the International Fuel Cycle Evaluation results by signing it. The 1978 Non-Proliferation Act directly contradicts Article IV of the 1968 Non-Proliferation Treaty by requiring the United States to embargo supplies and to impose other economic sanctions on countries practicing reprocessing. Countries dependent on American fuel thus had to agree to choose the least efficient fuel cycle and not to practice breeding if they wanted to receive even natural uranium from the United States.

The ineptness of this law was well illustrated in June 1980, when President Carter again had to break his broken word by approving shipment of nuclear fuel to India, which as a nonsignatory of the Non-Proliferation Treaty had developed its own reprocessing plants. "A cut-off," said a State Department explanation,[9] "could reinforce the perceptions of many countries of the unilateralism of U.S. non-proliferation policy and that the U.S. cannot be counted on as a reliable supplier."

But whatever the legal niceties of the act, one effect was unmistakable: reprocessing and enrichment plants mushroomed all over the world. After a brief hesitation, the rest of the world simply went forward on its own, for the new U.S. policy made no sense: even if plutonium were the main danger, the best way to get rid of it is to reprocess and burn it. But more important, as British astronomer Sir Fred Hoyle pointed out, under the terms of the 1978 U.S. Non-Proliferation Act, "a nation that built itself a centrifuge to obtain weapons-grade U-235 would find that the United States, by supplying the [3.5 percent] enriched uranium, had assisted the weapons proliferation, to the extent of removing three-quarters of the uranium-238."[10]

Typical of this period was the British Windscale Enquiry, named for a major extension of the Windscale reprocessing plant, which became a kind of international tribunal of nuclear power in general, and of reprocessing in particular. After several months of hearings, during which special weight was given to the testimonies of American antinuclear activists (Prof. Albert Wohlstetter for the Friends of the Earth and Dr. T. B. Cochran for the National Resources Defense Council), Judge Parker fully vindicated reprocessing in a detailed report that has become an internationally recognized legal document on reprocessing, and on the broader issues of proliferation, terrorism and civil liberties.[11] In spite of high-level U.S. intervention, the British government immediately proceeded with the construction of the reprocessing plant.

Similarly, Japan and Germany went ahead with plans to build large reprocessing plants, shipping their spent fuel rods for reprocessing to France in the meantime. West Germany, resisting U.S. pressure, sold an enrichment plant to Brazil, which now has complete fuel-cycle capability of its own. India, Pakistan, South Africa, and Israel eventually obtained fuel supplies, though they have not signed the Non-Proliferation Treaty, and have not even fully accepted International Atomic Energy Agency (IAEA) safeguards.

In February 1980, the International Fuel Cycle Evaluation presented its 20,000 pages of reports produced at a cost of $100 million by seven hundred experts from more than fifty countries. Its conclusions, pointedly ignored by evaluation initiator Jimmy Carter, were that no fuel cycle was totally free from proliferation risks, that breeder technology posed no greater proliferation risks than other technologies, and that the solution to the problem was institutional.

The IAEA Safeguards

At present, the main international approach to nonproliferation are the IAEA safeguards; bilateral and regional (European, South American) agreements also essentially conform to them.

The International Atomic Energy Agency, headquartered in Vienna, was established in 1958 under the auspices of the United Nations as a result of Eisenhower's "Atoms for Peace" initiative. One of its purposes is surveillance of nuclear facilities by on-site inspections to detect diversion of fuel from peaceable purposes. The IAEA requirements, known as "safeguards," include inspection of inventories and records, and monitoring the flow of nuclear materials—in countries that accept the safeguards.

These safeguards have been highly successful in countries that do not plan to manufacture nuclear weapons anyway; for other countries, their value has ranged from dubious to nil, for both the choice of inspectors and their authority are limited.

A list of inspectors is submitted in advance to an inspected country, and that country can reject any inspector whose nationality might cause difficulties. Beyond that, it is customary for one of the (usually two) inspectors to be a citizen of the country inspected, or of one that is its ally. (This is carefully cultivated by the USSR, which itself agreed to inspection of power reactors only in March 1983 and which has nonproliferation in its satellites firmly under control, but which keeps in touch with nuclear developments in the outside world via its East Bloc inspectors.) Thus a state bent on diverting nuclear material need not fear impartial inspection without prior warning.

The authority over what to inspect is also severely limited. In order to protect commercial data, the production and movement of raw and refined uranium ore (yellowcake) are not subject to safeguards—which would be quite a loophole if the entire system were not already flawed. A more serious deficiency is that the IAEA is authorized to detect actual or suspected diversions that *have* taken place, not those that are *about* to take place. Thus, after Israel destroyed the Iraqi research reactor at Tamuz in June 1981, two IAEA safeguards inspectors dispatched to the destroyed reactor in November 1981 found "no evidence of noncompliance" by Iraq, noting only that the stores of natural and depleted uranium were found in their original condition and that "this material will become subject to IAEA safeguards only when it is used, i.e., chemically purified.

Now depleted uranium (deficient in U-235) has no use other than acting as the fertile material to be irradiated for plutonium production, although natural uranium can be used for the same end; but tons of it just lying around next to a reactor with 20 percent enriched uranium, which would act as the irradiator, apparently do not violate the safeguards.

This method of plutonium production, wrote the deputy director of the IAEA Safeguard Department, in an article that has no official standing,[12] "might" require additional cooling and the introduction of conspicuous hardware which, together with notification of modifications of the reactor, would not have escaped detection.

Unless, of course . . . There are several possibilities, one of which is the right of any country to withdraw from its previous acceptance of IAEA safeguards on three months' notice, assuming that a government about to embark on the project of throwing a nuclear bomb on its neighbor would strictly adhere to all subtleties of diplomatic etiquette.

"The system was, and is," says longtime member of the IAEA's Board of Directors Bertrand Goldschmidt, "more a matter of politics than of policing, despite its foundation in technical inspection and surveillance. . . . The system is completely ineffective in the case of a government that is determined to break its nuclear pledges and close its borders to international inspection."[2]

This brief account of the IAEA, limited to the safeguards aspect, would be unfairly one-sided without mentioning the excellent work the IAEA does in enlightening the public on radiation, waste disposal, decommissioning of power reactors, and other subjects that have been turned into scares by antinuclear agitators.

Needed: A More Fundamental Approach

If all these shortcomings of the IAEA safeguards could be eliminated (and they could not, unless a certain loss of sovereignty were forced on the non–weapon owning states), it would not seriously reduce the threat of nuclear armament by a government determined to do so, for these safeguards are purely technical. Even then, they are limited to the electric power industry, on the false assumption that this is the only (or even the most likely) sector that might be abused for weapons production. As in other fields, technical solutions are limited because they do not strike at the heart of the problem. Similarly, the roots of drug abuse would not be touched by monitoring all syringes, and grievous harm would come from prohibiting the use of them.

However, that is an imperfect analogy, for proliferation is not as yet a case of individual criminals engaged in an illegal activity, as drug pushers are. If and when ultracentrifuges or laser enrichment make homemade uranium-235 bombs a credible threat (and the scaremongers now barking up the plutonium tree will falsely whine, "We told you so!"), this will pose problems not fundamentally different from those faced today by Interpol or other organizations cooperating in international crime detection and law enforcement.

The real problem with proliferation is that the potential "criminals" are governments, and the "cops" are other governments that differ from their counterparts in that they are more powerful and can make their own rules. This is a fundamentally unstable situation, unlikely to be cured by superficial checks of fuel rods in power plants.

The history of governments trying to police one another is a history of sorry failures—from the Geneva conventions at the turn of the century (which met to outlaw war but ended by drawing up the rules on how to wage it) to the present United Nations.

To the contrary, a government bent on its one and only job—ensuring the security and civil rights of its citizens—might not attempt the futile task of keeping nuclear weapons from tyrannical and aggressive governments; like a physician who finds shots against malaria more workable than hunting down all of the world's mosquitoes, it might rather take steps to defend itself and its like-minded allies from nuclear attack.

Ideally, the replacement of a retaliatory national strategy by a protective one, as suggested, for example, by General Daniel Graham's *High Frontier*[13] and more recently proposed by President Reagan, would give the free world a degree of protection from nuclear attacks that cannot be achieved by policing all possible sources of nuclear raw materials. And by its moral superiority (of limiting itself to the defense of civil rights and security) it might hasten the internal collapse of the slave empires with which the United States now seeks accommodation, thus striking at the real root of proliferation.

This is admittedly an ideal, not immediately realizable prescription; but certain measures on the road to that goal, specified in the recommendations that follow, are feasible—even "politically feasible."

Free-Market Issues

Beyond the proliferation issues discussed above, foreign trade poses no problems peculiar to nuclear fuels, reactors, or other facilities. A government that accepts the principle that its only job is to protect the civil rights of its citizens should not interfere with this trade except where issues of security are involved.

The prime example of such an exception is trade with the Soviet Union, which should be kept to a minimum—not to punish the USSR or (primarily) to deprive it of nuclear technology with possible military applications, or even to pressure it by economic sanctions, but as a simple policy of peace: to force the Soviet leadership into reallocating resources from their military machine to looking after their people.

This principle is not peculiar to nuclear goods and services; it is the principle that should have been applied to the Yamal gas pipeline, for example. In the nuclear field, it can be seen at work. The Soviets were denied direct U.S. assistance to their nuclear program, and as an economic sanction this was quite ineffective. Today the Soviets have an entire nuclear industry in place that is in some cases (breeders, reactor mass production) more advanced than that of the United States. They offer enrichment services to any Non-Proliferation Treaty member country, payable in a toll of the supplied uranium, and they even export nuclear power plants—to Libya and Cuba, for example. But the billions invested in the Soviet electric power program are, in part at least, billions withdrawn from the war machine.

Apart from such security aspects, a government dedicated only to the maintenance of liberty would not interfere in the free trade of nuclear goods and services beyond protecting it from witch hunts and superstitious zealots; it would not subsidize an export-import bank, but would merely express its thanks to Japanese or French taxpayer for gifts in the form of goods from subsidized industries.

Such a government dedicated exclusively to protecting the security and civil rights of its citizens is, of course, entirely imaginary; real governments spend most of their time attempting to make everybody live at everybody else's expense.

A real government cannot be expected to offend a multitude of lobbies simply by abolishing the export-import bank. A government cannot be expected to outrage the media by privatizing all nuclear services, but it can and should induce private industry to take over reprocessing[14] by guaranteeing, under penalty of full compensation (or however much is needed to recover the investment), that government policy in this field will not change again over the next ten years.

This, incidentally, is a free-market policy badly needed for *all* ventures of the nuclear industry. Much of the industry's plight is due to what it politely calls "regulatory uncertainties"—the astronomic costs of unpredictable, retroactive government regulations, most of which are intended to appease the antinuclear lobby. (See Chapter 9.)

Recommendations

The first step in any successful policy must be to acquaint people with the truth, however unpopular it may be; accordingly, such a policy must clearly state the following:

• There is no way to stop the eventual dissemination of technical knowledge.

- No technical precaution can for long prevent a determined country from producing a nuclear weapon.
- Prohibiting nuclear electric power would not prevent, or even significantly impede, nuclear weapons proliferation; it would, however, sacrifice the thousands of lives now lost to less safe power sources.
- The most effective protection against nuclear terrorism and blackmail is moral strength: a refusal to give in may cost lives today, but will save many thousands tomorrow. (This should be proclaimed as U.S. policy *before* any actual incidents take place—and in hard-nosed terms—to convince blackmailers that their projects will meet certain failure.)
- The only effective protection from large-scale nuclear attack is military strength—if it is strong enough to *protect* (rather than retaliate), it is automatically strong enough to deter.

In the field of foreign relations, the United States might evaluate its option of withdrawing from the Non-Proliferation Treaty, which many regard as both immoral and ineffective; however, as long as it is a party to the treaty, the United States cannot at the same time stand by the 1978 Non-Proliferation Act, which directly contradicts the treaty and enhances proliferation in effect. The act should be amended or repealed.

The IAEA safeguards are deficient in the political rather than the technical field; in particular, if inspections are to be effective, the inspected country should have no warning of an impending inspection, nor should it have the right to reject any of the inspectors. The custom of the inspector being a citizen of the inspected state or its ally should be abandoned. However, the United States should make it clear that such improvements, while welcome, will not remove the fundamental flaws of the system.

Apart from proliferation issues, foreign trade poses no problems peculiar to nuclear fuels, reactors, or other facilities. The United States should deny goods and services to the Soviets whenever this forces them to reallocate resources to peaceable purposes. But the United States should not otherwise obstruct the free exchange of nuclear technology or fuels; in particular, it should offer enrichment services (and when it has them, reprocessing services) to all countries.

Finally, the United States should vigorously promote and implement the policy that strikes at proliferation in the free world at its very roots: maintain a principled, determined, and credible military posture vis-à-vis the Soviet Union and other potential aggressors so as to make it unnecessary for smaller countries—in particular Israel, South Africa, and Taiwan—to seek nuclear weapons for their defense.

11
ELECTRIC UTILITY ISSUES

Rene H. Males

Summary*

Technological progress has given the electric utility industry a competitive edge. As a result, the industry has enjoyed sustained and, until recently, rapid growth. While regulated, the industry has thrived, finding new markets where electricity has proven more economic, cleaner, or more convenient than alternative ways of providing the same service for consumers.

The 1970s, however, brought challenges that have radically changed the industry's course. Real price increases for the industry's fuels and capital equipment, on top of inflation, and additional costs incurred to meet new social and environmental considerations, created financial problems for the "electric companies." Regulators responded sluggishly, and the industry's financial plight worsened. Moreover, a philosophical-political change from competing in energy markets to husbanding scarce energy resources cut off potential new markets for the electric utility industry while curbing sales efforts in existing markets. The forces affecting these changes in social direction also caused construction schedules to lengthen. Finally, building plans made during a time of rapid growth in consumers' electricity needs have turned out to be more than is now needed to meet current loads.

The 1980s may restore growth in electricity sales, albeit at a slower pace compared to the halcyon decades of the 1950s and 1960s. Moreover, adjustments in consumers' electric rates are catching up with earlier cost increases. Inflation appears to be slowing to more manageable levels. In addition to brightening

* I would like to acknowledge the substantive contributions and editorial assistance of Dominic M. Geraghty of EPRI and Robert G. Uhler of NERA, without whose help writing this chapter would have been impossible.

business conditions, the industry has technical gains in prospect. These will help utilities meet new problems with greater ease by regaining the industry's older structure of declining real costs.

Major challenges, however, need to be met. Acid rain, for example, is an environmental issue that will require innovative solutions, both technical and political, to avoid imposing unnecessarily large costs on consumers. Also, absorbing the remaining increased costs will cause short-term "rate shock" for some companies, which could abort the industry's financial recovery. Diversification and deregulation have been proposed as more sweeping solutions to the utilities' long-term problems.

Whatever this decade brings, the electric utility industry has a better price advantage over competitive fuels (primarily oil and natural gas) than it had in the 1960s. Further, a strengthening economy and a growing need to modernize the nation's industrial sector provide opportunities for greater electricity sales with little need to build and finance additional generating capacity.

Introduction

Ironically, much of what has been written about the electric utility industry has focused on its monopoly characteristics and the countervailing regulatory institutions. In fact, electric utilities have been broadly influenced by each era's technologies and have been specifically shaped by competition with alternative energy sources in each utility's service territory. Generally, neither energy nor electricity per se is the product desired by consumers. Rather, people want primary services such as light, motive force, or heat; electricity is an intermediate service capable of satisfying those needs. From this point of view, electricity competes in energy markets for the end users' dollars.

Five Aspects of Competition

Specifically, an electric utility faces competition in five ways: (1) other *forms* of energy—primarily oil, natural gas, coal, and solar—can furnish the primary services the consumer desires; (2) moreover, electric users themselves and even other electric utilities, in certain situations, can provide electricity—*enterprise* competition; (3) *generic* competition exists because people can substitute other goods or make capital investments in more efficient energy-conversion devices or in insulation to obtain the same primary service but with less electricity; (4) even though many consumers cannot rapidly move from one electric utility service territory to another, *geographic* competition occurs in the industrial sector in that manufacturers can locate, expand, or concentrate production in areas having attractive electricity prices; (5) finally, there is *general* competition for consumers' expenditures; for example, how people allocate their incomes—buying air-conditioning service versus paying for a cabin in the mountains.

An electric utility has a monopoly only in the sense that the company receives an exclusive right to provide electric service in a given geographic area. For this right, however, the electric company must agree to allow its prices to be determined by some form of regulatory agency. Typically, the state law that creates the regulatory function stipulates that the utility should be allowed a "fair return" on its investment, which provides the electric service.[1] In return, the utility must agree to offer service to all consumers in the geographic area. The exact extent of this obligation to serve is somewhat unclear. Such legislation, however, is generally clear in another respect: a utility may not discriminate among similar customers. The question of what constitutes discrimination and what is merely appropriate differentiation based on the cost of service to various customers is one of the central arguments in most rate hearings. In these adversarial proceedings, all parties have the opportunity to submit evidence and to argue about what constitutes "fair and nondiscriminatory" rates. Appointed or elected state officials (and sometimes the courts) have the final word on rates: the regulated electric company does not.

History of the Electric Utility Industry

Historically, competitive forces have created challenges for and offered opportunities to the electric sector of the energy industry. The sector's history also shows how technological developments have changed the competitive situation of electricity vis-à-vis other fuels or other forms of energy service. Thus, in the past, competition and technology have shaped the growth of the electric utility industry. Those same forces influence the industry's present situation and will provide the options for future developments.[2]

The Early Years

Just over one hundred years ago, Thomas Alva Edison perfected the incandescent electric light bulb, a device designed to compete with gas lamps, wax candles, and oil lamps. But before such electric illumination service could be sold, Edison also had to develop a practical generator of electricity (a "dynamo," in his term, which mechanically converted other forms of energy to electricity) and to build a reliable distribution network.[3]

Such early electric systems were physically created and legally formed as local lighting companies that purveyed their services over distances of several blocks in densely settled city locations—"free enterprise" in the best sense of the term. This electric lighting service was better than that available from gas works or other alternatives. Electricity was less expensive due to its higher efficiency in converting fuel (thermal energy) to light. Moreover, electric lighting was safer as well as more convenient: there was no odor, smoke, or soot.[4] Automatic water heaters and washing machines were still at least fifty years in the future.

As the technology of electricity distribution improved and power could be sent over longer distances, these small businesses coalesced into larger systems. Eventually, each utility would offer service within a whole municipality. A company was either a city-owned "department of power" or an investor-owned firm that operated in one or more towns. Unfortunately, the fledgling industry's relatively narrow market—the lighting load—had a regular diurnal pattern that left each utility's electricity-generation capacity unused much of the time (i.e., during daylight hours). Fortunately, the pervasive need for motive power by the nation's growing industrialized economy led to the development of electromechanical equipment that reversed the electric generation process. These devices converted electricity back to motive power and opened new markets for electric companies. Motors were used to drive elevators and a host of factory machines, replacing steam, waterwheels, and belts. Residential electric motive applications (e.g., fans) were also developed to complement the six-day eight-to-six electric loads of the commercial sector.[5]

By the 1920s, much-improved electricity-generating technologies and the engineering know-how to transmit electricity at higher voltages became available. These scientific and commercial developments led to a further agglomeration of smaller electric firms into even larger systems. At the same time, the electric utility business became increasingly capital-intensive, requiring large investments in generation, transmission, and distribution relative to revenues. This led to the joining of electric companies in different areas of the country into holding companies. These large corporations provided financial economies on a scale to match the industry's burgeoning capital requirements.[6]

By the early 1930s, some 75 percent of the electricity-generation business in the United States was controlled by sixteen holding companies—a concentration of market power that ultimately collapsed. Financial abuses, along with the Great Depression, sounded the death knell for these organizations as President Franklin D. Roosevelt and Congress enacted the Public Utility Holding Company Act of 1935. Today, the electric utility industry is regulated primarily at the state or municipal level, although the federal government continues to play an important role.

In the first two decades of this century, utility managers hotly debated the advantages and disadvantages of regulation. On one hand, a public utility commission could grant the advantages of a protected regional service territory and the power of eminent domain.[7] On the other hand, regulation meant that utility owners had to give up the sole right to set prices and dictate conditions of service. Moreover, a commission could review management policies and actions. This balance—uneasy at best—between utility managers and state regulators has been defined and redefined in a number of watershed commission decisions and court cases.[8] By the 1920s, utilities in most states were regulated in one way or another. Along with the limitations imposed on the financial structure of the industry, the 1930s also ushered in a federal regulatory agency, the Federal Power Commission (now known as the Federal Energy Regulatory Commission). The enabling legislation, the Federal Power Act,[9] focused on the regulation of interutility sales of electricity and sales across state lines.

During this early era, technology and markets affected one more element of the industry's institutional structure. Eager to extend electric service to rural areas, Congress passed the Rural Electrification Act of 1936 (REA).[10] REA provided financial aid and a federal mechanism for the creation of rural electric cooperatives. Extending electric service to dispersed farms, of course, was considerably more expensive per customer than servicing urban consumers. Many utilities, having their roots in America's towns, had offered rural service only on advance payment of line-extension costs. By 1950, either through the cooperatives, municipal or district systems, or state or privately owned utilities, electric service was available to all but the most isolated households.

Also in the Roosevelt years, the federal government took an activist role in the power-generation aspects of the electric utility industry by establishing the Tennessee Valley Authority (TVA) in 1933[11] and the Bonneville Power Administration (BPA) in 1937.[12] TVA was originally created to develop the resources of the Tennessee River basin, including hydroelectric power–generation in conjunction with navigation and flood control. Following the full development of its hydropower resources, TVA built a comprehensive power production system by adding fossil-fueled and nuclear generating plants. In so doing, TVA became the country's largest electric utility, with approximately 5 percent of the nation's generating capacity. More than half of TVA's electric energy is sold to municipal and cooperative systems, with the balance going to industrial consumers and federal agencies.[13]

In contrast, the BPA was created as a marketing agent for the power generated at thirty-three federal dams in the Pacific Northwest. BPA has grown as an energy broker rather than as a power producer. As a middleman, it operates the nation's largest network of long-distance, high-voltage transmission lines, which serve as the main electric power grid for all interconnected utilities in the region. There are several other federal electric systems: Southwestern Power Administration, Southeastern Power Administration, Alaska Power Administration, and Western Area Power Administration. Combined, the federal systems represent about 12 percent of the total U.S. generating capacity.[14]

Post-World War II Developments

Except for rural electrification and the federal power programs, the period after the great surge of the 1920s was one of stability and consolidation for the electric utility industry. During World War II, electricity, like other goods and services, was made available preferentially to support the war effort.

During the period 1950 to 1970, the utility business enjoyed its "golden era." Sales boomed, increasing on average 7.5 percent annually. Because of economies of scale and better technology, the cost of electricity decreased slightly, going from an average of 1.8 cents per kilowatt-hour in 1960 to 1.7 cents in 1970 (Figure 11-1). Generating units increased in size

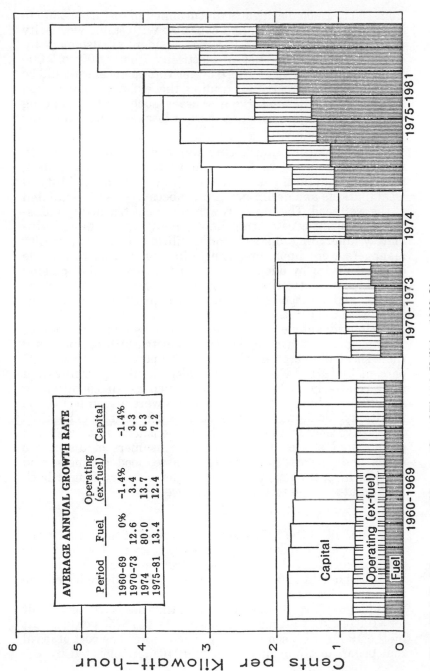

Figure 11-1. Cost Components of Investor-Owned Electric Utilities, 1960–81.

from 150 to 1,000 megawatts and declined in cost from $300 to $100 per kilowatt of installed capacity. With stable electricity prices and rising personal incomes, residential customers increased their average use of electricity from 3,000 to 7,000 kilowatt-hours per year. These larger volumes of sales could be served without greatly expanding the existing distribution systems (other than to serve new areas, such as the expanding suburbs). Fuel costs remained more or less constant at 30 cents per million Btu (mbtu).

During this time, marketing was a major focus of utility managers. This led to fierce competition with other fuels, particularly natural gas, which was growing in availability in local markets and declining in price because it was regulated at the wellhead. The rapid growth of the electric utility industry also meant constructing new generating plants and ultimately upgrading distribution facilities. Moreover, utility managers were improving the technologies used to provide electric service by adopting supercritical coal units, placing distribution networks underground, developing nuclear plants, and incorporating computer-supported management systems in the utilities offices.

Taking advantage of declining real costs and new end-use technologies that favored electricity, electric utilities increased their share of energy markets from 14 percent in 1950 to 23 percent in 1970.[15] Not only were electric utilities saturating their relatively exclusive markets, such as air conditioning and lighting, but the power companies were also entering the traditional domains of other fuels, such as space heating. Aggressive energy-marketing battles ensued using sales and promotional strategies familiar to consumers of disposable goods and services such as toothpaste, food products, and movies. In particular, the natural gas industry fought back with what it called total energy service.

Crises of the Seventies

The decade of the 1970s saw the reversal of these exhilarating trends and the dimming of the industry's golden era. The price of electricity climbed steadily from a low of 1.5 cents per kilowatt-hour in 1969 to 5.5 cents in 1981. This reflected the underlying increase in the utilities' costs of doing business. For example, coal-fired generating units that once had been built for $100 per kilowatt had increased in cost to $750 per kilowatt by the 1980s, with some costing much more. The cost of fossil fuel began to climb in the early 1970s. Accelerated by the

OPEC oil price rise, it surged past $1.90 per mbtu and was still headed upward at the end of the decade (Figure 11-2). Even so, the share of primary energy delivered in end-use markets in the form of electricity increased to 33 percent by the end of the 1970s.

Nuclear technology, which had been developing commercially since the mid-1960s, suffered even more dramatically than coal-fired systems. Early nuclear units had been brought on line for an estimated $150 to $200 per kilowatt. The exact cost, however, is difficult to determine because of subsidies by manufacturers competing for a foothold in what appeared, at that time, to be the major future electricity-generating technology.

Groups hostile to the technology intervened in an already unsure regulatory process while ever-increasing demands for more stringent safety features stretched the time to construct nuclear units from four years to more than ten years. In addition, extra equipment and greater safety margins drove the cost of such units to $1,500 per kilowatt with some of the longest-delayed and most problem plagued units costing much more (see Chapter 9).

During that period, the electricity business changed in two dramatic ways. First, as the nation recognized that it depended significantly on vulnerable and costly imported fuels, marketing of all energy forms became unacceptable. Instead, conservation became the watchword and, in many cases, was mandated by law or required by regulation. The growth in electricity sales fell from 7 to 3 percent per year because of higher electricity prices, two severe recessions, and the cessation of industry marketing—and, in many cases, demarketing, (i.e., the active promotion of conserving electricity). The second change was the industry's deteriorating financial health. The stock market reflected this as investors accorded utility stocks lower prices while the bond market mirrored the industry's weakness with down ratings and higher cost capital.

Today, some analysts attribute the utilities' financial problems to OPEC; other observers cite the institutional structure of the industry; some claim that utility managers have been complacent; some experts focus on the shortcomings of regulation while still others believe unreasonable pressure groups have created the electric utilities' plight. Each of these causes has contributed to the current situation. But four factors, all relatively unforeseen if not unforeseeable by utility executives in the early 1970s, appear to be crucial: (1) rapid and chronic

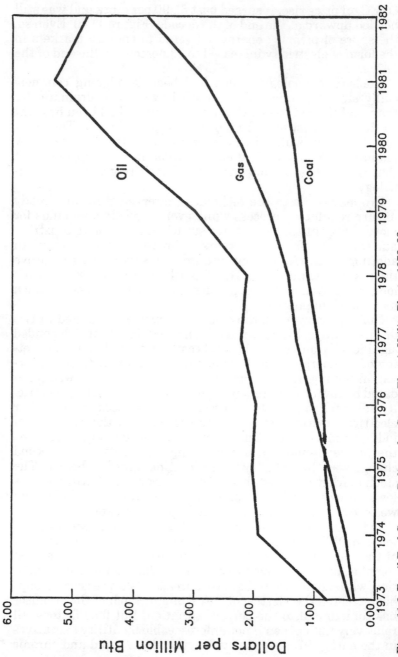

Figure 11-2. Fossil Fuel Costs as Delivered to Steam–Electric Utility Plants, 1973–82.

inflation, (2) real cost increases for utility equipment, (3) real cost increases for fuel, and (4) inadequate regulatory response.

Inflation

During the golden era, the average annual rate of inflation was 2.5 percent. In the 1970s, it was 8 percent per year. Moreover, the effect of inflation on the cost of many of the major components of an electric utility system was even more striking. For example, in 1965, a coal-fired generation station could be built for $100 per kilowatt. Today, reproduced to the same standards, the plant would cost about $350 per kilowatt because many of the elements of the plant's cost—such as construction labor, concrete, and engineering charges—have all gone up much faster than the average rate of inflation.

The utility business, of course, is not the only sector that is affected by inflation. But an electric company is different from other firms because it cannot raise prices of its own volition to reflect inflation. Instead, it must go before a regulatory commission to do so. Moreover, most of the time an electric company can raise the price of electricity only after the deleterious consequences of inflation are evident in the firm's financial accounting statements (i.e., after profits have already been adversely affected).

Real Cost Increases for Equipment

Cost increases for utility equipment over and above the rate of inflation have occurred for a number of reasons. Again, using the $100 per kilowatt generating plant as an example, to build it would now cost at least $750 per kilowatt. The additional costs beyond the $350 per kilowatt noted above reflect factors other than inflation. For one thing, very different generating units are being built today. New coal-fired plants include equipment to control sulfur oxide emissions, means for minimizing the use of water, facilities to improve ash disposal, capabilities to burn diverse fuels, and so forth. Moreover, constructing such generating stations now takes almost twice as long as it did during the 1960s. This delay adds considerably to costs, primarily because of the carrying charges on the capital invested during the construction period.

An even more dramatic factor has been the increased cost for nuclear facilities, as noted above. The actual cost of the early nuclear units is hard to estimate because the plants were constructed under turnkey contracts and were often subsidized by equipment manufacturers. A cost of $150 to $200 per kilowatt would be in the right range. Today, such units cost $1,500 to $2,000.

Real Cost Increases for Fuel

The price of oil increased from $2 per barrel in 1972 to about $35 per barrel by 1980. As major oil buyers, utilities had to pay these higher prices or switch to lower-cost fuels. Utilities with little hydro capacity or companies far from coal fields and close to the coasts or to major oil-producing areas, generally use oil as their major fuel. Overall, oil now represents just over 10 percent of utility fuel consumption, with most of this concentrated in the Northeast, California, and Florida.

Other fuels have also increased in cost. Coal, which is used to generate about 50 percent of all electricity, has risen in price since 1970 from 40 cents per mbtu to $1.60 today. Natural gas costs have gone up by a factor of 10, from 30 cents to $3.00 per mbtu. Uranium, the fuel for nuclear units, increased from $6 per pound of yellow cake to $45, but is now available at around $30. The only "fuel" that did not increase in cost was falling water. Hydroelectricity, however, provides only 12 percent of the electricity generated in the United States.

In 1970, fuel costs accounted for 21 percent of the utilities' total expenses (Figure 11-3). Today, the fuel bill is almost 50 percent. Most of these costs were passed on directly to electricity consumers through fuel adjustment clauses. These rapidly increasing costs became a concern to regulators and the focus of attention of consumer groups. In some jurisdictions, such cost pass-throughs have not been allowed or have been substantially restricted.

Inadequate Regulatory Response

As noted above, electric utilities, in contrast to nonregulated businesses, must subject their price changes to public regulatory bodies for approval. In quasi-judicial hearings, usually at the state level, regulators determine electricity rates, which reflect costs and include what they consider a "fair return on investment."

During the golden era, when electric utility costs were declining, such regulation functioned well. Spreading the cost reductions among rate-payers was easy. But as utility costs began going up, the system worked less well and problems arose. The commissions found it difficult to apportion the higher costs among increasingly vociferous and often conflicting consumer groups—the poor and the wealthy customers;

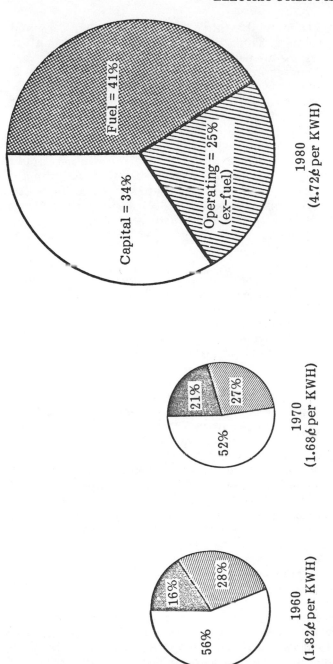

Figure 11-3. Cost Components of One Kilowatt-Hour of Investor-Owned Electric Utilities: 1960, 1970, 1980.

residential, commercial, and industrial ratepayers; new and existing consumers. Thus, as costs increased, so did politics. State commissions are sensitive to the political processes that create them; and approving large rate increases, of course, is "strong medicine" for an electorate to swallow. It is not surprising, therefore, that utilities' finances were squeezed as market forces pushed up utilities' costs faster than the commissions allowed the companies to hike their rates.

As inflationary pressures grew, utility rate hearings became an increasingly important forum in which consumers could vent their anger at inflation—the disease that was rapidly eating away people's earnings and savings. In response, politically attuned commissions, whether elected or appointed, whittled down the size of the rate increases, constraining the amount of cost increases that the utilities could pass on to ratepayers.

The utilities' profits were particularly prone to such political squeezing. As one of the most capital-intensive of all industries, pinching an electric company's allowed rate of return—its profits—has a pronounced effect on stockholders. The electric utility business requires a $2.50 investment to support a dollar of annual revenue. This is about the same as the telephone industry, but much larger than the 40¢ per $1 of revenue for steel and the 25¢ per $1 of revenue average for all manufacturing.

In addition, because of inflation, utilities' interest costs—the cost of debt capital—has jumped from 5 percent in 1965 to as high as 18 percent in recent years. Yet through most of the 1970s, the allowed return on equity capital was hardly increased. And even in the latter part of the 1970s, when the *allowed* rate of return was raised by commissions from 13 to 14 percent, the *actual* earned return remained between 11 and 12 percent (Figure 11-4).

This sequence of events—resulting in low returns—triggered investor disfavor, and electric utility stock prices fell. During this period, most utility stocks sold well below their underlying asset or book value (Figure 11-5). Nevertheless, to finance the expansion of the generating plant to meet consumers' growing requirements, utilities had to sell stock. But every time they did so, they traded away part of the existing stockholders' equity in the electric company.

Even worse, many utilities did not receive sufficient cash revenues from rate-payers to cover the companies' dividend payments—a situation made possible by the peculiarities of

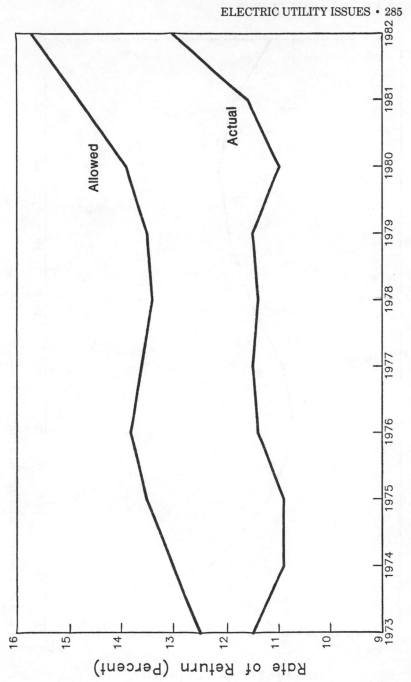

Figure 11-4. Allowed Versus Actual Rate of Return for Electric Utilities, 1973–82.

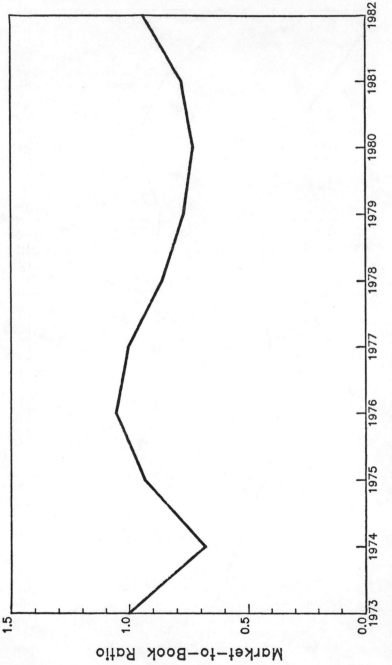

Figure 11-5. Market-to-Book Ratio for Electric Utilities, 1973–82.

regulatory accounting. In some states, the reported earnings of a utility can be much more than its actual cash earnings. Such companies, however, must borrow or sell more stock not only to pay for new construction but also to pay stockholders dividends—a practice analogous to the farmer who continues to sell his seed corn while losing his ability to buy more.

While some utilities today would need a boost in *earnings* to make their securities competitive, the required increase in electric *rates,* ironically, would be small—on the order of 10 percent. But that would be 10 percent on top of what is already perceived as the whopping past annual increases that were needed to keep up with the nation's inflation and the industry's rising real costs. This leads to another irony. When the economics of the utility business changed at the end of the 1960s, regulators delayed taking action. When they did finally respond, the rate increases permitted were smaller than needed to maintain, let alone restore, the companies' financial health. With persistent inflation and continuing cost increases, utilities got further and further behind and were forced to request ever larger rate increases. These in turn became even more irritating to consumers. Consequently, regulators found it ever more difficult to respond realistically to the utilities' financial needs.

Current Issues for Electric Utilities

The events of the 1970s left their imprint on energy markets generally and on electric utilities specifically. During that decade, the most important event was the dramatic change in price of energy relative to other goods and services. The real price of electricity rose, although its rise was less steep than those of other major energy forms, such as gas and oil. Thus, today electricity is in a better competitive position. Over the next decade, total growth of U.S. energy consumption is estimated at something less than 2 percent per year. Given electricity's relative edge over other fuels, it is projected to grow at a more rapid rate, slightly less than 3 percent per year.

Plant Construction

In the early 1970s, the utilities' commitments to build generating facilities were based on a high prospective demand for electricity—consistent with past experience but proven wrong by events. Responding to the much slower growth in demand—the result of higher electricity prices and conservation programs—utilities delayed their construction plans and canceled units. Moreover, the utilities' planning process per se became more difficult because the time it takes to build a nuclear plant in the United States increased from about five years in 1970 to ten years or more today. Similarly, coal-fired station construction times also lengthened from three years to eight years. Such stretching out was due, in part, to the retroactive imposition of rules and regulations after original licensing requirements had been established.[16]

By the early 1980s, generating reserve margins had risen to over 30 percent for many utilities, whereas, traditionally, a target of 20 percent had been considered adequate to provide for unexpected outages, deratings, or maintenance.[17] As a result, capital costs have been higher than they would have been if perfect foresight had been possible. Such reserve margins have made it unnecessary to order additional new capacity. Only in a few regions of the country—those with above average growth in electricity demand or with the need to replace obsolete capacity—is there any generation planning and construction activity.

New plant planning is complicated by several factors. First, there is uncertainty about the rate of growth in electricity demand. The second factor is the effect of self-generation by individual customers. Such power production has been favored by the Public Utility Regulatory Policies Act (PURPA)[18] and the Fuel Use Act,[19] because cogeneration units can use primary fuel more efficiently than central station power plants can. When electricity and heat are required simultaneously, there is less "waste heat." Nevertheless, the amount of cogeneration actually built by nonutility firms (principally industrial firms) and the amount of self-generation used by these firms have not increased substantially in the last decade. There is great uncertainty, however, about the proportion of electrical energy that this form of generation will provide over the next decade. A third uncertainty involves the effect on generating capacity requirements of load management (either time-of-use rates or load controls). The future course of fuel costs, particularly oil and gas, is a fourth unknown. Such costs affect the economics of existing generating units, whose output—electricity—competes in end-use markets with these other fuels.

Electricity Growth

The decline in the rate of growth of electricity demand in the late 1970s and early 1980s has not been completely analyzed.

Nevertheless, the increasing price of electricity was certainly a major factor in dampening kilowatt-hour sales. Slower economic growth, characterized by two severe recessions, was another known cause. Moreover, government actions and utility programs (many of them mandated by law[20] or by state regulation[21]) induced consumers to conserve beyond what higher prices alone would have accomplished.

Lower-than-anticipated utility revenues, of course, accompanied weaker-than-planned energy sales. Coupled with the rapidly increasing fuel costs, the higher capacity-related expenditures discussed above and higher carrying charges due to the increased cost of money forced utilities to ask regulators for large rate increases. This, of course, pushed prices up still higher and further depressed sales. On the other hand, today the economy is recovering and this should stimulate electricity sales.

Conflicting Goals

Recently, electric utilities have been torn more than ever between two conflicting goals: providing consumers an essential service in a reliable way at a reasonable cost while at the same time maintaining the companies' financial viability. In the past, with economies of scale, low inflation, and steady growth in sales, furnishing such service did not infringe on the financial goals of the companies. However, reliability has a cost, and with the declining financial performance of utilities, making the trade-off between providing reliable service and preserving financial health has become more difficult.

This problem confronts managers of both investor-owned and debt-financed utilities. Briefly, for the investor-owned firm, a financial manager's aim is to earn the allowed rate of return or exceed it, thereby maximizing the wealth of the company's stockholders, subject to the constraint of a given service-reliability level. This decision rule is meaningless for a totally debt-financed company such as a rural cooperative whose rate-payers are also the company's owners. However, the goal of efficiency is a surrogate for the profit-maximizing rule. So no matter which type of firm is involved, operating the company to maximize economic efficiency is a generally applicable rule.

Today, there is an additional question: Over what period should economic efficiency be maximized? For example, should a large investment be undertaken that will lower costs over the long run but raise rates in the short run? Similarly, should a maintenance expense be deferred to enhance a utility's

profitability in the current year? As the financial situation of
utilities has worsened, this period has shortened. Thus, decisions based on financial expediency in the near term could lead
to undesirable long-term effects—decreased reliability of service and higher costs. Utility managers face this dilemma
because they operate an enterprise for a coalition whose members—owners, creditors, customers, employees, and society—
have different stakes.

The Profitability/Sustainability Matrix

A utility's financial goals can be illustrated with a profitabil-
ity/sustainability matrix (see Figure 11-6). Profitability is defined as the difference between the utility's earned return on
investment and the company's cost of capital. If the difference
is positive, the investment is profitable; if negative, it generates a loss. Sustainability is the difference between the cash
funds generated by the utility (plus the cash it can raise
without changing its credit rating) and the cash required to
operate the firm (plus the amount needed to build new facilities). If the difference is positive, the enterprise is sustainable;
if not, the utility will eventually run out of money. Sustainability, of course, is an important measure of credit-worthiness. In
Figure 11-6, the vertical axis denotes profitability. If return on
equity is greater than the cost of equity capital, the utility is
profitable (i.e., above the horizontal axis). The horizontal axis
itself measures sustainability. If cash funds generated are
greater than cash used, the company is sustainable (i.e., to the
right of the vertical axis).

In the 1960s, utilities were both profitable and sustainable
(in the top right-hand quadrant). In the 1970s, however, utilities moved into the not sustainable but still profitable quadrant because of the companies' increased reliance on AFUDC.[22]
More recently, the further financial deterioration of utilities
has pushed them, as defined in this illustration, into the
nonsustainable and unprofitable quadrant of the matrix. Utilities have failed to earn their cost of capital.

Although they are in this unenviable quadrant today, new
financial strategies could move the electric companies toward
the upper right-hand quadrant. For instance, making cost-
saving investments (with constant revenues) would push utilities towards profitability. Similarly, ending the practice of
paying dividends out of capital (to maintain dividend levels)
would help sustainability. Allowing utilities to recover their
cost of money during construction (called construction-work-

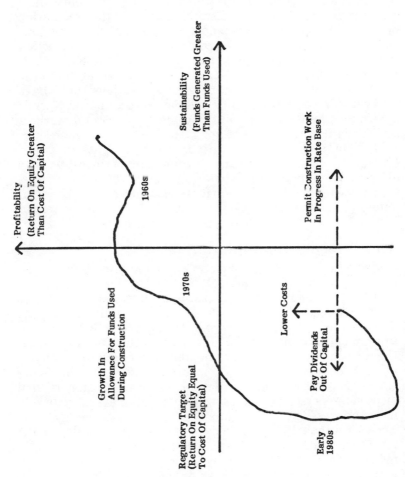

Figure 11-6. Profitability/Sustainability for Utilities.

in-progress, or CWIP) would also move utilities toward sustainability. Regulatory policy, of course, has a effect in terms of this matrix and in the real world.

The matrix also illustrates "ideal" rate-of-return regulation. By definition, this means that utilities would just earn their cost of capital: no more, no less. Also assuming that regulation provides for the utility's cash flow needs and neglecting transaction costs, such a utility would be located at the intersection of the two axes. This does *not* imply a market-to-book (M/B) ratio of one. Companies not earning their cost of capital in the short-term but with good earnings expectations could have M/B ratios greater than unity and vice versa. Market price is determined by the assessment in competitive capital markets of *all* earnings, both present and future.

The profitability/sustainability matrix highlights an important point: in the long run, utilities cannot invest in *any* alternative, if, because of constraints on revenues and costs, that investment does not earn its cost of capital. Unfortunately, present regulatory policies about allowed rate of return and cash flow often impose such constraints.

The Macroenvironment

Macroeconomic forces, such as economic growth and inflation, are essentially noncontrollable factors that greatly affect a utility's earned return and cost of capital. For instance, an electric company is particularly sensitive to inflation because its regulated rate-setting mechanism fails to reflect fully and in a timely way the effects of inflation on the utility's costs. If the inflation rate is greater than productivity improvements, a utility's operating costs will increase and erode the firm's earnings. Moreover, high inflation decreases consumers' real disposable personal income, making rate-payers generally more price sensitive and anxious to conserve energy. This causes revenue losses to electric companies. Finally, in the recent past, the utilities' cost of capital has also risen, usually faster than was anticipated by the regulatory process. Thus, with costs increasing and revenues decreasing, utilities' earned returns were falling at a time when the cost of capital was rising.

Although this macroenvironment is basically noncontrollable, two local decision-makers can ameliorate its consequences. First, regulators can offset the adverse effects of inflation by using "forward test years," revenue "attrition" clauses, "trended rate base," expedited hearings and decisions, and so forth. Second, utility managers can make investments that are less inflation-sensitive—by choosing such options as smaller, shorter lead-time generating plants, less capital-intensive equipment, and captive coal mines. In addition, utility managers can file rate increases as soon as the need arises—and pray for speedy, adequate relief.

Because of their double-leveraged structure, electric companies' profits are very susceptible to falling revenues. First, utilities' high financial leverage (the proportion of equity to total capital) magnifies the effect on earnings of a fall-off in demand growth. Second, the companies' high operating leverage (the proportion of variable costs to total costs) also affects earnings—revenues from sales must first cover the utilities' fixed charges and operating costs. Only the residual is then available to pay a return to equity owners. Because electricity demand is closely related to regional economic growth, a recession will significantly affect a utility's earnings. This, in turn, would limit the firm's ability to finance new investments to ensure continued reliable electric service.

Competition

Facing a decrease in sales growth, an unregulated company might decide to develop a better sales strategy for its "core product." In a perfectly competitive market, such a firm would strive to produce and sell an amount of product that would equate the company's marginal costs with market price. In regulated markets, government officials are supposed to substitute their actions for the "invisible" forces that operate in competitive markets. In return for limiting competition (i.e., granting an exclusive franchise or monopoly), the regulator rather than the utility sets a price perceived to be fair. Moreover, the utility cannot restrict its output; it must serve the amount demanded by its customers.

Electric utilities do have competitors. Much of an electric company's "core product," which is like a commodity, can be replaced by other forms of energy and other commodities. For example, utilities have "generic" competitors (e.g., stores that sell sweaters to keep people warm); "form" competitors (e.g., gas and oil companies that sell Btu in gaseous or liquid form); and "enterprise" competitors (e.g., firms that make their own power—cogenerators). Such competition has siphoned off sales, eroded utility revenues and diminished company earnings. Worse, when electric utilities most needed a successful competitive marketing strategy, they were *prohibited* by law and discouraged by regulators from promoting their form of energy service. Such restrictive policies are slowly changing with the increased recognition that electric utilities face competition in energy markets.

Technical Innovation

During the 1950s and 1960s, electric utilities posted the largest productivity increases of all sectors in the economy. These gains were, in part, the result of synergism between federal research, which promoted basic sciences, applications research by the utilities' principal equipment suppliers, and demonstration activities by the electric companies. This successful interplay had run its course by the late 1960s. Utility suppliers were particularly affected by the impact of antitrust lawsuits, the irregularity of the major equipment business, and the huge development effort needed for the new nuclear technology. Moreover, prospects for other new technologies (e.g., coal gasification, fluidized-bed combustion, and direct current transmission) appeared to require major research-and-development (R&D) efforts with great uncertainty as to success, and even then, large risks in their commercial viability.

This unsettled environment and the threat of federal action involving electric utility R&D motivated the leaders of the industry to mobilize their research efforts and create the Electric Power Research Institute (EPRI), an organization dedicated to shaping the industry's technical destiny.[23] This approach appears sufficiently successful so that others are following the model. The gas industry, which also faces complex institutional problems and a need to apply resources to research, formed a similar organization—the Gas Research Institute. More recently, high-technology industries have duplicated this approach.

EPRI's work, along with efforts by the federal government at the Department of Energy[24] and continuing research by some of the industry's equipment suppliers, has begun to deliver commercial or nearly commercial products that respond to several of the industry's principal needs. These include: (1) generation options that can use coal or uranium while meeting society's increasingly stringent safety- and environmental-control requirements at reasonable costs; (2) application of techniques or hardware to do the basic utility jobs at lower cost; and (3) planning systems or equipment that will allow utilities to meet the uncertain future more flexibly and yet economically. Already, EPRI-developed technologies have yielded reported savings to utilities of almost $1 billion per year. Many of these savings represent the avoidance of higher-cost alternative responses to increased requirements. But some reflect reductions in cost from traditional approaches. Consumers, of course, are the ultimate beneficiaries of this research.

Outlook for the 1980s and Beyond

The outlook for the electric utility industry today seems brighter, compared with the experiences of the last decade. Inflation is now much lower, and interest rates have come down from the heights of the late 1970s. The high real rate of interest may also fall as investors perceive a declining default risk in capital markets. There is no assurance, however, that these lower inflation rates will continue indefinitely. To the extent that the pace of inflation remains close to productivity increases in the industry, utility problems would be eased. Specifically, slower inflation and lower interest rates would benefit the electric utility industry in terms of its carrying costs and its cost of capital. Such conditions would allow electric companies to refinance their high-cost debt, which was incurred over the last decade.[25] While common equity sold below book value in the past cannot be recouped, raising capital in the future via the sale of common stock would be on a par with the value of the business enterprise, and thus dilution could be avoided.

Economic growth is picking up, which should increase the demand for electricity. The resulting jump in revenues—given current reserve margins and little need for further capacity additions—should help utilities cover their high fixed costs of the existing plant investments, thereby lessening the companies' need for frequent rate increases.

Utilities also have learned much about the vulnerability of relying on traditional planning and operating concepts in a volatile, uncertain environment. For example, new activities started in difficult times, such as load management, have earned a permanent place in capacity planning. In addition, long-term supply plans now often include nontraditional options, such as drawing on energy generated by a utility's customers, relying on renewable resources for power generation, investing in conservation, and so forth. The financing and risk-sharing strategies that were developed in the 1970s will continue as future options.

In the decades to come, the electric utility industry will face a number of key opportunities. First, the financial condition of utilities should improve because of the changing macroenvironment. This should help electric utilities' cash flow substantially. Second, new technologies should become available to meet new requirements and to reduce costs. Third, the companies' ability to manage loads should help lower costs. Finally, new or increased uses of electricity, where its intrinsic form has a high value to the customer, provide another opportunity for astute utility managers.

There are two potential structural changes which, although speculative, could transform the electric utility industry: deregulation and diversification. As unlikely as such revolutionary changes would be, other more modest modifications in regulation will take place. Both the regulators and the regulated acknowledge that the present system, when confronted by major difficulties, does not work satisfactorily. Can something better be devised?

Utilities, of course, still face major problems. The most critical is the continued national goal of reducing ecological and human health risks from economic activities. A possible requirement may be cutting SO_2 and NO_x emissions—the precursors of acid rain. There are, however, numerous other environmental, legislative, and regulatory matters that could severely affect utilities.

A second, less pressing problem involves greater participation by the public in utility decisions. This has led, for example, to extensive public hearing requirements to obtain approvals for constructing major utility facilities. In the same spirit, expanded utility operations in such areas as credit relations with consumers or helping rate-payers manage their energy needs have been mandated by regulatory bodies.

A Window of Opportunity?

Because of the fall-off in electricity demand growth after 1974, many electric utilities now have sufficient near-term generating capacity—a legacy of the ambitious building plans that were implemented in times of much higher demand growth. Completing these generating plants was a major factor in crippling the finances of the industry. Today, the industry asks: "Is there a window of opportunity for utility executives?" If load growth remains relatively slow, utility financial conditions could improve considerably, creating a favorable window that could last as long as five years.[26] During this period, internally generated cash should be plentiful. How should these earnings be used? Broadly speaking, there are two alternatives: financial strategies and investment strategies. In the financial strategies, assuming an internal financing target of 40 percent, the excess cash could be used to improve short- and mid-term stock performance. In the investment strategies, the excess cash could be used to acquire revenue-producing assets. More specifically, the alternative strategies might include:

- *Financial Strategies*
  Increase dividends
  Repurchase stock at market price
  Repurchase debt
- *Investment Strategies*
  Build for the future; begin construction of baseload plant in anticipation of demand growth spurred by real price reductions
  Invest in load management devices
  Diversify into independent and unregulated enterprises

Financial strategies could provide some short-term benefits, but these could evaporate when new plant construction begins in the early 1990s. Even the investment strategies, while better than the financial ones, might not sustain improvements over the long term. A "build for the future" strategy might produce a stable pattern of financial indicators. Yet, with all the strategies, once any serious construction is necessary, the financial performance of a utility quickly deteriorates. Regulation must also be improved concomitant with the improved planning by utilities.[27] This combination offers the best opportunity to put electric utilities on a sound financial footing.

New Technology

There has been widespread research, testing, and demonstration of new electricity-related technologies. These will eventually emerge as commercial equipment and products. The introduction and penetration of such items into the marketplace could reverse the recent upturn of the industry's cost trends (Figure 11-7). Prior to the 1970s, electric companies enjoyed decreasing average costs as output expanded. Today, however, utilities are reluctant to commit themselves to untried technologies—understandably, because the reliable provision of an essential service must take first priority. In addi-

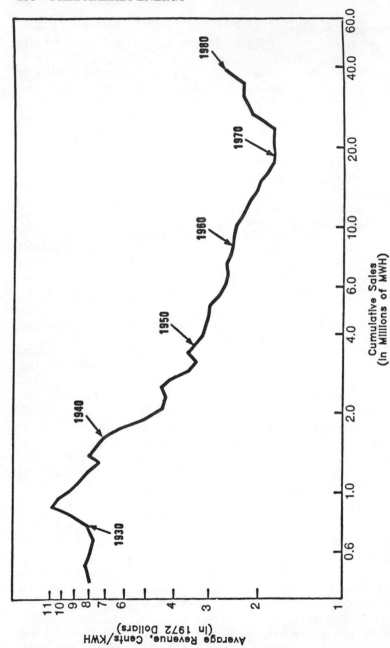

Figure 11-7. Electric Utility Industry Costs.

tion, the optimum time for implementing a maturing new technology is unclear. Moreover, there are uncertainties and risks associated with the cost of the technology and its performance. These would affect the adopting utility's return on investment and that system's reliability of service.

In brief, there is little incentive for electric utilities to take big chances because regulation limits the firm's return on investment to levels commensurate with the average risk of the traditional utility business. Regulation, however, does not limit the downside risk of financial loss associated with a "lemon" technology.

Partly because of this nonsymmetrical treatment of risk, utilities have considered forming consortia for the commercial testing of new technologies.[28] Each utility in a group would provide a part of the investment, one "host" utility would test the technology while all of the companies would benefit from the information gained about operating performance. Instead of one utility shouldering the risk of losing the entire investment, each utility would take the chance of losing only a part of the required capital spending.

Demand-Side Planning

One of the first nontraditional types of investments made by utilities was in the area of demand management. Utility planners began to look for ways to decrease or shape their customers' demands (both peak and average) because the construction of new power plants had higher costs than provided for in current rates. By dampening peak demand or by leveling demand over the day and year, less generating capacity is needed and a better capacity-utilization factor results. Reducing average demand not only lowers both short-term operating expenses and the long-term need for capacity but also affects revenues. In the 1970s, voluntary conservation by consumers grew, and lost revenues hurt the industry. Ultimately, however, better capacity utilization does benefit both consumers and utilities. Various types of load management options—an earlier name for demand-side planning—were implemented to influence loads, such as direct utility control of consumer electricity use at peak times, rate structures that encouraged off-peak, electricity use, and storage technologies on the customers' premises.

With little previous experience in such programs, utilities had to determine which investments represented an "optimum" choice. The industry quickly discovered that a good deal more information was required about consumer behavior and how utilities might affect it. Over the past few years, various approaches and incentives have been tried, and while partly successful, more data and analysis are needed to understand the potential of this investment option from the utility's and the consumer's points of view.

Demand-side programs can lessen the need for taking on the risks of building long lead-time power plants, and thus are likely to become an important strategy for reducing the total costs of utilities. Yet, consumers' behavior is being modified. In brief, balancing investments in traditional generating supply options with investments in demand-side activities for the mutual advantage of the disparate stakeholders in the electricity business will be a major challenge for utility managers in the 1980s.

Marketing

In recent years, electric utilities have developed *integrated* marketing plans, including a major new effort under the auspices of the Edison Electric Institute.[29] This is in contrast to the halcyon pre-1979 era when marketing was really a sales strategy. In those days, with declining marginal costs, the more electricity a utility produced and sold, the happier everybody was. A key element of this sales strategy was the utilities' emphasis on product quality—service reliability and adequacy. Stressing these selling points continued even after more recent legislation prohibited utilities from explicitly promoting the sale of electric energy and electric appliances.

Today, marketing embraces and integrates substrategies, such as sales of existing electric services, introduction of new services, promotions, rate design, load management campaigns, and advertising. Further, marketing must include image-setting of the utility and its public role, positioning the company's various services, market segmentation, and all interactions with existing and potential consumers and other stakeholders.

Most utilities have embraced this integrated concept, some explicitly and others through incremental programs. Many utilities, for example, are segmenting their market by service quality. Not all customers need (or are willing to pay for) conventional high-quality service. To satisfy such customers, utilities offer interruptible service at a lower rate. In addition, some utilities "demarket" in segments where revenues do not cover costs but promote their core service in profitable market segments. In general, many marketing programs are aimed at telling a utility's story better—marketing the electric company as a value-adder—explaining that the company is part of an essential industry, one closely linked to the productivity of the nation's economy.

National Marketing Strategy

In the late 1970s, when the ban on utility advertising and sales promotion went into effect, electric utilities' sales forces virtually withdrew from direct contact with their customers. This eliminated many crucial two-way personal communication channels. In time, this led to increasing consumer dissatisfaction with the "anonymous" big utility. Simultaneously, consumers' most overriding concern became the high cost of electricity. Reacting to this deteriorating relationship between companies and their customers, the investor-owned electric utilities trade association developed a national marketing strategy based on two concepts:

- *Focus on Value/Cost:* Electricity performs many essential everyday tasks well, at relatively low costs; consumers need to be reminded of its value for the money.
- *Cost Control/Customer Options:* Electricity consumption can be positioned as controllable by offering consumers different levels of service reliability, load-control devices, and information; if usage can be controlled, so can cost.

Many utilities face a dilemma in sales promotion. Utilities with more than adequate capacity question whether they should try to invigorate their core business. If done carefully, in terms of attracting higher load-factor customers, promoting energy-efficient appliances, and so forth, the increased use of electricity should benefit customers and utility owners alike. As another approach, a few utilities have executed mutually satisfactory off-system sales contracts to neighboring utilities. In light of the improved outlook for the economy, the electricity business has the potential to reward investors, especially given the utilities' better approaches to planning and marketing— approaches developed out of necessity during the turmoil of the last decade.

New Uses of Electricity

Beyond the traditional uses of electricity, there are significant new uses in the development stage or in the early application phase. Perhaps the most important are in the industrial sector where new technologies to make U.S. products more competitive are being adopted. These new electricity applications include lasers for heating, welding, and cutting; automatic control or robotization to speed production and improve quality; and, air-quality control for better product purity and to meet more stringent worker health requirements.

Similar opportunities for new applications of electricity exist in the commercial and residential sectors. The most attractive single market is heating, ventilating, and cooling (HVAC). Over the last two decades, electric heating has gained market share and now commands half of all new construction installations because of electricity's convenience, cleanliness, and low cost. Moreover, a new, improved electric heat pump is being developed to make further inroads into this market. Here, conservation arguments to reduce heat loss and health considerations to reduce indoor pollutant levels have stimulated an avid interest in electrical devices. In addition, systems to clean and filter indoor air or to increase the exchange of heat with fresh air could represent new electric loads of about the same magnitude as existing heating and cooling loads in efficiently designed new buildings.[30] Beyond the HVAC—related climate-control systems, there is a potential for the automation and computer control of existing manually run functions. Finally, the proliferation of both personal computers and electronic games could become a residential market for electricity equivalent to present TV loads.[31]

It is difficult to foresee all the marketing opportunities that new technologies could engender. For instance, electric vehicles, perhaps only for specialized short-distance hauls, could represent a new market segment in the 1990s. A number of utilities, several manufacturers, and EPRI are developing such vehicles, including their batteries and control systems. Electrification of transportation, however, remains mainly in the bulk movement of people and goods by existing rail facilities.

Diversification

Diversification, another possible move for electric utilities, could go in at least three directions: (1) they could combine into larger entities by horizontal integration; (2) they could integrate vertically; and (3) they could diversify into nonrelated businesses. As noted above, the thrust of the early development of the utility industry, was to combine into larger entities, at least until about 1930. Even so, the industry now consists of about 300 investor-owned firms; 2,000 municipals; 1,000 rural electric cooperatives; and 40 federal generating agencies, as well as a few state or public combined agencies.[32] Some leaders of the industry have suggested that a dozen or so regional utilities could serve the nation more efficiently. Such a move would be difficult, however, given the current status of the Public Utility Holding Company Act and antitrust laws.[33]

As for vertical integration, the second direction, utilities would have fewer institutional constraints (compared with horizontal mergers), although the approval of local utility commissions and the Securities and Exchange Commission (if the utility comes under the Holding Company Act) would be necessary before buying into related businesses upstream (e.g., coal mines, rail car facilities) or downstream (e.g., retail appliance sales, customer repair services). Recall that the initial Edison Company was a vertically integrated company from generator to user appliance.

The third direction for diversification would be into nonrelated businesses—via the conglomerate route. During the recent period when utilities were trying to find solutions to their financial problems, much was written about this issue.[34] A few electric utilities have diversified into related businesses (e.g., oil exploration, coal mine purchases, and electrical equipment) and nonrelated businesses (e.g., real estate and telephone). Some of these companies have done very well financially, perhaps because they diversified at a time when earnings were good and the business environment stable. When financial troubles arose, other electric utilities, observing the good fortunes of the diversification pioneers, developed and implemented similar plans of their own.

Today, the industry is divided on diversification. Some companies argue that they should "stick to their knitting," and other firms assert that gradual diversification out of the utility business is the only sensible course to follow. There are, of course, many institutional barriers to diversification, among them nervous regulators.[35] A related question is how the cost of capital to the diversifying utility might change as investors reassess the future risks of the company. If utilities go into lines of business whose risks are perceived to be greater than those of the electric industry, the returns required by investors might rise correspondingly, and the sources of capital might shift.

Deregulation

During the recent period of financial deterioration of electric utilities, some analysts blamed the regulatory process, claiming that it failed to respond to the rapidly changing circumstances of the economy and the corresponding changing needs of the utilities. Moreover, the success of deregulation in industries such as communications, air transportation, and rail transportation has kindled an interest in deregulating the electric utility industry, in part to enhance its financial condition.[36]

Critics of the industry's current structure who question its definition as a "natural monopoly" also believe that deregulation would stimulate competition in electricity markets. Because (it is alleged) the present situation does not serve consumers well in terms of promoting economy and efficiency and did not serve investors well in terms of a fair return, some argue that it is time to look at deregulation. As far as utilities are concerned, there is no unanimity of view with some believing that deregulation is the only reasonable answer[37] while others see substantial obstacles and no clear advantage.[38]

Deregulation proposals take many forms. Starting from the most complete, there are those who advocate the immediate opening of competition on all levels. Because of the complexity of this approach, however, most deregulators propose a step-by-step process, such as allowing bulk power service customers (either utilities or major customers) to select service from whatever generating source they wish. This approach is usually accompanied by subsidiary proposals to mandate transmission service on an open basis. More modest deregulation plans envision continued regulation of retail electric service while allowing competition at the generation and transmission level. Others would deregulate only additional (new) generating facilities, consistent with PURPA's requirement to buy electricity from all offerers owning certain types of generating facilities. Some approaches merely envision a more extensive interchange of electricity among existing generating sources, building upon a more active operation of power pools.[39]

Such deregulation schemes would involve major changes in historical relationships, links that have grown over time and that have created vested interests based on delicately balanced conflicting objectives. For example, deregulation would bring major changes in the regulatory sector. Complete deregulation would make that quasi-judicial administrative process obsolete. Of course, partial deregulation of the electric utility industry would still require lesser degrees of regulation. This, too, would pose difficulties for the regulator, as was found in the natural gas industry. Controlling the transfer price between the regulated and deregulated sectors is a thorny problem. In a different vein, the electric industry has grown with an uneasy truce between the investor-owned and noninvestor-owned firms. Under deregulation, this balance would be radically altered. Before any action, a careful analysis of the various deregulation scenarios should be undertaken to assess their impact on all stakeholders.

Although deregulation is not considered a panacea for the electric utility industry's ills, this particular debate continues. For example, there are major differences of opinion about financial risks, service reliability, supply adequacy, system coordination, market dislocations, and irreversibility. The potential benefits of deregulation, which also are debatable, include lower costs, improved quality of service, greater flexibility in plant siting, and economic efficiency.[40]

Stakeholders—Reconciling Their Different Objectives

A stakeholder is an entity whose well-being would be affected by utility decisions and whose own decisions would affect the utility. Over the past decade, the behavior and decisions of stakeholders have changed—from predictable and benign to uncertain and often deleterious to utilities. Consumers, for example, have reacted to increasing electricity prices by conserving—resulting, of course, in decreased sales growth and crimped revenues. Investors have demanded higher returns on their capital investments in electric companies because of a perception of higher risks. Regulators, responding to their constituencies, have made it difficult for utilities to earn their cost of capital.

Given the behavior of these external stakeholders, can a utility adopt a strategy to achieve a satisfactory trade-off between conflicting interests and objectives? The simple answer is no. The best strategy for one stakeholder is probably *never* the best for another entity.[41] But often these differences reflect conflicting short-term objectives. In the longer view, many of the divergences in short-term interests can be at least partially reconciled.

Different strategies will cause different trade-offs depending on each stakeholder's discount rate. For example, because consumers' discount rates are often very high, such rate-payers would prefer low short-term electricity prices. As for utility owners, they might prefer low-risk capacity strategies, unless the riskier ones are linked to a higher return on investment.[42]

Conclusions

After many decades of robust growth—a long period characterized by satisfied investors and complacent customers—a series of problems in the 1970s jolted the electric utility industry. Inflation was particularly damaging because of its corrosive effects on the financial condition of regulated companies— the effect of so called "regulatory lag." In addition, real cost increases for both fuel and capital equipment exacerbated the electric utilities' woes caused by inflation. Moreover, the long lead time for completing large construction projects aggravated the industry's financial difficulties as energy sales growth slowed substantially—the effect of higher prices, recession, and conservation.

For the 1980s, the outlook is appreciably better. Today, rates more adequately reflect current costs. Consumers have adjusted, at least in part, to the higher prices while a more intensive use of existing generating facilities appears likely. Plant construction schedules have been scaled back and are more in balance with future consumer requirements. Moreover, macroeconomic conditions in mid-1983 are more favorable, with inflation well below double-digit levels, fuel prices stabilizing and capital equipment costs unlikely to escalate.

This improved environment offers electric utilities certain opportunities. For example, selective marketing could help utilities use their capacity more effectively, particularly during nonpeak periods. Utilities also need to reconfigure their generating capacity mix to reflect today's relative fuel and equipment costs. The present mix mirrors past conditions when fuel prices were radically lower and environmental constraints were less binding—making a significant portion of today's mix uneconomical. Improvements in current technologies and the availability of new technologies appear to be adequate to meet such needs. Another opportunity involves improving the price structure of electricity. Basically, costs must be recovered over the range of a utility's services. While average prices are approaching "average marginal costs," differentiating rates by time of year or by hour of day would ensure a better tracking of costs. Finally, increasing sales represents an opportunity for many utilities because the extra revenues from carefully expanded energy sales should more than cover marginal costs.

An alternative to exploiting these opportunities would be changing the institutional structure of the industry, perhaps by means of partial or complete deregulation. This would involve difficult transitional problems and appears inconsistent with the dual private and public nature of the electric utility industry. Another tack would involve diversification. The horizontal or vertical integration of electric companies might lead to economic improvements but could create regulatory problems. Diversification outside the electricity business might yield financial stability to the corporation but would do little to change the performance of the underlying—and still regulated—electric utility company.

Finally, coming full circle, this chapter began by highlighting the industry's history—as driven by competitive market forces and technical innovations. In this regard, the future of the electric service business should be no different. Recent events have actually improved electricity's price position vis-à-vis other energy forms. Moreover, technical developments both on the user side and on the generation, transmission, and distribution side of the electric utility business are encouraging. This would mean even more cost-effective applications of electricity for the production of goods and services as well as a more efficient provision of electric energy for the comfort and convenience of consumers.

Forecasting has often been done by looking in the rearview mirror. If the electric utility industry does that and the next decade merely repeats the past one, the outlook will be grim. Fortunately, there have been a number of dramatic changes in the macroenvironment, in the awareness of regulators and government about the utilities' problems, and in technology. If this trend continues, there is every reason to believe that a significantly improved future is possible, particularly if the industry chooses to create and seize its opportunities, pursuing them with Edison's boldness.

12
ADVANCED ENERGY TECHNOLOGY

Richard S. Greeley

Summary

The United States can be supplied with energy for thousands of years from one or more of five different energy resources using advanced technology: (1) synthetic fuels from coal and oil shale; (2) nuclear fission breeder reactors; (3) nuclear fusion reactors; (4) solar energy; and (5) geothermal heat. To ensure the orderly development of these advanced technologies, long-range policy needs to be established to

- define the rationale for, and place reasonable limits on, any federal support to energy research, development, and demonstration (RD&D) programs relative to GNP and other elements of the federal budget;
- encourage more private involvement in energy RD&D, and particularly in the transition from development and demonstration to commercialization;
- replace stop-and-go funding of energy projects with a deliberate, long-term energy RD&D program of decreased federal involvement as each technology progresses from research and development through demonstration to commercialization.

The energy program recommended includes continued support to the Synthetic Fuels Corporation but at a lower rate of financial commitment for stretched-out projects. Increased electric utility involvement in nuclear breeder and fusion programs is needed to ensure that commercial interests and decisions are introduced at the earliest possible moment. Continued tax credits for solar, geothermal, and conservation

programs will provide fo
within an expanding marketp

Backgrou

Overview

Following the energy crisis of 1973–74,
tablished Energy Research and Development A
(ERDA) initiated over thirty programs in advance
technologies.[1] These programs fell within six major
fossil fuels, nuclear fission, nuclear fusion, solar energy, g
thermal energy, and conservation and storage. The individual
programs are shown in Table 12-1. Federal funding of most of
these programs has continued for nearly a decade and a number of major demonstration plants have been constructed. Yet
none of these technologies provides a significant proportion of
U.S. energy supplies today.[2]

It is probably too soon to expect otherwise. New energy
sources typically take thirty to fifty years to become commercially important. However, all of these programs were initiated
with the understanding that federal support would continue
through the RD&D process. Following the second oil crisis in
1980, Congress authorized the Department of Energy (DOE) to
engage actively in the commercialization of many of these
technologies including synthetic fuels, solar energy, and geothermal energy.[3]

The primary rationale for initiating the programs represented by the first five fields was that each involved a huge, if
not limitless, energy resource. The estimated size of these
resources is shown in Table 12-2. The figures given represent
the potential resources within the United States that could be
recovered and used under reasonable future economic conditions, based on extrapolation of known or inferred geologic
strata (for fossil and nuclear fuels), or economically viable
systems (for solar and geothermal energy). Much larger resources have been identified in some texts as "speculative
resources."[4] Clearly, future U.S. energy consumption substantially greater than the current seventy quads per year could be
supported for many years, or perhaps indefinitely, by either
fossil or nuclear fuels, together with major contributions from
solar and geothermal energy.

Conservation and storage technologies were also identified
for RD&D by the federal government because of the widespread understanding that the United States was extremely

d be devel-

creative financing
ace for these technologies.

the newly es-
ministration
d energy
fields:

(Pressurized)

Improved

(High Temperature ... ...nology)
(Combined Cycle Power ...ant)
(Fuel Cells)
(Magnetohydrodynamics)

2. Nuclear Fission
Improved Current Technology (Light Water Reactors)
Advanced "Burners"
Breeder Reactors

3. Nuclear Fusion
Magnetic Confinement
Inertial Confinement (Lasers or Ion Beams)
Hybrid (Fusion-Fission Reactors)

4. Solar Energy
Direct
    (Solar Heating and Cooling of Buildings)
    (Agricultural and Industrial Process Heat)
    (Solar-Thermal Electricity)
    (Photovoltaic Electricity)
Biomass
    (Direct Combustion)
    (Conversion to Ethanol—Gasohol)
Wind
Hydroelectric (particularly "low-head" hydro)
Ocean-Thermal Electricity

5. **Geothermal Energy**
Vapor-Dominated (steam-electricity)
Liquid-Dominated (hydrothermal-electricity)
Low-Temperature, Non-electric
Geopressured Brines—Methane
Hot, Dry Rock

6. **Conservation and Storage**
Residential
Commercial
Industrial
Advanced Batteries
Compressed Air
Flywheels
NOTE: Tidal Energy and Ocean Wave Energy also received some attention, but never became independent programs.

Despite the promise of these resources, major questions remain as to the need and rationale for continued governmental support for advanced technologies and the proper role of government in demonstrating and commercializing them. The answers to these questions will be based in part on candid assessments of each technology in order to determine which resources are likely to form a significant portion of U.S. energy supplies in the future.

The following sections of this chapter are intended to provide a further basis for answering these questions. First, the chapter gives a brief description of the major advanced energy technologies being developed in the United States, and their current status. Second, the primary issues currently surrounding the technologies are presented. Third, the constituencies supporting each major field of energy technology and their political bases are described. Fourth, estimates of the underlying economic data are given to indicate the potential for commercial realization. Finally, the options available to the United States for future federal support to each technology are listed, and recommendations are made for a coherent, long-term energy RD&D program.

Description and Status of Advanced Energy Technologies

Table 12-1 is a list of many of the programs in advanced energy technologies undertaken by ERDA in the early 1970s and continued by the Department of Energy following its

Table 12-2
Resource Potential of Advanced Energy Technologies[a]

Energy Source	Potential U.S. Resources (Quads-10^{15} Btu)[b]
1. Synthetic fuels—fossil	20,000[c,d]
2. Nuclear fission	
Current technology	500
Advanced "burners"	10,000
Breeders	80,000
3. Nuclear fusion	very large
4. Solar energy	essentially unlimited
1% falling on U.S.	600 per year
Current utilization	5 per year
Future use anticipated	10–20 per year
5. Geothermal energy	
Basic resources	very large
Current utilization	< 0.1 per year
Future use anticipated	5–10 per year

[a] Source: Western Hemisphere Energy Resources, (McLean, Va.: Mitre Corporation, 1980), MTR-80W230.
[b] One "Quad" = one quadrillion (10^{15}) British Thermal Units
  = 1.055 exajoules (10^{18}) joules
  = roughly one trillion (10^{12}) cubic feet of natural gas
  = roughly one-half million barrels of oil per day for one year
  = roughly 50 million tons of coal
  = roughly, the heat input to generate 10 gigawatts for one year
[c] Potential U.S. resources of conventional oil = 900 quads
[d] Potential U.S. resources of conventional gas = 1,200 quads

Current U.S. consumption of energy is about 70 quads per year.

creation in 1977. Most of these programs are still in existence, although many have been cut back appreciably in recent federal budgets.

Synthetic Fuels

The term "synthetic fuel" has come to include many resources, including: coal liquefaction and gasifications[5]; oil from shale, tar sands, and heavy oil deposits; natural gas from tight formations and geopressured brines; and alcohols from grain, wood or other organic material (biomass). Figure 12-1 illustrates the many ways to turn coal into synfuels. The recognition that these processes were nearly ready to be commercialized by the late 1970s, but were financially and technically high risk even for large energy companies and consortia, led to

There are many ways to turn coal into synfuels

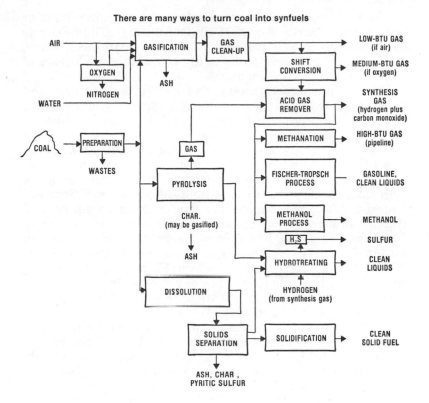

Figure 12-1. Processes for Converting Coal to Synthetic Fuel.

Three primary processes may be used to initiate coal conversion: gasification with air (or oxygen) and steam; pyrolysis by direct heating; or dissolution in an organic liquid.

The product from gasification is either a low-heat-content gas if air is used, or a medium-heat-content gas if oxygen is used. Both products can be used directly for combustion in a gas furnace. The medium-Btu gas can undergo "shift conversion" to adjust the ratio of hydrogen to carbon monoxide to the proportion desired in a synthesis gas. This latter interim product may be further processed to produce a high-heat-content gas of pipeline quality, gasoline, methanol, other clean hydrocarbon liquids, or a wide variety of synthetic organic products.

The products from pyrolysis are a gas which can be treated and used as above, a char which also may be gasified, and liquids which can be treated with hydrogen to produce clean hydrocarbon liquids and a sulfur by-product.

The product from dissolution is a liquid which can be hydrotreated as above or evaporated to produce a clean, sulfur- and ash-free, solid fuel.

Source: *Chemical and Engineering News* 57: 35 (August 27, 1979), 22.

the formation of the Synthetic Fuels Corporation (SFC) under the Energy Security Act of 1980 (P.L. 96-294). The SFC was established to provide price supports, loan guarantees, loans, and other financial support for the large-scale production of fuels from all of these resources.

The SFC has issued three calls for projects since 1981 and had selected two from the first round of over sixty submissions, one oil shale project in Colorado, and one coal liquefaction project in Wyoming.[6] The first project being constructed by the Union Oil Company of California is due to begin operation in late 1983. The second was withdrawn because a major partner dropped out, but has been resubmitted in the third round. In December 1982, three projects from the second call were announced: 4,600 bpd (barrels per day) methanol from peat in North Carolina; 4,000 bpd oil recovery from tar sands in California; and 6,000 bpd from heavy oil deposits also in California. Forty-six submissions were received in February 1983 from the third call (of which 17 were new projects), representing the following types of processes in 14 states: 10 coal liquefaction, 9 coal gasification, 13 shale oil, 11 tar sands, 3 miscellaneous coal and peat.[7,8]

The SFC has tentatively earmarked $1 billion for tar sands and heavy oil; $3 billion for oil shale; and $6 billion for coal projects. An initial emphasis in first- and second-round submissions on alcohol from wood, grain, and wastes has been shifted, since tax credits for gasohol make many projects economically viable without the SFC guarantees or support.[9]

The SFC thus is continuing to nurture the infant synthetic fuel industry through the period of building the high-risk "pioneer" plants. Presumably, the next generation of plants will be fully commercial, and their product will be competitive with conventional fuels. Otherwise the technology will have to be put in mothballs until oil and gas prices increase again.

Improved Recovery

In the meantime, a variety of RD&D programs supporting the secondary and tertiary recovery of oil and the extraction of gas from tight deposits have been terminated by the Department of Energy. These activities are now considered mature and available for commercial development and use. The experimental in situ gasification of coal, on the other hand, is still being supported by DOE. This program attempts to obtain a low to medium heat content fuel gas from controlled burning of underground coal seams, thereby avoiding underground or surface mining of the coal.

Improved Combustion

The need to burn coal without emission of sulfur oxide pollutants led to experiments and demonstrations starting in the late 1960s with fluidized bed combustion in which limestone is added with the coal to capture the sulfur. The federally supported R&D projects ran into various technical difficulties and achieved only limited success. One large demonstration project in England under the aegis of the International Energy Agency but with major U.S. funding is still under way. Despite the problems with the federal programs, atmospheric fluidized bed units are offered commercially in the United States by several manufacturers and a few are operating.[10]

Coal forms suspensions or slurries with oil, methanol, water, and even carbon dioxide (under pressure). The ability to prepare coal slurries and then to store, transport, pump, and burn the mixture like oil is an attractive concept and has been attempted in several demonstrations. Stable coal–water mixtures containing up to 75 percent coal, which can be handled and burned like No. 6 fuel oil, are at the point of being commercially available.[11] Problems remain with the ash and sulfur content of the coal when coal–water mixture is used in boilers designed for oil or gas.

Improved Efficiency

Many attempts have been made to improve the efficiency of combustion or of the electricity-generation process. The conventional fossil fuel or nuclear power plant converts only about 35 percent of the energy available in the fuel to electricity.

One approach is to increase the temperature of combustion to improve the thermodynamic efficiency, which depends on the difference in absolute temperatures of the steam or gas entering and leaving the turbine. This approach requires the development of improved turbine blade materials that can stand the higher operating temperatures. Federal support for turbine blade experiments was eventually stopped when it became apparent that the turbine manufacturers could well continue the R&D themselves.

Associated with these developments was the concept of combining a high-temperature gas turbine, operated using synthetic gas from an on-site coal gasifier, with a steam boiler heated by the exhaust gases from the turbine. A large "combined-cycle" demonstration plant was planned during the 1970s by the Commonwealth Edison (CE) Company of Chicago to be funded jointly by CE, the Electric Power Research Institute (EPRI), and the federal government. The project was eventually terminated before construction started because the estimated cost escalated until it outweighed the benefits of demonstrating an essentially proven technology.

Fuel cells that convert hydrogen gas and atmospheric oxygen to electric power and water in an electrolytic cell can approach 75 to 80 percent efficiency. Experiments are under way using natural gas or coal as the source of hydrogen. The Gas Research Institute, a private research organization established by the natural gas industry in 1979, is supporting the natural gas experiments.

The magnetohydrodynamic (MHD) process uses a high-temperature flame to produce an ionized gas that flows through a channel surrounded by magnets to generate electricity directly at high efficiency. The MHD Research Institute was established by Congress in Billings, Montana, in 1979 to develop the concept further. Joint experiments with the USSR were terminated following the Soviet invasion of Afghanistan.

Nuclear Fission

Previous chapters[12] have dealt with conventional nuclear reactors, their costs, and their regulatory and safety problems. A major portion of the DOE budget continues to be devoted to improvements in current nuclear technology and to the liquid metal–cooled fast breeder reactor (LMFBR) demonstration project, the Clinch River Breeder Reactor.

The so-called breeder actually is a nuclear power plant that transmutes nonfissionable uranium or thorium within the reactor into fissionable plutonium or uranium. More fissionable material is produced in each plant than is consumed in generating power. Figure 12-2 illustrates the uranium and thorium breeder concepts. The advanced reactor concepts such as the high-temperature gas reactor (HTGR) and the molten salt reactor (MSR) based on the thorium-uranium cycle have been terminated (except for the HTGR commercial demonstration plant at Fort St. Vrain, Colorado, which continues to operate and generate power).

Improvements to current technology are intended to extend the life as well as the safety of nuclear fuel assemblies. Additional funds are budgeted for developing safe technologies for radioactive waste disposal. RD&D on reprocessing of spent fuel assemblies to recover uranium and plutonium was suspended by President Carter in 1977, but may shortly be resumed. The efficient use of fuel in conventional reactors as well as the basic breeder concept is dependent on reprocessing.

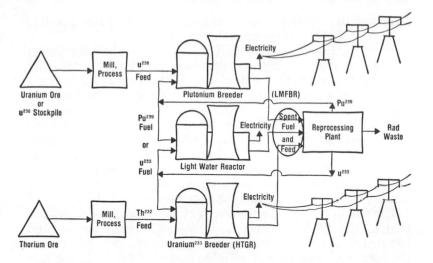

Figure 12-2. Nuclear Breeder Reactor Fuel Cycles.

Either uranium 238 or thorium232 can be used as "fertile" feed material in a breeder reactor to produce fissionable fuel through exposure to neutrons within the reactor. The term "breeder" means that more fissionable material is produced than is needed to keep the breeder reactor itself operating.

The fuel produced in a plutonium breeder—plutonium239—or in a uranium breeder—uranium233—is recovered by reprocessing the exposed ("spent") feed material. The plutonium239 or uranium233 may be used as fuel in a separate light water reactor or to start another new breeder, as well as in the breeder reactor in which it was produced. In either case, the primary product from each of the reactors is electricity.

Notes:

A plutonium "breeder" consumes U238:   U238 + neutrons → → Pu239 (fissionable fuel).

A uranium233 "breeder" consumes Th232:   Th232 + neutrons → → U233 (fissionable fuel).

A light water reactor consumes U233, U235, or Pu239:   U233, U235, or Pu239 → energy + rad waste (fissionable fuel).

These programs were initiated in the early 1970s when the use of nuclear power was increasing rapidly and a shortage in future supplies of uranium was envisaged. Improvements in current technology were needed to stretch uranium reserves 10 to 15 percent until advanced burners could multiply fuel efficiency by a factor of ten or more. Eventually breeder reactors would multiply fuel efficiencies by a factor of 60 to 100.

As the nuclear power programs slowed in the late 1970s and no new reactors were ordered, the urgency for improvements in fuel utilization efficiency disappeared. However, as a long-term program, the breeder in the form of the Clinch River Breeder Reactor (CRBR) withstood all attempts to kill it, until the October 21, 1983, Senate vote. Ground clearing had started on the site, and major components were being fabricated. The estimate of the cost to complete the CRBR was $2.4 billion for a total cost of about $4.5 billion, assuming initial operation in 1989.[13]

Some 753 electric utility companies had pledged funds totaling about $250 million for the CRBR. Initially the program was jointly managed by ERDA and a utility executive group. Difficulties with this management structure were encountered, and changes were made by Congress in 1978 to provide more direct federal management and control.

The CRBR would have been only one step toward demonstration of the LMFBR concept. Full-scale demonstration and commercialization are still in the distant future. In addition to the technical questions of reliable and efficient operation and the economic questions of the availability of uranium ore, enrichment costs, and the relative costs of conventional burner reactors, basic policy questions must be answered (see Chapter 9 and Chapter 10).

Nuclear Fusion

The promise of essentially unlimited energy resources from the fusion of the hydrogen isotopes deuterium and tritium and the challenge of "taming" the hydrogen bomb led to the magnetic confinement program during the late 1950s. Basically, a plasma of intensely hot hydrogen nuclei is confined and heated further within a magnetic field until the nuclei fuse and release energy. When more energy is released than is needed to confine and heat the plasma, "break-even" has been exceeded, and power for external use can be generated.

Despite early hopes for rapid success, the program has not yet achieved break-even. Currently, the Tokamak Fusion Test Reactor (TFTR) is under construction at Princeton, New Jersey; this machine is expected to come close to break-even. Plans for a Fusion Engineering Center, to speed the engineering and development of power producing reactors, were pressed in the late 1970s but have been deemphasized lately awaiting the results from the TFTR. The budget for the magnetic fusion program is $500 million for fiscal year 1983, and the same level is projected for 1984.

Inertial confinement fusion uses the energy in high-intensity laser beams, focused on a pinhead-sized glass sphere containing a deuterium–tritium mixture, to compress, confine, and heat the mixture to fusion conditions. Beams of ions are also being used in experiments at the national laboratories. Again, hopes for rapid progress have dimmed, and the budget for inertial confinement fusion for civilian power production has been cut back appreciably from about $200 million in 1980. Considerable research on laser and ion beams for military applications continues, and there may be advances that will contribute eventually to the civilian power program.

Some thought has been given to a hybrid fusion-fission machine, which might be easier to develop than a pure fusion device. The hybrid would use the fusion reaction, either magnetically or inertially confined, at less than break-even conditions to produce large numbers of neutrons. The neutrons would be captured in a surrounding blanket of nonfissionable thorium (Th 232) or uranium (U 238), which would undergo nuclear reactions to produce fissionable uranium-233 or plutonium-239. The fissionable uranium or plutonium would then be removed and used in power-producing nuclear plants[14] (see Figure 12-2.)

The hybrid concept is being studied, but development awaits further progress on fusion technology as well as fundamental decisions on reprocessing and recycling of irradiated fuel such as plutonium in nuclear power plants.

Solar Energy

Starting with a $1.2 million appropriation to the National Science Foundation in fiscal 1971, the federal budget authorization for solar energy increased dramatically at ERDA and then at DOE to exceed $1 billion in 1981.[15] A Solar Energy Research Institute was established in Golden, Colorado, in 1977, as were four regional centers in Massachusetts, Georgia, Minnesota, and Oregon. The program was explicitly directed by the Congress to assist in the commercialization of solar applications. President Carter announced on June 20, 1979, a goal of obtaining 20 percent of U.S. energy from the sun by the year 2,000. Currently we get about 5 percent from the sun (4.8 percent from hydroelectricity, 0.2 percent from wood combustion, and an additional small percentage from roughly 500,000 buildings with some form of solar collectors).

The energy from the sun can be converted to useful power in many ways. The federal program was developed in seven branches:

1. SHACOB—Solar Heating and Cooling of Buildings (both active and passive)

2. AIPH—Agriculture and Industrial Process Heat
3. Solar-Thermal—conversion of focused sunlight to electricity using a high-temperature boiler
4. Photovoltaic—direct conversion to electricity using photo cells
5. Biomass—conversion of organic material to electric power (through combustion) or to fuels (including gasohol) and chemicals
6. Wind—primarily large machines for producing electricity
7. OTEC—Ocean Thermal Energy Conversion to electricity

Large-scale hydroelectric power was considered fully commercial, but a "low-head hydro" program was initiated in 1977 under the geothermal energy branch to administer tax credits and loans for small-scale municipal and industrial systems. Emphasis was placed on renovating old dams and generating facilities.[16]

The RD&D program launched by NSF and expanded vigorously by ERDA and DOE developed several major demonstration projects. These included grants for solar collectors on roughly 15,000 homes and commercial buildings, a five-megawatt solar thermal "power-tower" facility in New Mexico, large purchases of solar photovoltaic cells, and five large wind energy conversion systems for production of electricity.

Currently, the federal budget includes continued strong support for the photovoltaic program. Most other demonstration programs have been terminated or are being concluded. Off-budget items include the Solar Bank, established to provide low-cost loans for the capital cost of residential and commercial solar energy systems. The loan program is being implemented by the U.S. Department of Housing and Urban Development (HUD) following a court decision in July 1982, which found that administration attempts to end the program were not considered within the will of Congress.[17] Other off-budget items include an estimated $390 million in tax credits for solar energy. These credits include active solar collector installations on residences and commercial and industrial buildings, alcohol fuel subsidies, and low-head hydro installations. The tax credits expire by law in 1985.[18]

Geothermal Energy

The Pacific Gas and Electric Company currently produces over 1,000 megawatts of electricity from a large dry-steam resource at The Geysers, north of San Francisco, with another 1,100 megawatts under construction or planned for completion by 1990.[19] No other dry geothermal steam reservoir has been found in the United States.

A federal RD&D program was launched in 1974 to exploit a number of high-temperature water (hydrothermal) resources in the West for electricity generation, and lower temperature water resources throughout the country for building and process heat. A research program was also initiated to obtain steam by pumping water through fractures created in hot, dry rock. The Geothermal Energy RD&D Act of 1974 (P.L. 93-410) provided loan guarantees to stimulate the development and demonstration of these resources, and the Energy Security Act of 1980 (P.L. 96-294) included geothermal resources under depletion and intangible drilling cost tax provisions similar to oil fields.

To date, one 50-megawatt commercial project is under construction at Heber, California.[20] There are numerous known geothermal resource areas (KGRAs), but questions concerning their capacity to provide energy for twenty years or more and technical problems with corrosion and sediment formation in pipes and heat exchangers have delayed or prevented additional projects. Of thirty-seven prospects identified in 1977 for potential developments by 1990, with a total ultimate electrical generating capacity of roughly 30,000 megawatts, fewer than 25 percent may be economically competitive with electricity from fossil or nuclear plants.[21]

R&D advances were estimated to bring an additional 25 percent into competitive range. The tax credits of P.L. 96-294 and the 15 percent investment credit of P.L. 96-223 would render some 70 percent of the capacity cost-competitive. Without a major breakthrough in technology or discovery of additional resources, the best that could be expected from U.S. geothermal resources would be about 20,000 megawatts, or the equivalent of two quad per years.[22]

The anticipated large low-temperature resources have proven disappointing in the East and have been tapped for limited industrial, home, and farm-building heating in the West, a total of 0.02 quad by the end of 1982. The postulated extremely large geopressured brine resources underlying much of the Gulf Coast of Louisiana and Texas (some reports indicate a total of 50,000 quads of energy from methane contained within the brines) have been tested with two limited wells but do not appear to be competitive yet with conventional natural gas.

Conservation and Storage

Increasing fuel prices in 1973–74 and again 1979–80 were expected to result in many individual and corporate actions to conserve energy. Although some energy experts doubted the ability of Americans to save energy, consumption dropped by 4 quads from 1973 to 1975 (74.6 to 70.7 quads), rose as real prices decreased from 1975 to 1979 (70.7 to 78.9 quads), and then dropped as prices increased again from 1979 to 1982 (78.9 to 70.9 quads). Some of the recent decreases in energy consumption are due to the recession. However, the real nature of energy conservation is indicated by figures on energy consumption per constant dollar of GNP. In terms of quads per 1972 GNP dollars, the index Q/GNP has fallen steadily since 1973, from 59.5 then to 48.1 in 1982, about 20 percent.

There are three basic aspects to energy conservation:

1. Use less: lower the thermostat, turn off lights, travel less, etc.; these are one-time measures and can be expected to fluctuate with the real or perceived price of fuel as well as other economic conditions.
2. Decrease heat leaks: insulate buildings and pipes, stop steam leaks, etc; these measures, as long as they are maintained, provide continuing savings that accumulate with time to lower the Q/GNP ratio.
3. Use heat more efficiently: install heat exchangers and recuperators, improve engines and generators, improve vehicle aerodynamics and decrease weight, cogenerate electricity and process heat for industry (or district heat for nearby residential and commercial buildings), store off-peak power for use during on-peak hours, etc.; these measures are primarily technical and can be expected to decrease the Q/GNP ratio further and perhaps dramatically over time.

The sharp increases in fuel prices caused immediate responses in conservation types 1 and 2 and slower responses in type 3. In addition, many industries switched from high-priced oil to cheaper fuels. Industry set a goal in 1974 of 20 to 30 percent energy savings by 1980, and that goal has essentially been met. The auto industry, spurred in part by legislation requiring improved gas mileage up to 27.5 miles per gallon (mpg) as a fleet average by 1985, has achieved steady improvements in mileage performance. The national average auto efficiency increased from a low of 13.1 mpg in 1973 to over 15.5 mpg in 1981.

ERDA initiated many programs in 1974 to encourage energy conservation.[23] Besides a major advertising and public relations campaign, including awards to business for energy savings and a campaign to label the energy efficiency of appliances, ERDA undertook RD&D, and DOE continued research programs in advanced technology for industrial and electric utility use: improved materials for higher temperature combustion chambers and for increased heat transfer, superconducting magnets for generators, and more efficient electrical transformers and motors.

ERDA and DOE supported demonstrations in cogeneration in which an industrial firm constructed a boiler to produce more steam than needed within the plant and sold either excess steam or excess electricity to the electric utility. In some cases the electric utility sold excess steam (or the exhaust steam from its turbogenerators) to a nearby industrial facility or distributed the steam throughout a neighboring commercial and residential district for heating. This "dual use" of steam from a turbogenerator raises the average total energy efficiency of an electric power plant from roughly 35 percent for just electricity generation to well over 60 percent, and decreases by a corresponding amount the waste heat emitted directly by the power plant to a cooling pond or stream.

The federal government also attempted to force the conversion of large power plants from gas and oil to coal. Although this fuel-switching was not intended to conserve overall energy, it was aimed at decreasing oil imports and the use of presumably scarce natural gas. Under the Energy Supply and Environmental Coordination Act of 1974 (P.L. 93—319), the Federal Energy Administration and later DOE were empowered to force power plants to switch to coal. After considerable legal wrangling and protests by environmentalists about increased pollution from coal, about twenty facilities did switch to coal, either voluntarily or under compliance with an FEA-DOE order. Natural gas has been "backed out" of most utility boilers and essentially no new oil base-load utility boilers are now permitted under the Power Plant and Industrial Fuel Use Act of 1978 (P.L. 95-620). Some oil and/or gas-fired generators for supplying peak-hour electricity are still allowed.[24]

ERDA and DOE were concerned also about the inefficiencies that arise because of the daily cycle in electricity demand and the consequent need for peaking units. RD&D was supported in energy storage technology including compressed air in underground caverns, flywheels, and advanced batteries. The latter two technologies have application to electric automobiles as well as to electrical generation. Research on batteries continues, but the planned demonstration projects were abandoned.

Issues

General

The basic issues for each of these advanced technologies are twofold:

1. What is the future potential for the technology, given a realistic appraisal of conventional oil/gas/coal/uranium resources, prices, and availability to the United States?
2. What is an appropriate federal role in supporting its further development and commercialization?

The government's crystal ball is not particularly good. In the first national energy plan, "Creating Energy Choices," published by ERDA in 1974, each of the more than thirty technologies were planned to become commercially available and provide or save significant amounts of energy by 1985. Despite high oil prices (which in 1979 exceeded by nearly three times the highest estimate made in 1974 for the year 2000) few if any of these advanced technologies can compete today with conventional oil or gas; none are likely to come anywhere close to meeting the optimistic projections for 1985.

It is now time—or even past time—to take a hard look at each technology and determine whether it is worth continued support, either public or private. For instance, wave power and tidal power never received program status or funds and were essentially dropped at the outset in 1974, although there are still calls to have the United States join Canada in a huge tidal power project at Passamaquoddy Bay. Are there other candidates for abandonment after a decade of RD&D?

What should be done about the technologies such as fusion and the breeder that have only very long-term prospects? Should we continue to pump half a billion dollars a year into each of these programs?

What should be done about the technologies that have proven to be *technically* feasible, such as most aspects of solar and geothermal energy, but are not yet economically viable? Should we let the market decide their future?

What should be done about the synthetic fuel technologies that require multi–billion dollar demonstration plants to pioneer their commercial feasibility? Should we let the market decide their future as well?

During the Nixon, Ford, and Carter years, each successive presidential energy message called for increased federal involvement and spending. Congress generally complied. The rationale was threefold:

1. To accelerate the development and commercialization of technologies to meet the energy crisis, even though private industry would eventually have commercialized them anyway
2. To subsidize the risk in commercialization because even the largest private companies would be unwilling to "bet the company" on certain large projects, however promising
3. To undertake development in cases where no single firm or consortium could recapture its investment in RD&D because there was no patent protection or where the potential for commercialization and return on investment was too far in the future.

ERDA was authorized in 1974 to conduct research, development, and demonstration of promising energy technologies.[25] The Department of Energy, which replaced ERDA on October 1, 1977,[26] was authorized to conduct RD&D and also "to place major emphasis on the development and commercial use of solar, geothermal, recycling and other technologies utilizing renewable energy resources."[27] Authorization to assist in commercialization of nuclear power had already been in place since the early 1950s. Similar authorization to assist in the commercialization of synthetic fuels from fossil sources was provided to the Synthetic Fuels Corporation (SFC), an autonomous federal organization. Therefore, by 1980, the federal government was deeply involved in commercializing all of the advanced energy technologies.

The Reagan administration entered office in 1981 with the premise that commercialization was properly left to private industry. Attempts were made to cancel or eliminate most such programs, including the SFC. Under strong pressure from various interests, the SFC, the Solar Bank and certain nuclear programs have been retained to date.

An additional general issue is the role of the national laboratories in conducting RD&D on advanced technologies. These laboratories were established in the 1940s under the Manhattan Project to develop nuclear weapons. They all have become multipurpose laboratories heavily involved in many, if not all, of the technologies described here. In some fields, such as nuclear fusion, they have unique facilities. In others, such as fossil and solar energy, they have developed substantial expertise and extensive facilities and instrumentation. They have been attacked, on the one hand, by some industrial executives for moving too far into demonstration and commercialization activities (properly the role of industry) and, on the other hand, by some university deans for moving too far into basic research (properly the realm of universities).

Clearly, a challenge exists to utilize the vast expertise and facilities of the national laboratories. To shut one or more of them down as having outlived their usefulness would be politically difficult and a waste of valuable human and technical resources. To continue to support them without clearly specifying their roles is to risk continued confrontation with industry and academia, and frustration for the scientists and engineers at the laboratories.

A further case in point is the future of the energy technical centers. These centers were originally established by the U.S. Bureau of Mines in Pittsburg, Pennsylvania; Morgantown, West Virginia; Bartlesville, Oklahoma; and Laramie, Wyoming. They were incorporated into the Department of Energy in 1977 and given primary missions in coal cleaning, coal slurries, fluidized bed combustion, coal conversion, secondary and tertiary oil recovery, coal liquefaction, and in situ coal gasification, respectively. The administration is currently attempting to transfer these centers to private operation and recently awarded the Laramie Energy Technology Center to the University of Wyoming. Most of the funds for their support, however, will continue to come from DOE, although closer ties to and support from industry and universities will be encouraged.

Technology: Specific Issues

Synthetic Fuels

The Synthetic Fuels Corporation, after considerable discussion within the Reagan administration, is proceeding with price supports and loan guarantees for the initial production of commercial quantities of synthetic fuels. The primary issue remains, however: Should the commercialization of synthetic fuels be left completely to private industry and the marketplace?

The earlier rationale for the SFC was that the United States needs a substantial synthetic fuel industry to provide a major source of domestic fuels as conventional oil and gas reserves and production decrease, and that the proposed plants still involve a degree of technological as well as considerable economic risk. Executives from some of the largest energy firms argued that federal support was not needed and that the synthetic fuels would be commercialized in due course.[28] Other executives argued that large firms and even consortia of large firms could not afford to invest several billion dollars in a plant and then not be able to sell the product.

Some of the urgency for developing synthetic fuels has disappeared with the current oil "glut" and the break in world oil prices. Yet, a lead time of ten years or so is required to implement large-scale production facilities. At a supported price of up to $67 per barrel for a 10,000-barrel-per-day facility, the SFC is "betting" up to $100 million per year per project that oil prices in the United States will rise substantially from today's price of about $30 per barrel.[29] More and larger projects that would be required to meet a national goal of 1 to 2 million barrels of oil equivalent per day—a small amount relative to current U.S. oil consumption of roughly 15 million barrels per day, but large enough to prove the technology and the economics—could cost the SFC in one year most of the $15 billion allocated for its total lifetime. It should be noted that the funds for the SFC incentives are "off budget," derived from the windfall profits tax (actually an oil excise tax) enacted as the Crude Oil Windfall Profits Tax of 1980.

Another major issue surrounding synthetic fuels is their environmental impact. This impact occurs in four areas:

1. Air and water pollution during construction and operation
2. Socioeconomic effects of "boom towns" associated with construction of the plants
3. Availability of water to supply the processes
4. Health effects, particularly potential toxicity of the synthetic fuel products

Federal involvement in research on these impacts started with their identification nearly a decade ago. The National Environmental Policy Act of 1969 (NEPA) requires the effects to be fully identified, described, and evaluated before initiating a major federal program. Each synthetic fuel project has had, or will have, an environmental impact statement (EIS) prepared. The draft EISs for the solvent-refined coal projects I and II, finally abandoned by DOE in 1980–81 for budgetary reasons, were sharply criticized and had to be extensively revised.

Despite a decade of study, questions remain in all four impact areas. Construction of the initial synthetic fuel plants with SFC support will provide an opportunity to measure and better define the impacts for future larger-scale plants. However, unless specific measurement programs are required through the NEPA process, the fundamental questions of the true severity of the impacts will not be answered.

Improved Combustion

Several firms are now offering fluidized bed combustion units for sale, and others are selling coal-water slurries as a boiler fuel. Still other firms are improving the combustion efficiency of conventional pulverized coal and stoker-spreader boilers. Thus, despite the failure of federally supported programs in these areas to achieve a single clearly successful project, private industry has taken over.

On the other hand, fuel cells and magnetohydrodynamic systems are still in the experimental stage. The Gas Research Institute (GRI) is supporting fuel cell RD&D.[30] It would seem that the Electric Power Research Institute (EPRI), established by the electric utility industry in 1971, could provide increased support to the MHD program.

The often competing and sometimes contradictory roles of the private industry research institutes, EPRI and GRI, versus the DOE national laboratories and energy technical centers are highlighted in the field of fossil energy. During the period of expanding RD&D budgets in the 1970s they all prospered. Now there are serious problems with supporting all of them in every field they have chosen to enter.

In the field of improved combustion, there are excellent opportunities for basic research at the universities, for applied research at the national laboratories and energy technical centers, and for demonstration and commercialization at the private research institutes and within private industry. The administration, speaking through the Office of Science and Technology Policy and the Department of Energy, must clarify the role of each participant and identify the appropriate level of funding to be expected over the next ten to twenty years.

Nuclear Breeder

The issues surrounding the breeder program center on four very different aspects:

1. The technology. Is the Clinch River Breeder Reactor (CRBR) sufficiently advanced to compete with France's Phoenix (operational) and Super-Phoenix (now under construction but suffering delays)? Is it sufficiently developed to be safe to operate?
2. The resource. Due to delays and cancellations in conventional nuclear power plants, no imminent shortage of uranium is now foreseen as had been projected when the CRBR project was initiated. Is there not time now to conduct more RD&D and delay an advanced breeder demonstration?
3. Fuel reprocessing and recycling. The breeder concept depends on reprocessing the fuel to extract the "bred" plutonium and then recycling the plutonium through additional power plants. Plutonium is widely feared as a cancer-causing chemical and also as a nuclear weapon ingredient. Can these fears be allayed?[31]
4. The developer. The breeder reactor is ultimately intended to be an electrical power plant. Should not the electric utility industry fund and conduct the demonstration and commercialization programs?

The support for the CRBR is undoubtedly political. Yet proponents argue that, unless the United States supports a major breeder program, it will ultimately have to buy the technology from foreign sources. To cancel CRBR would delay our entry into the competition by at least another decade.

Proponents also argue that the CRBR represents well-developed and safe technology, that eventually the enormous resource the breeder makes available will have to be tapped, and that reprocessing is required anyway to make efficient use of conventional nuclear fuels.

Recently, the 753 utilities that have pledged funds to the CRBR project formed a task force on methods of additional support. The task force recommended that a private breeder reactor corporation be formed that would provide up to 40 percent of the CRBR costs, subject to federal guarantees of project completion and reactor operation.[32] Increased private support is thus a current possibility. However, the breeder program will be primarily a federal responsibility for many years to come (if the program survives annual attempts in Congress to cancel it).

Nuclear Fusion

The primary issue in nuclear fusion is the level of funding apparently required to maintain technical progress—about

$500 million per year.[33] The timetable for achieving a commercial fusion power plant places the first ones in operation no earlier than 2010 to 2020. This timetable assumes that no fundamental technical problems arise during the program development cycle.

Must the United States invest in two nuclear technologies, the breeder and fusion, both of which promise essentially unlimited energy resources? Can the United States afford both at a total cost of nearly $1.5 billion per year for twenty to perhaps forty years?

Other issues that must eventually be addressed involve institutional and environmental questions. Currently envisioned fusion technology would involve huge power plants of 5,000 megawatts each. These would require connection to a very large utility grid to provide adequate power in reserve when the fusion plant is shut down for service. The largest power plants in service today are 1,200 megawatts each.

Fusion power plants will also generate radioactive components, although in quantities roughly 10,000 times less than fission plants. These components will require removal and disposal in high-level radioactive waste repositories. Handling large quantities of the reactor fuel, tritium, also involves some risk, since tritium is radioactive.

Continuing the present course of action implies not only a very large federal subsidy to the electrical utility industry, but also an emphasis on large electrical grids and a continuing commitment to radioactive waste handling and dispoal. However, making a choice now between the breeder and fusion is not appropriate, given the different levels of development and the uncertainties in future potential of each technology. The only choices are the level of support and the source of funds.

At a meeting in 1978 on the strategy for commercializing laser fusion power,[34] several industry representatives urged greater participation by private industry in the early stages of the program. Although the fusion program will be the responsibility of the federal government for the foreseeable future (because no private firm or consortium could envision a return on investment within a reasonable length of time), lack of private industry involvement will prevent or delay identification of opportunities for accelerating commercial applications.

Solar and Geothermal Energy

Each of the many individual technologies included within the overall title of "solar and geothermal" has specific issues. The primary question, however, is whether, or to what extent, each technology should be required to compete on its own in the energy marketplace.

Private firms offer active and passive solar heating systems for residential, commercial, agricultural, and industrial applications; photovoltaic systems for direct production of electricity from a few milliwatts to megawatts in size; wood combustion units for residential and industrial heat; biomass conversion to ethanol and gasohol; wind-driven electrical generators from a few kilowatts to a megawatt; small-scale hydroelectric plants in the same power range; geothermal steam-powered heating and electrical generating plants up to 50 megawatts; and many other innovative energy systems. Hundreds of firms are involved in manufacturing these systems, and thousands of firms provide installation and maintenance services.[35] It is, to a certain extent, the recognition of this widespread private commercial activity that led the Reagan administration to curtail sharply the federal solar and geothermal energy commercialization efforts.

On the other hand, many proponents of these efforts felt that solar and geothermal energy should receive no less federal incentives than do fossil and nuclear energy. In a study of historical incentive programs for energy systems over the period from 1933 to 1977,[36] the average federal incentive for coal, gas, oil, nuclear, and hydropower was estimated to be 15 cents per million Btu produced over that period. These incentives were combinations of federally sponsored RD&D programs, low-interest loans, loan guarantees, tax credits, depletion allowances, tax exemptions, and cash grants. Nuclear energy received $1.90 per million Btu in incentives between 1950 and 1977 whereas coal, gas, and oil received 9, 5, and 21 cents per million Btu, respectively, over the same period.

The argument was advanced that solar energy should receive the benefits provided in the National Energy Act of 1980 (the currently existing residential, commercial, and industrial tax credits plus the Solar Bank) plus a long-term incentive of 20 cents to $1.00 per million Btu to achieve parity with fossil and nuclear energy. Furthermore, because of national benefits to be derived from increased use of solar energy (such as lower pollution, decreased vulnerability to foreign oil supply interruptions, and increased jobs), an additional long-term subsidy of 70 to 90 cents per million Btu was recommended.[37]

These arguments have not been persuasive to date, but the current tax credits and the Solar Bank have received strong support in Congress. Whether these incentives, together with continuing high energy prices and the ready availability of solar energy systems, will stimulate significant use of solar energy by 1985 to 1990 remains an open question.

Conservation and Storage

Conservation and storage technologies face the same basic issue as solar and geothermal energy systems: To what extent should the marketplace prevail?

It is interesting to speculate what the result would have been today without the federal conservation and storage program from 1974 to 1982. Would overall energy use be any greater or less? Several projections in the 1970–1974 period forecast U.S. energy consumption by 1985 to exceed 120 quads. Actual consumption, even with prompt recovery from the recession, is unlikely to exceed 75 quads, nearly 40 percent lower. Perhaps 20 percent of this decrease is due to structural conservation (involving some capital investments), and will result in permanent improvements in energy use efficiency.

Were these improvements at least partially the result of the federal program or completely the result of higher prices and the elasticity of demand? There is no question that federal information and public relations programs alerted many people to energy-saving opportunities. The weatherization program for low-income families permitted many of those who could least afford the capital outlays to share in the energy savings. The mandatory auto efficiency standards forced the initial engineering, at least, of higher miles-per-gallon cars. But the overriding pressure to cut energy costs came from the rapid increase in fuel prices.

A remarkable marketplace response to the potential for energy conservation is the recent surge in firms offering "shared savings" plans. These "energy management" firms will arrange an energy audit of a commercial building or industrial facility, and will then contract to guarantee 15 to 20 percent or more savings in fuel costs. A fraction of the fuel costs saved, usually 50 percent, is shared with the energy management firm. The owner of the building or facility puts up no money and at the end of the contract, usually five years, retains all of the energy conservation equipment and 100 percent of the future fuel-cost savings. The energy management firm and its investors generally receive a return on their investment of 30 to 50 percent or more. The energy-conservation measures undertaken by the energy-management firm include normal attention to insulation, heat leaks, and so forth, together with computer controls to optimize the performance of the heating, ventilating,and air-conditioning system. Even after a decade of energy conservation, new energy-management firms are finding a rapidly growing marketplace.[38]

This technique of third-party financing is also being offered for many other energy systems, including coal and wood-fired boilers and gasifiers, solar collectors, and a wide range of non–energy related equipment. The advantages of this creative financing stem from the recent tax acts, which provide additional tax credits and rapid depreciation benefits to investors. Cash-short and debt-limited commercial and industrial firms can take advantage of the new tax provisions through various leasing and shared savings plans. The net cost to the federal government for all forms of energy tax credits in fiscal 1984 is estimated at $5.6 billion, which is $1 billion less than in 1983. Many states also permit similar tax credits.

Constituencies and Political Bases

General

Each of the advanced technologies has a constituency and a political base. This constituency will frequently advertise in the newspapers and on the air, hold conferences, publish newsletters and books, and above all, lobby for support in Congress. The constituency will generally be represented by one or more professional or trade associations in Washington, but the political base will center around the states containing the energy resource or facing similar energy problems. Often the political base will reflect liberal versus conservative viewpoints where there is no dominant energy resource location (e.g., nuclear, solar, and conservation).

During the 1970s, most constituencies and their political support were mutually supportive, aiming to stimulate federal support for development of all energy resources. Although the environmentalists and those professing a "soft path" of geographically distributed solar technologies attacked synthetic fuels and centralized nuclear power, the growing federal energy budget provided almost every interest with a major advanced energy technology program.

At the beginning of the 1980s it became clear that not all programs could continue to be supported and that decisions would have to be made to abandon some and cut others. The ability to make and sustain these cuts will depend on an understanding of the particular constituencies involved.

Specific Technology Constituencies

Synthetic Fuels

The coal lobby is represented by the National Coal Association, the once-powerful United Mine Workers, and a number of groups whose members believe that we must use our vast coal resources much more intensively. Their political bases are in Appalachia, the Midwest, and the mountain states. They have been joined by the oil- and gas-producing companies, many of which have bought coal reserves and/or coal-producing companies. Their national associations press for federal support for synthetic fuel development.

Nuclear Breeders and Fusion

The electric utilities and many technologists, particularly those in the national laboratories, have lobbied very hard for a long-term program, starting with conventional nuclear power plants in the 1950s and now envisioning breeders in the 1980s and 1990s, and fusion in the twenty-first century. Their political base is broadly distributed among those who feel the United States should maintain its lead in high technology.

Solar, Geothermal, and Conservation

The solar lobby arose from consumers hit by high energy prices and environmentalists concerned about fossil and nuclear power. They were joined by many technologists and by entrepreneurs who envisioned a new growth industry. The political base tended to center in the liberal Northeast rather than in the Sun Belt.

The recession and current oil glut have quieted many of those who offered quick solutions during the energy crisis. Most constituencies recognize the need for control of spending on federal energy programs and argue only over the extent of the cuts.

These cuts have been substantial, although not as deep as proposed by the Reagan administration. The total energy budget outlays decreased from $8.85 billion in fiscal 1982, the last Carter budget, to an estimated $7.87 billion in 1983, to a proposed $7.1 billion in 1984, a 20 percent cut. Most demonstration programs (except the breeder) were terminated. The energy R&D budget at DOE has decreased from $4.83 billion in fiscal 1982 to $4.43 billion in 1983 to a proposed $4.26 billion in 1984. Deep cuts are proposed for solar and conservation R&D in fiscal 1984; solar and other renewable resources down 60 percent from $252 million in 1983 to $102 million; and conservation down 75 percent from $410 million in 1983 to $101 million in 1984.[39] Congress is unlikely to approve such deep cuts.

Economic Data

General

It is difficult to compare the various claims for advanced energy technology system costs since they are based on very different assumptions. Also, the cost estimates have risen dramatically so that comparisons rapidly became out of date. In a recent study by the Rand Corporation, actual costs of synthetic fuel plants (or the most definitive cost estimates based on detailed designs) were two to three times greater than initial or preliminary cost estimates.[40] Furthermore, solar energy, which has no fuel cost, and nuclear energy, which has a relatively low fuel cost, can be compared only with fossil technologies on a life-cycle basis.

Despite the uncertainties involved, it is useful to have some idea about the general magnitude and relative order of the costs of conventional and advanced energy technologies. Table 12-3 provides a detailed list of these cost estimates as of 1980.[41] Estimates for the nuclear breeder and fusion are not given, since they are too uncertain to be included at this time.

Conventional coal is still the cheapest fuel, even considering the costs of environmental controls on the boiler. Natural gas, which in 1980 was a bargain as a clean fuel, has risen in price rapidly as a consequence of the Natural Gas Policy Act of 1978, and now many industrial firms are switching back to oil. (Few users of natural gas can switch easily to coal because of space limitations, the problem of handling the coal and ash, and environmental regulations.) World oil prices, as noted above, have fallen since 1980 to about $30 per barrel landed in the United States. Heavy oil, which may be recovered using steam injection, is considered a "synthetic fuel" by the SFC and may be competitive even with the current price of conventional oil. Also, the Canadian tar sands are being processed to yield oil at competitive prices.

Otherwise, the synthetic fuels, alcohols from biomass, and solar heat are all still more expensive on the average than conventional fuels. The presumption that continuing escalation of world oil and natural gas prices would result in early commercial success of synthetic fuels and solar heat has not been borne out. On the other hand, it should be noted that

Table 12-3

Cost Estimates for Energy Technologies[a]

Fuel		Capital Cost[b,f]	Price[c,f]
Conventional Fuels	1. Utility Coal	$ 4–5	$ 7½ /bbl
	2. Natural Gas, Domestic	10–20	8½
	3. Oil, Domestic	10–20	31
	4. Natural Gas, Alaskan	10–20[g]	35
	5. Natural Gas, Imported	10–20[g]	22
	6. Oil, World (Mid-East)	2– 2½	34
Synthetic Fuels	7. Heavy Oil, Domestic	10–20	24–30
	8. Shale Oil, Domestic	20–30	35–40
	9. Tar Sands, Canadian	20–30	30
	10. Oil from Coal	30–40	40–60
	11. Gas from Coal	20–40	30–50
	12. Methanol from Coal	35–45	65
Fuels from Biomass	13. Ethanol from Corn	25–35	100
	14. Ethanol from Cellulose	40–50	150
	15. Methanol from Cellulose	40–50	90
Solar Heat	16. Solar Heat for Buildings	80–200	85
	17. Solar Process Heat	60–100	60

Electricity	Capital Cost[d,f]	Price[e,f]
1. Coal-Fired	$ 750	4 ¢/kwh
2. Nuclear	1,250	6
3. Wood-Fired	1,200	7–10
4. Hydroelectric	1,200	3–6
5. Geothermal	1,200–2,000	6–25
6. Solar-Thermal	2,300–12,000	6–10
7. Wind	1,500	6–10
8. Ocean-Thermal	2,500–5,000	10–20
9. Photovoltaic	2,500–18,000	25–100

NOTES:

[a] Source: R. S. Greeley, *Western Hemisphere Energy Resources* (McLean, Va.: Mitre Corp., 1980).

[b] Thousand dollars per daily barrel of capacity for 20-year lifetime of facilities.

[c] 1980 dollars per barrel or equivalent.

[d] Per installed kilowatt.

[e] 1980 cents per kilowatt-hour at the bus-bar (convert to 1980 dollars per barrel equivalent by multiplying by 5.8; assumes a heat rate of 10,000 Btu per kilowatt-hour and 5.8 million Btu per barrel).

[f] Conventional fuel prices are average purchase prices in 1980. Synthetic, biomass, and solar costs and prices are estimates based on experiments under way in 1980.

[g] Plus transportation facilities (ships or pipeline).

certain synthetic fuel projects may produce high-value products such as jet fuel, which may be competitive in price. Also, some uses of solar heat in certain locations are competitive in price today, such as replacing electricity in hot water and home heating units.

In producing electricity, conventional coal is still the cheapest fuel. However, on the basis of delivered price in cents per kilowatt-hour, coal and nuclear power plants are roughly equivalent. Many existing nuclear power plants are providing appreciably lower-cost base-load power than coal-fired plants. Synthetic fuels will probably not be used to generate electricity but will go to higher value uses such as transportation. Wood-fired industrial boilers and smaller utility boilers are being introduced in some areas at competitive prices. Solar energy used to generate electricity, including wind-driven generators, are generally the highest cost options, although competitive in selected applications in areas such as Hawaii and portions of the Southwest. High-quality geothermal reservoirs can provide electricity competitive with new coal or nuclear power plants in selected areas.

Conservation of energy may be considered an alternative fuel source. Many energy-saving measures are immediately cost-effective and save most of the cost of the fuel not used. However, installing heat recuperators and similar devices can involve considerable capital costs as well as operating and maintenance costs. Each investment in energy-conservation equipment must be analyzed on a life-cycle basis and compared with other available alternative fuels.

Representative Data

The following estimates provide an indication of the relative costs and yields of national programs in advanced energy technologies.[41] The total costs include both federal and private expenditures:

- Synthetic fuels. About $100 billion for 2.5 million barrels per day capacity (provides about 225,000 direct jobs during construction).
- Solar energy. About $500 billion (of which $30 to $40 billion would be federal expenditures) for 4.5 million barrels per day equivalent fuel saved (provides about 1 million jobs during implementation period).
- Breeder. About $4.5 billion for the Clinch River Breeder Reactor demonstration plant, and perhaps $80 to $100

billion to displace 1 million barrels per day equivalent fuel (providing roughly 160,000 jobs during construction).
• Fusion. About $500 million per year for twenty to forty years' R&D, which will increase later for demonstrations and commercialization.

Options and Recommendations

General Options

Clearly the power of the marketplace to stimulate energy conservation and creative financing options for conventional and advanced energy technologies must be recognized. The difficult choices have to do with identifying a clear national need to accelerate the development and demonstration of technologies that do not yet or cannot provide a marketplace return on private investment.

The following alternative options for supporting the future development and commercialization of each energy resource are listed to indicate the range of federal versus private industry activities that are possible.

Specific Technology Options

Synthetic Fuels
1. Continue Synthetic Fuels Corporation support to the construction of the "pioneer" plants. Provide federal assistance with R&D to lower costs of the products below current world prices.
2. Provide federal support only to R&D, possible site selection and "banking" for future plants, and solving environmental and water availability questions.
3. Cut out all federal support and allow the marketplace to dictate the type and pace of synfuel developments.

Nuclear Breeder
1. Continue federal support to the Clinch River Breeder and reprocessing technology.
2. Continue federal support only to R&D on new breeder concepts and on reprocessing technology. Delay any development or prototype activities until the breeder technology is more clearly needed.
3. "Leap-frog" the liquid metal fast breeder reactor (LMFBR) technology and put federal support on a hybrid fusion-breeder approach.
4. Abandon federal support for the breeder.

Nuclear Fusion
1. Step up federal support to fusion technology programs, including rapid implementation of the Fusion Engineering Center to solve the engineering development of the magnetic fusion concept.
2. Continue federal support to the R&D of magnetic and laser fusion at about the same level until the success of one particular approach indicates increased support is justified.
3. Shift the approach to emphasize a hybrid.
4. Cut back on engineering developments and emphasize research until scientific questions are answered more fully.

Solar and Geothermal
1. Continue federal support to the demonstration plants for the Thermal-Electric (Power-Tower), Photovoltaic, Ocean-Thermal-Electric, Wind, and Geothermal programs with stronger electric utility involvement. Also continue the federal tax credits for residential and industrial solar and the Solar Bank.
2. cut back federal support to providing information on solar technologies and to continuing valid areas of R&D such as the photovoltaic program.
3. Cut out all federal support and allow the marketplace to dictate the type and pace of solar/geothermal technology development.

Conservation and Storage
1. Continue tax credits and strong federal R&D including demonstration programs in advanced conservation and storage technologies.
2. Cut back federal support except for tax credits for residential, commercial and industrial projects, information programs, and low-income weatherization.
3. Cut out federal support except for existing tax credits to industrial firms.
4. Cut out all federal support including tax credits.

Recommendations

Long-Term View

We must recognize the long-term nature of energy technology development. The introduction of a new energy technology is a thirty- to fifty-year activity. Stopping and starting major energy programs is not effective in contributing to successful energy system development and commercialization.

Clearly, the market, acting through relative prices, will eventually shift our resource base away from depleting oil and gas reserves. There are four primary alternative resources that can be used: (1) coal and shale oil, (2) nuclear fission, (3) nuclear fusion, and (4) solar. Luckily, we do not have to shift abruptly or make a choice soon. Despite earlier fears, conventional oil and gas reserves will last many years longer. But plan we must for the era when the real costs of oil and gas escalate.

Synthetic oil and gas products fit into the existing energy transportation, distribution, and consumption infrastructure. A transition based on market prices from conventional oil and gas to synthetic oil and gas would appear to be a natural occurrence over time. There appears to be a financial hurdle in starting this transition because of high technical and economic risks. The Synthetic Fuels Corporation is playing a major role in clearing this hurdle. However, unrealistic target dates for early volume production and falling or steady oil prices could make the SFC task very expensive. A stretch-out in synthetic fuel projects is warranted, with provision for acceleration if oil and gas prices increase unexpectedly.

Nuclear power plants also fit into the existing electric utility infrastructure. If conventional nuclear power continues as a viable option, then either the breeder or fusion reactors will eventually be required to utilize the vast energy resources available in nuclear fuels. A decision between them does not have to be made now. The basic question is this: What is the minimum RD&D program that will permit steady technical progress to be made in both programs? Perhaps in the long run the breeder will become commercialized first and fusion will supplant it later on a cost-competitive basis. Again, there appears to be no national urgency about either program.

Solar and geothermal energy are already commercial entities. A decision could be made that solar provides national benefits in terms of lower pollution, more jobs and secure supplies and hence deserves a continuing national subsidy. In any case, following that decision, the marketplace can operate in a rational and predictable environment.

Similarly, conservation is generally low cost, provides jobs, and avoids using foreign supplies of oil. Again, a decision could be made to continue the tax benefits now being used so successfully. Essentially, however, the marketplace is working.

Within this long-term context, the following specific recommendations are made:

- Synthetic fuels: Continue the Synthetic Fuels Corporation but stretch out projects and lower the rate and prospective size of financial commitments.
- Nuclear breeder and fusion: Increase electric utility involvement and support in these programs to determine their scope and pace and to ensure that commercial interests and decisions are introduced at the earliest possible moment.
- Solar, geothermal, and conservation: Continue tax credits to provide for creative financing and a measure of national support for these resources.

Concluding Remarks

Despite seeming lack of progress in achieving either "energy independence" from imported oil or major contributions of energy supplies from advanced technologies, the United States has come a long way from the energy crisis of 1973–74, and even 1979. We understand better the nature of the plentiful energy resources available to us and the length of time any new technology requires to become commercialized. Most of the technology developments started so optimistically in 1974 are now well researched, and many can be turned over to private interests.

Given another ten years of development and operation of the pioneer plants, synthetic fuels will be ready technically to enter commercial markets. If oil prices fall and stay down for many years, the pace of commercialization will inevitably slow or stop. If and when oil prices should again escalate rapidly, synthetic fuels will be accelerated in production sufficiently in response to market forces to avoid a third crisis.

Similarly, advanced solar, geothermal, and energy-conservation technologies will need several years or more to reach the status of proven technology. As with currently available solar energy and energy conservation systems, commercialization will take place at a rate determined primarily by the price of competing fuels. Since the price of natural gas is now closely tied to the price of oil, the latter will control the pace of commercialization of these technologies as well as synthetic fuels.

The nuclear breeder and fusion are both considerably more than a decade away from being technically feasible systems

As they near commercial development, they will have to compete with coal-fired power plants and with conventional nuclear reactor "burners." In the long term, they may either complement or compete with each other.

Therefore, a "natural" evolution of advanced energy systems is likely to occur. Synthetic fuels from coal and oil shale will eventually supplement conventional oil and gas. Solar, geothermal, and various energy-conservation systems will be utilized wherever they are economically viable. The nuclear breeder will be adopted when the price of fissionable uranium warrants utilizing "fertile" materials now unused. Nuclear fusion will become viable either as a potentially less hazardous system than the breeder or simply as a cheaper system. Conventional nuclear reactor burners may be continued in service using fuels generated by the breeder or a hybrid fusion breeder.

This evolution from one energy resource base to another will take many years. Federal support for research and selective development projects will be appropriate throughout the years since private interests would in many cases not be able to recapture their investments, given the time scales involved. However, the federal role must be well-managed and sharply circumscribed, with private firms encouraged to participate at the earliest possible moment in the research as well as the development, demonstrations, and commercialization.

Above all, research must be continued on advanced energy technologies. For instance, better catalysts for more efficient and cheaper production of synthetic fuels are urgently needed. A remarkably clearer understanding of catalytic action is currently being achieved. Biological methods of synthetic fuel production are extremely attractive. The new bioengineering firms are close to major advances in this field. Advanced solar energy systems, particularly low-cost photovoltaic cells with long life and high efficiency could be a revolutionary development for decentralized electric power production. The direct production of hydrogen from water using sunlight is another dramatic possibility.

Steady support over the long term, at whatever level of federal funding is deemed appropriate relative to the gross national product and other budget items, is necessary to ensure the orderly, efficient development of the abundant energy resources available to the United States.

EPILOGUE: ENERGY AGENDA FOR THE GOVERNMENT

Richard N. Holwill

Introduction

For almost a decade the word "energy" seemed inseparable from the word "crisis." Hysteria about the problem led first to price controls, then to a Federal Energy Office, a Federal Energy Administration, and finally a fullblown cabinet-level Department of Energy (DOE). With decontrol of domestic crude oil prices and the advent of an international oil glut, the crisis subsided. The challenge now is to end the harm being done by the institutions created during the panic of the 1970s and create a structure that will minimize the problems that might result from any future supply shortfall.

To achieve this goal, political capital should not be squandered on dismantling DOE, particularly not if abolishing the department means merely transferring its functions elsewhere. The priorities should include: (1) ending price control on those fuels that remain under regulation; (2) establishing an emergency policy that works toward resolution of the crisis instead of complicating the problem; (3) recognizing those functions that should legitimately be given to the federal government; (4) and ensuring that future energy questions can be decided in a market free of federal intervention. The elimination of DOE as a long-term goal should not be abandoned, but it should be put into perspective.

Decontrol Natural Gas Prices

The Reagan administration has presented legislation freeing the market for natural gas early in the 98th Congress. This legislation is designed to increase freedom in the market and to avoid the partial-control option proposed by those who want to cap natural gas prices either at the Btu equivalent of crude oil or at the price of residual fuel oil. Moreover, the legislation is free of extraneous clauses allocating incremental prices to industrial users or second-guessing the market by attempting to specify priority uses for gas. Language is included in the bill to prevent pipelines from abusing the monopsony they have over isolated producers, or the monopoly that they have over consumers. Finally, the legislation invokes the extraordinary power of the federal government to abrogate those preexisting contracts that were negotiated prior to the introduction of a truly free market. (See Chapter 2 for a somewhat different view of what is required.)

Achieving a free market for natural gas will not be an easy task. Misinformation abounds about supplies, price, price trends, and the structural nature of gas markets. Most important, producers, pipeliners, utilities, and consumers are so accustomed to price controls on natural gas—which have existed for more than five decades—that they never discuss natural gas in broad economic terms. Rather, the debate focuses on an essentially irrelevant point: the wellhead price of gas.

Interfuel Competition

The debate should actually begin at the point of consumption: the burner tip. Competition with other fuels takes place there, not at the wellhead. Gas competes in a more complex economic environment than imagined. In most markets, the wellhead price of gas is now lower than the price of other space-heating fuels as a direct result of price controls. While international oil prices increased dramatically over the past few years, gas prices remained constrained, and gas consumers remained protected.

The debate on gas prices should not be distorted by talk of escalation in gas prices. Prices will rise; the burner-tip price may approach the price of No. 2 fuel oil (two-oil). It will not likely exceed that price. To put this into perspective, remember that two-oil is an alternative fuel for heating homes. In other words, natural gas users may begin paying a price roughly equivalent to that now paid by those consumers who use oil for heating purposes.

Natural gas must compete directly with two-oil, which is generally priced between the per-gallon prices of crude and gasoline; the markup over crude varies, but tends to be in the 15 to 20 percent range. Natural gas must also compete with residual fuel oil, known as resid or six-oil. Resid is, as the name implies, a waste product that remains after all gasoline and other refined products have been distilled from crude. It is a simple combustible used, like gas, in industrial boilers. Refiners will sell resid at a discount below the per-gallon price of the crude from which it is made simply to get rid of it. The price tends to run as much as 20 percent below that of crude oil, depending on sulfur content. During periods of high gasoline demand, refiners unavoidably produce more resid. While they are raising the price of gasoline to clear the market, they are also cutting the price of resid to avoid ullage problems—that is, the problem of having no room left in their storage tanks. As a consequence, resid prices may fall dramatically below crude costs during periods of supply–demand disequilibrium for gasoline or distillates. If natural gas prices are tied to resid prices, they would be pulled down during periods of high gasoline demand, thus making the sale of resid and the production of gasoline more difficult.

This somewhat academic discussion of refined product pricing is important to the natural gas question because gas competes both with low-priced resid and with higher-priced two-oil. Since space-heating markets are highly seasonal, pipelines and utilities have excess capacity in nonwinter months. Utilizing this capacity allows them to amortize their equipment more efficiently. All consumers benefit from the off-season sales of natural gas because overhead costs and, therefore, prices set by public utility commissions are lower.

Pipelines that concentrate on maximizing sales prices to distribution companies serving home heating customers do so at the jeopardy of their off-season markets where the alternative fuel is resid. The market-clearing burner-tip price cannot be projected with certainty. However, given the structural dynamics of interfuel competition, that price will fall somewhere between the prices of resid and two-oil, and will quite likely vary from utility market to utility market depending on the relative mix of space-heating and industrial consumers in each market.

Other Features of Decontrol

Critics of decontrol often argue that because consumers must receive their gas through a fixed distribution system, they have no recourse if the price of gas exceeds the bounds described above. Where industrial users have burners capable of using a variety of fuels—gas, resid, two-oil, and, occasionally, coal—residential and other small users do not have these options. Moreover, building code restrictions often require a lined chimney flue for oil but not gas, a fact that can make conversion prohibitively expensive.

It is futile to argue that burner-tip gas prices would never exceed two-oil prices; to attempt to do so is to accept the logically impossible challenge of proving a negative. In response to those who would protect against this possibility with price controls, we suggest an alternative: the decontrol legislation should include language allowing utility customers to demand common-carrier treatment from their pipeline if, upon review by the Federal Energy Regulatory Commission (FERC), the facts of a case demonstrate that a pipeline has abused its monopoly power by contracting for exorbitantly high-cost gas. If the consumers win their case, they would be free to arrange for new gas supplies on their own. They would be obligated to pay the pipeline and utility companies a cost-of-service charge for delivering their gas and would pay producers directly for the gas used. Admittedly, this system could lead to excess litigation unless strict standards were included in the legislation to ensure that this feature functions only as a safety valve.

A similar system should be provided for producers who often have difficulty getting pipelines to take their gas. In this case, the producer would use the provisions of the law as an entry valve into a market where pipelines can have a monopsony. The producer would be responsible for providing gas to an existing spur or trunk line, and for marketing the gas to consumers anywhere in the country either on a direct sale or barter basis. He would be obligated to pay transportation charges to the pipeline companies that move his gas on a common-carrier basis, and would pay rates reviewed by FERC in cases where sufficient alternatives are not available to create a competitive market for transportation services.

The effect of this scheme would be to put the wellhead price into its proper perspective. It would be a function of the burner-tip price, minus transportation charges and might vary drastically throughout the country. This concept is not radically different from the crude oil production business where the producer is accustomed to receiving a "wellhead netback price," which represents the price a refiner is willing to pay, minus collection and transportation costs.

This system encourages efficient competition. Unlike the cost-of-production rules used by the old Federal Power Commission, natural gas would no longer be a consolation prize for not finding oil. Unlike the multivintage system imposed by the

Natural Gas Policy Act (NGPA) of 1978, natural gas from the deep, tight, and high-pressure formations that qualify for uncontrolled prices would no longer be favored by producers. Gas exploration would focus first on the lowest-cost, most easily found gas, and then on gas that is closest to major markets. The transportation-cost factor would give a boost to exploration in the vast Eastern Overthrust Belt where geologic age and metamorphic forces clearly indicate that gas is the preferred exploration target. In sum, true decontrol will lead to the efficient and aggressive exploration of many new potential gas provinces and will tend to leave the most difficult deep zones until those days in the future when prices and technology justify such high-cost exploration on more rational grounds.

Incidentally, the high prices from gas in those deep zones—often two to three times the Btu equivalent price of crude oil—would not be justified in a free market. Pipelines have paid those prices only because they were able to "roll them in"—that is, average them—with lower-cost vintages of gas. In some cases, pipelines have, either through choice or contract pressure, shut in lower-cost gas in favor of gas qualifying for higher prices under the NGPA; this, too, would be eliminated by a system where the wellhead price is determined by the burner-tip price, minus transportation and other costs.

One problem remains: over the past few years, pipelines and utilities, forbidden to negotiate on price, have negotiated gas contracts on terms of sale and have put a high premium on security of supply. As a consequence, many contracts contain clauses specifying unrealistic prices upon deregulation. Other contracts—primarily those for decontrolled gas—have willingly paid unreasonable prices for natural gas. Despite appearances of partial decontrol, these facts make it clear that markets were distorted to the point that rational decisions could not be made. Therefore, the transition to a decontrolled market will require the use of the extraordinary federal power to abrogate contracts. This power has been used only rarely. The last use came when the nation switched from the gold standard. Congress felt it necessary to wipe clean the contractual slate and thus force renegotiation of contracts in which the conversion of dollars to gold was essential. This change in the natural gas market is of almost as great a magnitude.

One final remark about natural gas decontrol: some proposals link decontrol to some form of windfall profits tax. However, such a tax will constrain those supplies of gas that, from a cost standpoint, represent the frontier of exploration. We should encourage exploration along these frontiers, not inhibit it. In the coming year, taxes should be discussed only in terms of the revenues needed to run government and not as penalties for those sectors of society that Congress decides are sufficiently unpopular to bleed for a few extra dollars.

Emergency Energy Policies

As its next priority in energy, the administration must address the concerns of those who claim that the existence of a multi-billion dollar, cabinet-level Department of Energy is justified by the possibility of another interruption in oil supplies. The key to the abolition of DOE is the development of an emergency energy program that does not rely on a government bureaucracy.

The elimination of price and allocation controls was the first step in the development of such a policy. The administration must now work to develop a clear legislative statement that indicates to all that allowing a market-clearing price is the centerpiece of an emergency policy. Such a statement is needed primarily because of a general cynicism that assumes that this policy will not be long-lived. (See Chapter 3.)

Encourage Private Oil Stockpiling

As a direct consequence, business and private consumers in the United States have not taken the prudent steps that private firms and individuals in Europe and Japan have taken to protect themselves against sudden disruptions in the distribution of crude oil or refined products. Consumers in many other countries rely on decentralized private sector storage facilities for gasoline, distillate, and resid under policies generally termed "dynamic storage." The term reflects the fact that the storage life of refined products is limited, and the consumer must therefore continually use products before they become clogged by sediments or lose vital components through evaporation.

Although private storage capacity has grown, it has not been encouraged; U.S. government officials continue to term such storage facilities as "hoarding capacity." This bias implies that, if a new round of allocation controls is ever imposed, the allotment of fuel will be reduced to an individual or firm that has maintained a reserve against an emergency. Before adding storage capacity, a prudent person will consider the possibility that the extra supplies will be allocated away by government fiat and may well decide that the extra security is not worth the investment.

The new legislative statement should recognize the multiple benefits of additional decentralized product storage capacity by assuring those who add capacity that the emergency energy policies of the U.S. government will not punish the prudent. The laws should allow for immediate expensing of capital costs associated with such facilities and tax credits for individuals who maintain extra storage capacity, particularly for home heating oil. In most cases, home gasoline storage may not be feasible for safety reasons. However, businesses could certainly maintain this capacity for their fleets and even for employee atuomobiles. The law should allow maximum flexibility for storage, subject only to local safety regulations.

Obviously, a law such as this could be changed in the event of an emergency. Those investing in extra gasoline, heating oil and resid storage would, nonotholess, have an upper hand in the event of a legislative retreat if Congress is forced to change the law because a change disallowing private storage could be stopped. Moreover, those maintaining this storage would be able to argue that they acted in good faith and should at least receive an exemption.

The government role in the operation of this dynamic storage system would be minimal. It might publish information about maximum shelflife of products or perhaps explain how such capacity could be used by consumers to make off-season purchases of out-of-season fuels. However, no major government action or expense is required or even warranted.

Some will still object, calling this policy an authorization to hoard; such claims are illogical. Clearly, the runner who failed to condition himself for a marathon could not logically claim that the winner cheated by preparing for the race. By the same token, the imprudent consumer cannot complain if someone else is ready for an emergency.

Why Price Rationing Is Fairer Than Other Schemes

Others have traditionally complained that any policy that relies on prices to clear the market is unfair to the poor. They call this "rationing by price." Price rationing is, nonetheless, better than any other form yet devised, particularly if you look at the dynamics of the last two supply shortfalls. In 1973 and 1979, the United States was the only major country to experience gasoline lines; it was also the only country to constrain prices through controls or to attempt to allocate supplies to users based on a bureaucratic anticipation of need or priority. (See Chapter 3.)

Those who argue for price controls must endorse some other form of rationing. The effective policy during the period of gasoline lines, rationing by inconvenience, ignored the value of time and the amount of resources wasted as workers killed time in gasoline lines. Those who argue for a coupon rationing system ignore the huge cost, incredible security problems, and the potential for counterfeiting.

Those who argue for price controls also ignore the fact that higher oil prices will spur development of alternative sources of energy faster than any form of government subsidy. In the best of times, government grants, loan guarantees, and purchase agreements can be given only to a few of the thousands of potential entrepreneurs who would experiment with alternatives. By contrast, higher prices would enable every backyard inventor to expect a better price for his alternative fuel or new energy-saving gadget and, thus, would be even more of a spur to innovation. The contradiction in current policy is obvious: historically, most innovation has come from unknown individuals working alone, but most current government technology support is going to established companies that are tinkering with projects that bureaucrats decided might possibly work. Had current technology-development policies been the vogue in the last century, a synthetic fuels corporation would very likely have overlooked Thomas Alva Edison while giving a loan guarantee to a candlemaker to develop an improved wick, or a grant to a carriage company to test high-mileage hay for horses.

The serious problems associated with allowing prices to clear the market cannot be ignored by those who endorse a free market in an emergency. The federal, state, and local governments must cover higher fuel costs to provide basic services to ensure the health and safety of the population. Those on fixed incomes cannot adjust as quickly to higher prices as those who are able to insulate their homes or invest in more efficient furnaces.

These problems are not insurmountable. The Windfall Profits Tax (WPT) on domestic crude oil is already generating billions of dollars annually. In the event of a crude-oil shortage, price increases would generate a dramatic increase in government revenues. As an integral part of its emergency policy, the government should be prepared to recycle these revenues: first, to states and localities to offset higher fuel prices for emergency services; second, to hospitals and other institutions operating in the public interest; and third, to those in genuine need of assistance with heating bills. In this way, the market would allocate supplies, and the revenues from the tax would protect those who need assistance to survive. Perhaps the simplest and quickest recycling scheme would be to lower the withholding rate for federal income tax and to raise the rates for those on public assistance. This system would be far more efficient and equitable than any administered by a government bureaucracy.

The SPR and Existing Emergency Programs

With the exception of its information function and the Strategic Petroleum Reserve (SPR), DOE should get out of the emergency planning business. (See Chapter 3.) As discussed earlier, the market can clearly handle pricing and allocation of supplies. Moreover, if the country is willing to commit itself to a free market, the concept of the SPR should be entirely reevaluated in the cold light of its high costs, considerable risks, and marginal benefits.

The cost of the SPR is high. If it is topped off at 750 million barrels as planned, the cost—assuming an average nominal cost per barrel of $35—would exceed $26 billion. The implicit annual carrying cost would be $5 billion. The security of the storage facilities cannot be controlled perfectly. Oil could be hijacked from within the salt domes using slant drilling techniques and only a little ingenuity.

The greatest danger, however, may be the contamination of the crude oil by sabotage. If the oil in the reserve were spiked with tetraethyl lead, it would poison all of the refinery catalysts with which it came in contact. This could force the shutdown of a substantial portion of U.S. refinery capacity and would do more damage to U.S. energy balances than any so-called embargo.

Selective Oil Embargoes Never Worked and Never Will

Presumably, the United States would begin to draw some supplies from the SPR when and if another (selective) oil embargo is mounted against this country. This plan, however, is based on two false assumptions: first, that the embargo in 1973–74 and the Iranian revolution resulted in supply shortfalls: and second, that an embargo could be mounted in the future. The simple truth is that oil is a fungible commodity. Just as a grain embargo against the Soviets will not be effective unless the United States completely withholds grain from the international market, no oil embargo can work unless oil producers are willing to leave the oil in the ground. In such an event, all oil consumers would suffer equally as prices for the remaining production began to climb.

Despite the gasoline lines in 1973 and 1979, the empirical evidence bears out this point. Crude oil imports did not fall during the Arab oil embargo or the Iranian revolution. In each case, however, price controls compounded the problems and exogenous factors contributed to supply disruptions. In 1973, prior to the crisis, petroleum inventories were unusually low because of an extraordinarily long strike by Japanese seamen the previous year. Thus, stocks of refined products were inadequate to satisfy demand. In 1979, West Coast gasoline reserves were extraordinarily low as a consequence of a longer-than-normal driving season and an imbalance in demand for refined products.

The 1979 case is particularly interesting. The key to the 1979 shortage was actually an abundance of hydroelectric power, which contributed directly to a slowdown in demand for residual fuel. Resid—too viscous to be moved long distances through pipelines—could not be easily shipped from the West Coast to the East where it was in high demand. As resid storage facilities filled up, refiners were forced to limit total refinery runs to the amount of resid that could be sold. Gasoline production suffered, and gasoline lines developed. In fact, they began to develop before the Iranian situation resulted in any change in crude oil deliveries to the West Coast. When those changes did take place, Iranian crude was replaced by crude from other producers.

In both 1973–74 and 1979–80, had government not intervened in the marketplace, no gasoline lines would have developed. Not only did price controls prevent prices from climbing to a market-clearing level, but allocation controls were imposed through which states "set aside" some quantities of fuel for preferred users. This set-aside directly contributed to still longer gasoline lines. (See Chapter 1 and Chapter 3.)

Provided that in the future we allow prices to clear in situations such as these, the Strategic Petroleum Reserve makes little sense. If we assume that oil producers might again attempt to impose an embargo on the United States *without limiting production,* oil now flowing to the United States would move into other markets; the oil now sold to those markets would be freed to come to the United States; and the SPR would not be used.

A General Embargo Simply Raises Prices

If we assume that oil producers would shut in production, then all oil consumers would be equally affected as prices rise. In this case, massive diplomatic pressure would be mounted by all oil-consuming nations on all oil producers. Prices would simultaneously rise, and in all likelihood, many producers would begin to do as they have done in the past—cheat on deliveries by allowing oil companies to lift far more crude than they should under production ceilings.

At the present time, excess worldwide oil capacity outside of OPEC could offset even a major cutback. In fact, most other producers would jump at the chance to increase revenues by marginally increasing production. Of course, at some point in the future, this might not be the case, and a shortfall could occur that would drive up the price.

But as world oil prices climbed, new oil and natural gas supplies would be brought on stream; numerous uneconomic gas fields are now available for production once prices hit certain levels. Moreover, with each supply-demand-induced hike in oil prices, new energy technologies become economically viable. In such a situation, oil prices would drop back gradually over a period of years, although perhaps in fits and starts.

In any of the above situations, or even in the event of an invasion of the Persian Gulf by a foreign power, widely dispersed product storage would offer greater security than an easily sabotaged SPR. Moreover, as the crude would already be in product form—that is, gasoline, distillate heating oils, and resid—it would have the added advantage of not requiring additional refining; given the fact that in certain circumstances refining can be a bottleneck of some proportions, this consideration is not insignificant.

In reevaluating the Strategic Petroleum Reserve, we must begin to consider the fact that we may not need it in an emergency. In fact, we may never need it except as a hedge against price disruptions. (See Chapter 3 and Chapter 4.) If we are unwilling to use the SPR in these situations, we will probably never use it. Finally, we must face up to the toughest question: Is this marginal increase in security worth $26 billion plus $5 billion each year after it is filled?

Exporting Alaskan Crude Oil

As a part of a legislative statement on emergency energy policies, the administration should press for authority to allow the exportation of Alaskan oil and gas. The present congressional ban on exports is logical only if one assumes that price controls will be reinstituted in the event of an emergency. The ban has caused harm in several ways. Its reversal would create substantial benefits for all concerned.

The harm done by the ban includes the creation of an oil glut in Alaska and California and, therefore, a reduction in marginal production as well as a slowdown in exploration activities. Because the ban necessitated shipping excess Alaskan crude through the Panama Canal to the East Coast—at great cost—tax revenues and royalties to the governments of the United States and the state of Alaska are lower than they otherwise would be. Most critically, the ban has shown our trading partners that the U.S. commitment to free trade is less than total, a fact that contributes to Japanese cynicism about U.S. objections to their trade practices.

Lower transportation cost and greater political stability afforded by U.S. production would offer a number of benefits to the Japanese. Legislation allowing exportation of crude oil to Japan could even be tied directly to improved U.S. access to the Japanese market through the elimination of nontariff barriers. The sale of crude to Japan could also be contingent upon the Japanese providing a like amount of oil to the U.S. East Coast from their production allotments in the Middle East; if, in a time of crisis, the Japanese failed to deliver, Alaskan supplies could be reclaimed, thus, protecting the interest of U.S. consumers.

Liquefied natural gas (LNG) pricing is sensitive to the length of a freight run. The freight run from Valdez to Japanese ports is short when compared with the distance from Japan to other LNG sources. LNG exports might, therefore, offer a profitable outlet for gas reserves in Alaska. At present, these supplies are shut in due to wrangling over the many issues associated with the overland pipeline to the lower forty-eight states. If the ban is removed, the costs and benefits of gas exports could be compared to those connected with the costly line through Canada.

In the final analysis, the development of a hydrocarbon trade between the United States and Japan would benefit all parties with little downside risk. The Japanese would be able to get crude at lower cost and the U.S. balance of payments would be improved. Providing a new outlet for oil and gas supplies now in surplus would spur exploration activities both in Alaska and on the West Coast. (See Chapter 5.)

All of these benefits, however, depend on assurances of a free market even during an emergency. This should be high on the Reagan administration's agenda.

Nuclear Power and Market Forces*

With regard to the current dispute over the risks and benefits of nuclear power, one fact is clear: it will be a key question in the energy debate for the remainder of the decade. The question of nuclear power will become increasingly important because much of America's electric power-generating base is nearing the point at which it must be replaced. As utility managers make plans for the 1990s and beyond, they should be able to consider nuclear power options. As environmentalists consider the implicit dangers of abandoning nuclear power— either inadequate power or added pollution from fossil fuels— they, too, will most likely reconsider the nuclear option. As a consequence, it must be discussed in a different context than other government energy questions.

Can Nuclear Power Compete Economically?

The ultimate question remains the same: What steps are needed to free the market? In this case a second question is in order: Can nuclear power survive in a free market? It is not a creature of the market; it was developed by government design and has never been fully privatized; it is subjected to the strictest regulations. Research and development continue to be largely government-controlled and -directed functions. It has been the target of a concerted, almost irrational, effort to distort the dangers associated with it. Simply put, the market-place is so confused that we cannot tell if nuclear power would survive if put to a competitive test against other sources of electric power even though it has proved competitive in other countries (e.g., Japan).

The challenge in nuclear power is to begin removing both

* The author wishes to thank Henry Sokolski for his valuable assistance on nuclear policy questions.

the barriers to nuclear power and the subsidies that it received. These must be stripped away carefully. To remove all of the barriers and none of the subsidies would result in a surge in supply and serve no useful purpose. Conversely, to remove the subsidies but leave the barriers would be to kill the industry without even giving it an opportunity to live or die on its merits. Nuclear power must not be forced on people against their wishes through government subsidies. By the same token, a vocal minority cannot be allowed to deny this power source to a majority of consumers by using regulations to make nuclear power unnecessarily expensive (see Chapter 8).

The ultimate test of nuclear power will be in deregulated competition to serve local markets on a cost basis. The concept of deregulating power generation independently of electric power transmission or distribution has been around for some time; with the development of interlocking regional grids, this concept is now becoming feasible. Because the fuel cycle cost of nuclear power is relatively low, an existing nuclear facility is able to produce incremental power with only a marginal increase in cost. In fact, many facilities can generate 800 to 900 megawatts for only slightly more than the cost of 500 Mw. Nuclear power should do quite well in a deregulated environment.

In the coming year, the administration should press for legislation to allow for experimentation with power deregulation of this type in one or more of the regional power grids. A large-scale experiment is needed as a first step toward general deregulation of electric power generation and to determine the actual need for new power facilities in the event of total deregulation. If this idea is not advanced, power companies might make their decisions on the next generation of power plants without benefit of what might be the greatest potential cost saving to consumers—competition. Moreover, discrete decisions made without reference to the potential of competition could lead to an overbuilding of generating capacity (see Chapter 11).

An experiment of this type would demonstrate to consumers that nuclear power can be inexpensive, and demonstrate to the nuclear industry that profits can be made through efficiency. At present, consumers see nuclear power only in terms of the extraordinarily high construction costs and resulting high base rates. The industry, meanwhile, is reduced to selling expensive "back-fit" items to keep business alive in a market devoid of new plant orders. Finally, if successful, these experiments could demonstrate that, in a free market, utilities would have no incentive to build extra cost into facilities only to increase base rates.

Streamlining Regulations and Licensing

The second step would be to streamline the regulatory process while simultaneously asking utilities to cover their own insurance costs. The delays in licensing have been well documented; the effect these delays have on cost is clear. Most, however, tend to forget the most important fact: the ad hoc rulemaking process through which plants are now licensed often reduces the safety of new plants (see Chapter 8). Because no item can be standardized, construction flaws cannot be ironed out in the manufacturing process.

Continued review of construction progress leads to continual changes in design. Many of these changes have little or nothing to do with safety in high-risk vessels; the aim of the design change is to obtain an incremental increase in the safety of very low-risk equipment. Not only does the incremental increase in safety pale in comparison with the increase in cost, but the result is not always an increase in system safety. When a system design is integrated and balanced, changes at one point can place pressure on other points in the system unless they, too, are changed. Late design changes can often have negative effects unless an entire system is redesigned.

Except to note that licensing reform should focus on preclearing certain reactor components and preselecting sites for construction of smaller light-water reactors, our purpose is not to outline specific regulatory reforms. Rather, we propose that linkage should be built into licensing reform. In coordination with reforms proposed by the nuclear industry, the Price-Anderson Act should be phased out. Price-Anderson, passed originally in 1957 to provide insurance to the nuclear industry in the early days when risk could not be assessed, has outlived its usefulness. Today, it is essentially only a subsidy that, by limiting nuclear accident liability to $560 million, provides utilities with below-market insurance.

Linking licensing reform to Price-Anderson repeal offers several benefits. First, utilities would be forced to make decisions based on the true cost of nuclear power. New plants would be judged based on true economic realities including insurance cost levels. The requisite level of insurance should be determined by local public utility commissions, which are responsible to local voters and rate-payers and will, therefore, be sensitive to the cost of unrealistically high insurance requirements. The costs, however, would not include the artificially high construction costs that result from licensing delays. Safety would be improved because the insuring companies would have an incentive to work with the utility to improve design and operational safety. This has been the case in numerous industries; the nuclear industry would likely be no exception.

Second, this linkage offers political benefits. The very people who have opposed nuclear power licensing reform have supported Price-Anderson repeal. They believe that no rational insurance company would insure a nuclear power plant. Without insurance, the industry would, they argue, be forced to shut down.

If insurance cannot be provided and shutdown occurs, nuclear power will have failed the most basic test in a free society: nuclear utilities were unable to offer reasonable assurance that they could be responsible for their liabilities. The more likely result would be far different. Despite the reluctance of the insurance industry to provide coverage, either existing insurance companies would fill this void in the market or some other party would begin to offer such a service. The industry and its opponents would be shortsighted if they discounted the ability of a free market to respond to a demand for insurance of this type.

The Nuclear Waste Question

Nuclear waste disposal is the second major area of contention in the debate over nuclear power. Utilities first raised the issue with dubious claims that they were running out of space in spent-fuel storage ponds; some even threatened to close if they were not allowed to shift fuel from one facility to another. Seizing on these claims, the opponents of nuclear power began trying to block any proposal for high-level nuclear waste disposal. As a direct consequence, the question of long-term nuclear waste storage has been badly distorted.

Without regard to any particular piece of waste disposal legislation passed by Congress, many solutions are technically feasible now. For example, above-ground storage in dry casks can extend the capacity of on-site storage. In a technical sense, the problem is not critical; in a political sense, it is severe. Until the public is confident that the long-term problem is solved, antinuclear forces will be able to play on the legitimate fear of the layman.

Responding to this political problem, the nuclear industry has pressed for a permanent solution to waste storage. Technically, its solution offers some advantages; economically, it is troublesome. The solution involves reprocessing spent fuel, extracting the useful elements, and suspending the waste in a vitreous glass medium. This process reduces the volume of high-level waste by a factor of five, but it increases the volume of medium-level waste. Moreover, the cost of geologic storage will not necessarily be lower than the cost of other storage media.

Unfortunately, even this solution is not viable unless reprocessing spent fuel is economically justified by the sale of the recovered uranium and plutonium. Both the Ford and Carter administrations examined this question and decided to defer a decision on reprocessing until such time as the economics became attractive. To date, the economics have not changed sufficiently to justify an investment in reprocessing facilities. Clearly, the U.S. government should not push this option on the private sector with promises of subsidies.

Waste disposal is an important issue. Proceeding with the development of a long-term storage facility is desirable. A geologic storage facility should offer several features: it should be retrievable because at some point in the future, this waste might be a resource. Most important, a storage facility must be managed by the private sector, which—unlike the national laboratories—has an incentive to move quickly, efficiently, and safely. Environmentalists should favor a long-term geological repository because it is essential to accepting foreign spent fuel, a policy that many environmentalists advocate on nonproliferation grounds.

If the public is allowed to make rational decisions about individual future nuclear power plants based on their relative cost and relative safety features, the future of nuclear power is not dim. It is clearly environmentally safer than coal and less subject to price shocks than oil-fired facilities. In a competitive market, nuclear power could offer substantial cost savings. If, with regulatory reform, the cost of nuclear power is not exaggerated, it should have a chance to survive. However, if the industry is saddled with additional cost burdens such as those implicit in overregulation or those associated with costly investments in unneeded reprocessing or enrichment facilities, its chances for survival in a truly free market will be reduced.

Rationalizing Enrichment

Similarly, extra cost should not be built into the system for enriching uranium unless the market will bear that cost. At present, European competition has cut into the U.S. domination of the free world's enriched uranium markets. Nonetheless, the administration seems intent on pursuing construction of the Gas Centrifuge Enrichment Project (GCEP) at the Portsmouth, Ohio, enrichment facility. Completion of this project will add unnecessary capacity and will retard the development of more promising enrichment technology including the laser isotope enrichment process. Most important, its completion would preclude privatization of the enrichment business. Privatization should be the goal of this administration; the goal should not be a new government program that must be subsidized for decades.

Privatization of the operating enrichment system has not been pursued for several reasons. Some do not see the need for privatization. This need is real and important: as long as the government operates enrichment facilities, it will price enriched material at levels approaching subsidies or at levels that are set to generate extra revenues. Privatization had not proceeded because it had been discussed in terms of the sale of these facilities at their *replacement cost;* this assumption implies an increase in the price of enriched material beyond what the market can bear. Instead of pursuing the sale of these facilities, the administration should revive the leasing option. If leased, the cost could be based on the depreciated value and not the replacement cost of the facilities. The operator would, at least, have a chance to make a profit and would perhaps choose to make the investments needed to improve efficiency. He should also be given the option of renegotiating the onerous take-or-pay electric power contracts that now allow TVA and other utilities to charge the U.S. government for power it is not using and will not use for nearly a decade.

If, after the private operation of these facilities is established, the market indicates a demand for enriched material, then a decision can be made on the future of GCEP in light of other enrichment options. In the interim, it is imperative that research and development be subjected to economic tests and the realities of the marketplace. This does not happen now in large part because nuclear power was developed in the public sector. The impetus for new nuclear power programs comes from the public sector and from individuals accustomed to working in an environment where economic considerations do not always take precedence.

Role of Government in Nuclear Research and Technology Development

Pure research into high-energy physics and similar distant applications of nuclear power should be transferred to the National Science Foundation, which directs fundamental research in general and is well equipped to work with the universities now doing this research. Through this reorganization, the debate on the future of nuclear power could be moved from narrow considerations of discrete projects to the broader questions, which are more important in the final analysis.

The development of technology is a more difficult problem. As long as general revenues are dedicated to these tasks, Congress is obliged to oversee the process. On the other hand, if funds from utilities and potential vendors made up the pool of funds for applied research, Congress could step out of the picture. But this option has been precluded since many public utility commissions deny utilities subject to their jurisdiction the freedom to invest in any operation that is not directly in the interest of the rate-payers. Given the fact that applications of new technologies must be developed and the fact that the utilities and consumers will benefit from this technology, the logical solution is to impose a user fee for the development end of R&D. This fee should be in the form of a direct tax of some type on electric utilities and the nuclear power industry; the revenues would be dedicated exclusively to development in areas sanctioned by the industry.

Through a user fee of this type, a rational decision could be made on the future of projects such as the Clinch River Nuclear Breeder Reactor project and GCEP. At present, cynics can say that a particular project is being pursued only because a powerful senator fought for his home district or because it was promised to the voters of a particular state by a successful presidential contender. In the case of Clinch River and GCEP, if the industry is not willing to pay at least a substantial part of the cost either directly or through a user fee as described above, the projects should be killed. In all future technology development projects, the industry should control the flow of funds, and Congress should be kept out of the picture unless general revenues are involved.

Through a series of steps, as described above, the nation

could make a rational transition from what is now mass confusion over nuclear power to a market-oriented structure. This transition would be sufficiently gradual to ensure that potentially valuable projects have a reasonable chance to survive and that nuclear power itself has a fair chance to stand the test of a free market. Starting this process will also enable the administration to end an embarrassing paradox: at present, it gives the appearance of being in favor of a free market in all things conventional, but virtually socialist on nuclear power. In the long term, such a paradox will not serve the best interests of the administration, of nuclear power, or of a free market for other energy sources.

REFERENCES AND NOTES

Prologue: The World Oil Outlook, S. Fred Singer

1. For a more complete discussion and references, see S. Fred Singer, "The Price of World Oil," *Annual Reviews of Energy,* vol. 8 (Palo Alto: Annual Reviews, 1983).
2. See Appendix 2 of Reference 1.
3. S. Fred Singer, "A Geophysicist Looks at World Oil." Invited address to the Southern Economic Association, Atlanta, November 7, 1979.
4. S. Fred Singer, "Building Pressure for an Oil Price Drop," *Wall Street Journal,* January 11, 1980.

1. Energy Policy in a Free-Market Environment, S. Fred Singer

1. A thousand cubic feet (1MCF) of gas has a heat content of one mbtu, one million Btu, or about one gigajoule (one billion joules).
2. By 1982 prices for gas and oil averaged about $2 and $6 per mbtu, respectively.
3. For example, wellhead prices for natural gas may range from $0.50 to $9 per MCF, in the same state.
4. There were additional categories, such as Alaskan oil, "tertiary" oil (requiring costly recovery techniques), and some uncontrolled "new" domestic oil.
5. A case can be made, however, for deregulation of pipelines, and certainly of electric power generating facilities. (See Chapter 2 and Chapter 11.)
6. One controversial issue will undoubtedly be whether a windfall profits tax should be attached to the bill.
7. The changes being discussed include: (1) modifying regulations about land restoration (following strip-mining) to allow regional flexibility, to permit creation of level land rather than recreating the hilly contours where that is economically more useful, and to replace design standards by performance standards, thus improving cost-effectiveness; (2) permitting the burning of low-sulfur coal without the use of expensive flue-gas-scrubbing equipment; and (3) setting appropriate standards for ambient air quality, but leaving the implementation methods to the users to be achieved at lowest cost.
8. Some negative externalities are harmful and uncompensated side effects, such as accidents, noise, pollution, the congestion of roads in cities, and the need to build and maintain roads.

2. Regulation and Deregulation of Natural Gas, Connie C. Barlow and Arlon R. Tussing

Peter R. Merril, "The Regulation and Deregulation of Natural Gas in the United States (1938-1985)," Harvard University, John F. Kennedy School of Government, January 1981.

Natural Gas: Making the Transition to Competitive Markets, proceedings of Inside F.E.R.C. conference, New Orleans, April 11-12, 1983. (New York: McGraw-Hill, not yet published.)

M. Elizabeth Sanders, *The Regulation of Natural Gas: Policy and Politics, 1938-1978*. Philadelphia, 1981.

Arlon R. Tussing and Connie C. Barlow. "A Survival Strategy for Gas Pipelines in the Post-OPEC Era." Adapted from a Tussing speech to the Interstate Natural Gas Association of America, September 27, 1982. *The Public Utilities Fortnightly*, February 2, 1983.

Arlon R. Tussing and Connie C. Barlow. "The Rise and Fall of Regulation in the Natural Gas Industry." Prepared for the Third Annual North American Meeting of the International Association of Energy Economists, Houston; slightly different versions in *Public Utilities Fortnightly*, March 4, 1982, and *Energy Journal*, October 1982.

3. Emergency Management, Benjamin Zycher

1. Economists do not know the precise nature of this "demand" for redistribution. It may be that individuals value positively the income levels of others, or it may be that what is valued is the ability of each individual to enjoy some minimum "standard of living," defined as some minimum consumption level of "necessities." Alternatively, individuals, uncertain about the future, may support redistribution policies as a form of insurance against possible decreases in their own income. Another view is that redistribution is the outcome of competition within the democratic process for the right to use the confiscatory power of the state for purposes of gaining wealth. It is unnecessary to delve here into all of the issues and complexities surrounding the theoretical and operational definition of the minimum living standard and of the individuals entitled to assistance in achieving it.

2. Of course, to say that the market will behave "imperfectly" is to say nothing; neither government nor any other institution embodying human behavior and choice behaves in an ideal fashion. Therefore, the standard argument implies only that *a potential* exists for allocational improvement through public policy; there can be no presumption that such policy as *actually implemented* will result in *actual* improvement.

3. Of course, this is a normative proposition. Many believe that policy should sacrifice some wealth in order to achieve other goals; an example is provided by arguments favoring public subsidies for the arts. To say, as I do, that maximization of aggregate wealth is the appropriate goal is to say that the burden of proof rests with those who would use the democratic process, in effect, to spend other people's money.

4. Paul W. MacAvoy, "Quick, Put a Tax on OPEC's Oil," *New York Times*, March 14, 1982.

5. Exchange-rate effects, of course, will differ across nations as foreign spending on oil varies. This is a second-order effect, important mainly for the prices of imported and import-substitute goods. Reduced dependence might reduce "vulnerability" for a nation if it could prevent export of its own oil during periods of rising prices. Export controls are not relevant for the United States because the International Energy Agency sharing arrangement, discussed below, implicitly precludes the imposition of export controls during disruptions.

6. Although this goal may seem self-evident, most political systems have chosen to constrain the range of choices open to individuals. Examples are drug and vice laws, and various other restraints on individual economic behavior. Some commentators argue further that individuals are induced by "artificial" influences (e.g., advertising) to make "inefficient" choices. Why some factors affecting human choices are in some sense less legitimate than others has not yet been explained. See, inter alia, John Kenneth Galbraith, *The New Industrial State* (Boston: Houghton Mifflin, 1979).

7. For a brief summary of this history, see Benjamin Zycher, "Untying the Energy Knot," *Inquiry*, July 1983.

8. The disruption caused by the Iran–Iraq hostilities took place when price controls on crude oil and gasoline were being phased out. No gasoline lines developed because the controls were not binding at the time.

9. See, for example, Joseph P. Kalt, *The Economics and Politics of Oil Price Regulation* (Cambridge, Mass.: MIT Press, 1981).

10. Note that lobbying for a particular policy, like redistribution, is regarded in traditional thinking as a collective good. (I benefit from lobbying for my preferred policies whether I or someone else actually engages in the effort.) Therefore, concentrated interest groups have a natural advantage in terms of getting their message heard by government officials. See Mancur Olson, *The Logic of Collective Action* (Cambridge, Mass.: Harvard University Press, 1965). Although some concentrated groups may favor markets to actual policy choices, they are more likely to lobby for other nonmarket policies favorable to them. Competition among such concentrated interest groups is likely to lead to policies that will satisfy all such groups in part—but not to the outcome that is most efficient socially.

11. This belief cannot be correct, particularly for a sector as competitive as the oil refining industry, because, in general, price controls on inputs (such as crude oil) cannot reduce the market value of derived

outputs (such as gasoline). Indeed, by reducing the refinery production of gasoline, such controls must reduce gasoline deliveries and thus drive up the retail price of gasoline. More generally, price controls can reduce only *reported* prices; *true* prices cannot be reduced and usually will be forced up by the price controls because the controls reduce output. Hence, with gasoline price ceilings the sum of the legal maximum price on the gasoline pump and the value of time spent in the queue *unambiguously is greater* than the price that would clear the market in the absence of the controls at the retail level.

12. The Phase IV EPAA oil regulations can be summarized as follows:

1. Retail sales of gasoline, diesel fuel, and heating oil would be set at their cost to retailers on August 1, 1973, plus the actual markup for each on January 10, 1973. The minimum markups, however, would be seven cents per gallon. Retailers of heating oil would be allowed to adjust ceiling prices automatically at the beginning of each month to reflect increased costs of imported heating oil but not other costs.
2. Other retail refined product prices were to be controlled at cost plus the actual markup for each product on January 10, 1973.
3. Refinery prices for products were limited to the May 15, 1973, price plus increased import costs borne after May 15.
4. A two-tier pricing structure for domestic crude oil was established with the ceiling price on old oil set at the May 15, 1973, level by grade and field plus up to thirty-five cents per barrel in order to maintain historical differentials. Other crude oil was exempt from price controls.
5. Gasoline retailers were to post on the pump the ceiling price and octane rating of the gasoline.

13. The real reason, of course, was that large firms were the most visible, and the costs of enforcing compliance by small firms were certain to be enormous.

14. This argument, of course, confuses a change in costs with a change in wealth.

15. William E. Simon, *A Time For Truth* (New York: McGraw-Hill, 1978).

16. See Note 11.

17. See Note 11.

18. See the *Strategic Petroleum Reserve Drawdown Plan,* Amendment Number 4, U.S. Department of Energy, December 1, 1982. One rather innocuous provision in the plan is in principle inconsistent with market allocation of oil: the plan allows the energy secretary, at his option, to reserve up to 10 percent of the volume of any month's SPR oil sales for specified groups faced with extraordinary circumstances. They would receive the oil at the most recent competitive auction price. The provision is innocuous because no special allocation is needed if oil is sold at the market price. Anyone wishing to buy oil

can do so at the market price under any circumstances; that is the meaning of "market price."

19. Even if the range were limited, oil would be allocated on a roughly efficient basis because it can always be resold. As transaction costs become increasingly important, restrictions on the number of bidders also become more damaging; there is some evidence from the Mandatory Oil Import Quota Program that transaction costs in the oil industry are not trivial.

20. The rate issue is less important because the market can allocate SPR oil over time.

4. Restrictions on Oil Imports?
S. Fred Singer

1. See Paul W. MacAvoy, ed., *Federal Energy Administration Regulation* (Washington D.C.: American Enterprise Institute, 1977); H. A. Merklein and William P. Murchison, Jr., *Those Gasoline Lines and How They Got There* (Dallas: Fisher Institute, 1980); and Paul W. MacAvoy et al, "The Federal Energy Office as Regulator of the Energy Crisis," *Technology Review*, MIT (May 1975) pp. 39–45.

2. S. Fred Singer, "The Many Myths about OPEC," *Wall Street Journal*, February 18, 1977; S. Fred Singer, "Limits to Arab Oil Power," *Foreign Policy* 30 (Spring 1978).

3. Public Information Package for the IEA Allocation System Test Number 4 (AST-4) (Paris: International Energy Agency, April 1983).

4. S. Fred Singer, "Our Strategic (for Others) Oil Reserve," *Wall Street Journal*, Aug. 3, 1983.

5. Kenneth W. Dam, *Oil Resources* (Chicago: University of Chicago Press, 1976).

6. Douglas R. Bohi and Milton Russell, *Limiting Oil Imports: An Economic History & Analysis* (Baltimore: Johns Hopkins University Press, 1978).

7. S. Fred Singer, "The Marginal Cost of Oil" (Charlottesville: University of Virginia, Energy Policy Studies Center, 1977).

8. For a discussion of tariffs in general, see Bohi and Russell, and Douglas R. Bohi and W. David Montgomery, *Tariffs and the Economic Costs of an Oil Disruption,* Discussion Paper D-82B (Washington, D.C.: Resources for the Future, 1982); Douglas R. Bohi and W. David Montgomery, *Oil Prices, Energy Security and Import Policy* (Washington, D.C.: Resources for the Future, 1982). For a discussion of "optimal tariff," see for example Robert Stobaugh and Daniel Yergin, eds., *Energy Future* (New York: Random House, 1979); James Plummer, ed., *Energy Vulnerability* (Cambridge Mass.: Ballinger, 1982); William Hogan, "Import Management and Oil Emergencies," in D. Deese and J. Nye, eds., *Energy and Security* (Cambridge, Mass.: Ballinger, 1981); H. G. Broadman, "The Social Cost of Imported Oil: Theoretical Issues and Empirical Estimates," in F. S. Roberts, ed., *Energy Modeling IV: Planning for Energy Disruptions* (Chicago: Institute for Gas

368 • FREE MARKET ENERGY

Technology, 1982). For a critical discussion of the above, see Elena Folkerts-Landau, "The Social Cost of Imported Oil," *Preprint,* October 1982.
9. S. Fred Singer, *An Evaluation of Options of the U.S. Government in Dealing wih the OPEC Cartel and with Other Incipient Cartels.* Report prepared for Secretary, U.S. Treasury, Order Number ES-318, Washington, D.C., April 2, 1976.
10. A number of conservative economists have advocated the use of a tariff in the face of declining oil prices—for example, William Brown, *Washington Post,* Feb 14, 1982; Paul W. MacAvoy, *New York Times,* March 14, 1982; Myer Rashish, *New York Times,* March 14, 1983; Paul L. Joskow and Robert S. Pindyck, *New York Times,* May 1, 1983; Editorial, *Business Week,* April 11, 1983.

5. Export of Alaskan Oil and Gas, Stephen D. Eule and S. Fred Singer

This chapter was adapted from S. Fred Singer, Milton Copulos, and David J. Watkins, "Exporting Alaska's Oil and Gas," *Backgrounder No. 248,* Heritage Foundation, Feb. 22, 1983.

1. C. J. Cicchetti, *Alaskan Oil: Alternative Routes and Markets* (Baltimore: Johns Hopkins University Press, 1972).
2. As of March 1983, this condition has been met. The price for a barrel of oil, without quality adjustments, was $28.69 for domestic and $28.43 for imported. (When these prices are adjusted for quality using the Consumer Energy Council of America's $0.75 figure, the differential grows from $0.26 to $1.01.) See *Weekly Petroleum Status Report,* "Refinery Acquisition Costs of Crude Oil," Dept. of Energy, June 24, 1983, p. 17.
3. Marshall Hoyler, "Statement on Alaskan Exports," address before the Budget and Audit Committee, Alaskan Legislature, April 23, 1983.
4. During 1982, production of Alaska-California oil fields was 2.8 mbd while refinery demand was only 1.9 mbd. See *The Export of Alaska Crude Oil* (Cambridge, Mass.: Putnam, Hayes & Bartlett, 1983), p. 21.
5. Sohio, ARCO, and Exxon account for 93.8 percent of the 1.507 mbd Prudhoe Bay production with Amerada-Hess, Chevron, Getty Oil, Mobil, and Phillips Petroleum accounting for the remaining 6.2 percent. Or the 93,000 barrels per day produced in the Kuparuk field, ARCO lifts 79.7 percent, with BP Alaska, Sohio, and others producing 20.3 percent. *Export of Alaska Crude Oil,* p. 6.
6. *Export of Alaska Crude Oil,* pp. 22–23.
7. Ibid., p. 11.
8. A more efficient option than the pipeline proposals described would be to allow Alaskan North Slope crude transport on foreign-flag vessels. This would, however, involve repeal of the Jones Act.

9. The International Energy Agency's participating countries are Australia, Austria, Belgium, Canada, Denmark, West Germany, Greece, Ireland, Italy, Japan, Luxembourg, the Netherlands, New Zealand, Norway, Portugal, Spain, Sweden, Switzerland, Turkey, the United Kingdom, and the United States.

10. The 1974 experience has been discussed and documented in some detail by a number of authors. For example, see P. W. MacAvoy, "The Federal Energy Office as Regulator of the Energy Crisis," *Technology Review*, MIT (1975), 39–45; H. A. Merklein and W. P. Murchison, Jr., *Those Gasoline Lines and How They Got There* (Dallas: Fisher Institute, 1980); Chapter 3, this volume.

11. See S. Fred Singer, "The Many Myths about OPEC," *Wall Street Journal*, Feb. 18, 1977. It should be noted that President Reagan ousted Libyan diplomats from Washington, and Libya made no effort to institute retaliatory action in the oil market against the United States; the Libyans also know that embargoes do not work.

12. An exemption exists for Alaskan oil shipped to the U.S. Virgin Islands; this oil may travel in foreign-flag tankers.

13. Opponents of exports claim that Alaskan North Slope producers would, in a free market, no longer discount their West Coast refinery prices, thus leading to higher product costs on the West Coast. For instance, the Consumer Energy Council of America (CECA) report, *The Consumer and Energy Impacts of Oil Exports* (Washington, D.C.: CECA, 1983) "assume[s] that the discount is passed through" to consumers (p. 35), while the same publication asserts that "any reduction in transportation costs will result in increases in wellhead prices, not . . . in decreases in consumer prices" (p. 10). The question becomes, why are refiners thought to be more altruistic than producers? In fact, studies indicate that West Coast product prices will not increase with the lifting of the ban, but rather that the distribution of the economic rents will be altered. See M. Trexler, *Export Restrictions on Domestic Oil: A California Perspective* (Sacramento: Energy Commission, 1982); Robert Nathan and Associates, "The Financial Consequences of Exporting Alaskan North Slope Crude Oil" (Washington, D.C.: Robert Nathan and Associates, Inc., 1981). For a good critique of CECA's study, see Marshall Hoyler, "The Politics and Economics of Alaskan Exports," Center for Strategic and International Studies, Georgetown University, Washington D.C., October 1983, Appendix A.

14. The Consumer Energy Council of America contends in its report, ibid., that Mexico would not be able to expand production enough to meet the demand on the U.S. Gulf Coast, and the United States would thus be forced to purchase oil from other (OPEC) sources (pp. 26–28). But Mexico would not have to *increase* production levels that much. Instead, Mexico could *divert* some present production going elsewhere and so take advantage of increased netbacks due to lower transport costs to the U.S. Gulf Coast.

15. Arlon R. Tussing and Associates, *Alaska's Economy and the*

Merchant Marine Act of 1920 (The Jones Act) (Anchorage: ARTA, 1982).

16. The appraisal of undiscovered probable Alaskan North Slope reserves is on the order of a ten-year supply.

17. Recently, the sponsors of the Point Conception regasification facility have delayed LNG deliveries until 1990, reflecting the difficulty of marketing North Slope natural gas in California. These difficulties arise from: (1) the current surplus of natural gas in California (caused in part by the oil glut there); and (2) the California LNG Terminal Act, which permits LNG to be imported only from Indonesia and South Alaska.

18. For a pessimistic discussion of these options, see U.S. General Accounting Office, *Issues Facing the Future Use of Alaskan North Slope Natural Gas* (Washington, D.C.: U.S. GAO, 1983).

19. For a revealing account of the extremes with which this view manifested itself, see the discussion of the Carter administration's reaction to the optimistic Energy Research and Development Administration's 1977 MOPPS study of natural gas in W. T. Brookes, *The Economy in Mind* (New York: Universe Books, 1982), pp. 139–45.

20. This ANGTS figure includes the proposed pipeline and gas conditioning plant.

21. This is not a total estimate as it does not include costs for pipe insulation, socioeconomic impacts, highway repairs, geotechnical data acquisition, state and ad valorem taxes, and satellite communications system, as does the $40.9 ANGTS figure, nor does it include LNG storage or dock facilities. See U.S. GAO, *Issues*, p. 53.

22. The Governor's Economic Committee, *Trans-Alaska Gas System: Economics of an Alternative for North Slope Gas*, 1983.

23. This location would incur less environmental risk than Point Gravina, which would necessitate cutting through the Chugach Range; it would also shorten the travel time to Japan.

24. See U.S. GAO, *Issues*, p. 99. This estimate of LNG tanker costs appears to be far too high. Numerous LNG tankers are available on the world market today. Some are being sold for their scrap value while others are being converted to carry bulk commodities, such as grain. The current low cost of tankers should thus push this estimate downward.

25. The Governor's Committee suggests that the pipeline project be built in three stages, with the revenue raised after completion of the first stage being employed to finance the next two. The committee envisions that after completion of the first stage in 1988, some 4.8 million tons per year of LNG could be exported. The second stage, scheduled for completion in 1990, would produce 8.9 million tons. In 1992, completion of the third stage would increase yearly production to 14.5 million tons. Potential backers of the project include the North Slope producers, the state of Alaska, Japanese utilities, and the Japanese government. One metric ton (1 tonne) of LNG is equivalent

to 1,357 m³ (48 MCF) of natural gas, or 1.4 m³ (8.8 barrels or 1.21 tonnes) of crude oil. Its heat equivalent is approximately 50 million Btu.

6. Taxation of Energy Producers and Consumers, James W. Wetzler

1. A minor amount of revenue goes into trust funds for land and water conservation and boating safety.
2. Specifically, the tax on fuel consumed in noncommercial aviation, which funds the Airport and Airway Trust Fund, is 12 cents per gallon for gasoline and 14 cents for other fuels, and the tax on fuel used for commercial cargo transportation in inland waterways is 8 cents per gallon, scheduled to rise to 10 cents in October 1985.
3. See, for example, U.S. Congress, Office of Technology Assessment, *Gasohol: A Technical Memorandum*, September 1979, p. 13–17.
4. The revenue loss resulting from a tax credit often understates the amount of direct subsidy that would be needed to achieve the same incentive because many direct subsidies are included in taxable income, while a tax credit directly increases after-tax income. Recently, the Office of Management and Budget (OMB) has published estimates of the "outlay equivalent" of tax expenditures—that is, the direct subsidy that would provide the same incentive as the tax reduction. The outlay equivalent of the $800 million revenue loss from the energy conservation credits is estimated to be $1.2 billion. See *Special Analysis, Budget of the U.S. Government, 1984*, p. G26. The $800 million and $1.2 billion estimates include solar energy incentives, even though these are treated by OMB as production, not conservation, incentives.
5. One version of a passive solar credit came to grief when House and Senate conferees learned that some passive solar designs use indoor swimming pools as their thermal mass.
6. The Carter administration proposed extending the credits to both passive solar design and wood-burning stoves. There was some suspicion that its support for wood stoves was linked to the imminence of the New Hampshire primary in 1980.
7. The half-basis adjustment for a 10 percent credit will reduce depreciation deductions by 5 percent. At a 46 percent tax rate, this amounts to a tax increase of 2.3 percent. However, because the lost deductions would have been spread over five years, the present value of the tax increase is about 2 percent, or one-fifth of the credit.
8. In this context, however, it is important to note that the Synthetic Fuels Corporation, the principal source of direct subsidies for energy production, is an exception, as it is effectively exempt from annual review of appropriations because it had budget authority of $14.9 billion at the start of 1983.
9. It is interesting to note that gasoline consumption in 1982 was well below the target set in 1977 by President Carter.

10. To encourage Congress to act on his proposal, Carter imposed an oil import fee and ingeniously altered oil price control regulations to have the fee passed on to consumers as a 10-cent-per-gallon increase in gasoline prices. However, Congress repealed the import fee, overriding the president's veto by overwhelming majorities, the first override of a veto by a Congress of the president's own political party since 1952.

11. Imposing the tax at the refinery "tailgate" avoids certain distortions that would emerge under a tax imposed at the wellhead. For example, the same petroleum product can be produced in alternative ways—through expensive refining of low-quality oil or through less expensive refining of higher-quality oil. An ad valorem wellhead tax, by imposing a larger burden on the higher-quality oil, would distort this choice in favor of lower-quality oil; that is, the technique that adds more value to the product after the tax is imposed. This distortion would be undesirable unless there were some reason to want to discourage imports of high-quality oil more than those of low-quality oil.

12. See, for example, Stobaugh and Yergin, *Energy Future*, (New York: Random House, 1979) and other references in Chapter 4.

13. To the extent that the government has better information than consumers about how to conserve energy, it can try to communicate that knowledge. Note, however, that one argument for the residential energy tax credits is that they act as a form of advertising, making tax-conscious consumers aware of the energy-saving features of certain products by appealing to the widespread desire to avoid taxes.

14. However, short-lived equipment is written off over 3 years, and much property owned by public utilities is written off over 10 or 15 years.

15. The tax credit is 6 percent for property in the three-year class.

16. For a dollar's worth of five-year equipment by a taxpayer in the 46 percent top corporate tax bracket, for example, the 10 percent investment credit is equivalent to an immediate deduction of 21.7 cents (21.7 × .46 = 10), and the five-year write-off of 95 cents (100 − .5 × .10 = 95) is equivalent, in present value at a 10-percent discount rate, to an immediate write-off of 78.4 cents, for total first-year equivalent deduction of 100.1 cents. At discount rates below 10 percent, ACRS is more generous than expensing.

17. Percentage depletion is subject to a variety of ad hoc limits, which may affect particular independents. On any property, percentage depletion may not exceed 50 percent of net income. For any taxpayer, it may not exceed 65 percent of overall taxable income. Also, the excess of percentage depletion over costs actually incurred is subject to various kinds of minimum taxes.

18. The remaining 85 percent of exploration expenditures, however, is treated less favorably than development expenditures. It, too, is expensed initially, but when the mine begins production, the amounts previously expensed must be included in income. They may subse-

quently be deducted a second time through the depletion deduction if the producers uses cost depletion.

19. Some of this favorable income tax treatment is counterbalanced by a tax on coal production that funds the Black Lung Disability Trust Fund. The tax is the lesser of 4 percent of the price or $1 per ton for underground mines and 50 cents per ton for surface mines.

20. The eligible fuels include (1) oil from shale or tar sands; (2) gas produced from geopressured brine, coal seams, Devonian shale, or tight sands; (3) gas produced from biomass; (4) synthetic fuel produced from coal; (5) certain kinds of processed wood fuels; and (6) steam produced from solid agricultural by-products.

21. In 1980, for example, taxpayers with adjusted gross incomes above $100,000 received 35 percent of dividends and 40 percent of royalties.

7. Coal and the Clean Air Act, Lester B. Lave

1. Michael Grenon, "Global Energy Resources," in J. Hollander, M. Simmons, and D. Wood, *Annual Review of Energy,* 2 (Palo Alto: Annual Reviews, 1977) and Harry Perry and Hans H. Landsberg, "Factors in the Development of a Major U.S. Synthetic Fuels Industry," in J. Hollander, M. Simmons, and D. Wood, ed., *Annual Review of Energy,* 6 (Palo Alto: Annual Reviews, 1981). See also Chapter 12, this volume.

2. Institute of Gas Technology, *Energy Statistics, Second Quarter, 1982,* 5 (Chicago: 1982).

3. Committee on Nuclear and Alternative Systems, *Energy in Transition: 1985-2010* (San Francisco: W. H. Freeman, 1980).

4. Lester B Lave and Lester P. Silverman, "Environmental Pollution and Energy: An Economic Evaluation," *Materials and Society* 3 (1979); G. E. Dials and E. C. Moore, "The Cost of Coal," *Environment* 16 (1974); and National Academy of Sciences, *Coal As an Energy Resource* (Washington, D.C.: National Academy of Sciences, 1977).

5. Lester B. Lave and Linnea C. Freeburg, "Health Effects of Electricity Generation from Coal, Oil, and Nuclear Fuel," *Nuclear Safety* 14 (1973) and Leonard A. Sagan, "Health Costs Associated with the Mining, Transport, and Combustion of Coal in the Steam-Electric Industry," *Nature* 250 (1974).

6. Samuel C. Morris, "Health Risks of Coal Energy Technology," in C. Travis and E. Etnier, eds., *Health Risks of Energy Technologies* (Boulder: Westview Press, 1983).

7. Lester B. Lave and Gilbert S. Omenn, *Clearing the Air* (Washington, D.C.: Brookings Institution, 1981).

8. The calculation assumes emission of 0.4 pounds of sulfur dioxide per million Btu thermal coal (0.2 percent sulfur, 12,000 Btu per pound coal, or approximately 90 percent fuel gas desulfurization for eastern coals).

9. Michael S. Baram, *Alternatives to Regulation* (Lexington, Mass.: Lexington Books, 1982).

10. Allen V. Kneese, Richard U. Ayres, and Ralph d'Arge, *Economics and the Environment: A Materials Balance Approach* (Baltimore: Johns Hopkins University Press, 1970).

11. Howard K. Gruenspecht, "Differentiated Regulation: The Case of Auto Emissions Standards," *American Economic Review* 72 (1982).

12. Although the data are far from completely reliable, they suggest that emissions levels changed little from 1970 through 1978. For total suspended particulate matter, sulfur dioxide, volatile organic compounds (hydrocarbons), and carbon monoxide, the percentage decreases in emissions were 46.1, 9.4, 1.8, and 0.5. For nitrogen dioxide, emissions were estimated to have increased 17 percent. Measured average annual concentrations for 1973 to 1978 declined not at all for ozone, less than 10 percent for total particles, 20 percent for sulfur dioxide, and possibly not at all for fine particles and acid sulfate aerosols. Only carbon monoxide and benzo (a) pyrene showed significant improvements. The difference in apparent progress implied by the conflicting data for community air quality and emissions levels could be due to biases in older measurement methods, relocation of air pollution monitors to cleaner areas, and dispersal of pollution sources to unmeasured areas. For example, sulfate levels in northwestern cities have hardly declined at all, in contrast to the marked decline in sulfur dioxide levels. See Lave and Omenn, *Clearing the Air*.

13. B. Peter Pashigian, "Environmental Regulation: Whose Self Interests Are Being Protected?," working paper published by the Center for the Study of the Economy and the State, University of Chicago, August 1982.

14. Bruce A. Ackerman and William T. Hassler, *Clean Coal, Dirty Air* (New Haven, Yale University Press, 1981).

15. Lewis V. Perl and Frederick C. Dunbar, "Cost Effectiveness and Cost-Benefit Analysis of Air Quality Regulations," *American Economic Review* 72 (1982).

16. Cynthia Grace Praul, William B. Marcus, and Robert B. Weisenmiller, "Delivering Energy Services: New Challenges for Utilities and Regulators," *Annual Review of Energy,* 7 (Palo Alto: Annual Reviews, 1982).

17. A number of analyses have concluded that the benefits of stringent abatement of automobile emissions are less than the costs; see Lester B. Lave and Eugene P. Seskin, *Air Pollution and Human Health* (Baltimore: Johns Hopkins University Press, 1977).

18. U.S. Environmental Protection Agency, *Acid Rain Research Summary,* EPA Report No. 600/8-79-028, Washington, D.C.: 1979.

19. William J. Baumol and Wallace E. Oates, *The Theory of Environmental Policy,* (Englewood Cliffs; N.J.: Prentice-Hall, 1975) and Edwin S. Mills and Lawrence White, "Government Policies Toward Automobile Emission Control," in A. Friedlaender, ed., *Approaches to Controlling Air Pollution,* (Cambridge, Mass.: MIT Press, 1978).

8. Federal Leasing Policy, Walter J. Mead and Gregory G. Pickett

1. Convention of the Continental Shelf, Article 1, U.N. doc. A/C, 13/ L. 35 (1958).
2. For an analysis of this 1978 legislation see R. O. Jones, W. J. Mead, and P. E. Sorensen, "The Outer Continental Shelf Lands Act Amendments of 1978," *Natural Resources Journal* 19 (1979), 885–908.
3. S. L. McDonald, *The Leasing of Federal Lands for Fossil Fuels Production* (Baltimore: Johns Hopkins University Press, 1979), pp. 24–25.
4. W. J. Mead, P. E. Sorensen, A. Moseidjord, and D. D. Muraoka, "Additional Studies of Competition and Performance in OCS Oil and Gas Shelves, 1954–1975" (Final report, U.S.G.S. Contract No. 14–08-0001-18678, 1980), p. 45.
5. W. J. Mead, A. Moseidjord, and P. E. Sorensen, "Toward Efficient National Policies for Leasing Oil and Gas Resource," *The Energy Journal* 4 (1983).
6. S. N. Wilcox, "Joint Venture Bidding and Entry into the Market for Off-Shore Petroleum Leases," Diss. University of California, Santa Barbara, 1975, p. 92.
7. U.S. House of Representatives, 1977, Report of the House Ad Hoc Select Committee on the Outer Continental Shelf, H. R. report, no. 95-590 to accompany H.R. 1614, 95th Congress, first session, p. 64.
8. The only allocative efficiency consequence of annual lease rental payments is that it leads to earlier production than is optimal. The rental payment can be avoided or minimized by prompt exploration and immediate production. Current federal regulations cause any rental payments to cease as soon as royalty payments begin. The optimum time profile of production would occur without a rental and royalty payment requirement and with repeal of the five-year rule. Lessees would then be free to maximize the present value of their leasehold. Thus, they would choose between present and future production such that the value of their leasehold would increase with the opportunity cost of capital. Well-established economic theory indicates that choices made under this rule will also increase the general welfare if there are no significant externalities and if the opportunity cost of capital corresponds with the social time preference rate.
9. That is, the bureaucracy is a substitute for the *sequence* of competitive secondary markets for OCS leases. For a discussion of secondary markets under alternative leasing regimes, see G. G. Pickett, "An Option Valuation Model of Bonus Bidding and Profit-Share Bidding for Offshore Oil and Gas Leases," Diss. University of California, Santa Barbara, 1983.
10. K. W. Dam, *Oil Resources* (Chicago: University of Chicago Press, 1976).
11. Contractor's Agreement, Long Beach Unit, Wilmington Oil Field, California, Article 4.

12. Federal Register 45: 106 (1980), 36790.
13. Ibid., 36794.
14. Ibid., 36792.
15. D. K. Reece, *Leasing Offshore Oil: An Analysis of Alternative Information and Bidding Systems* (New York: Garland Publishing, 1979). For similar models and assumptions, see Department of Natural Resources, State of Alaska, Division of Minerals and Energy Management, "A Report to the State of Alaska, J. S. Hammond, Governor 1979; G. Martin, Department of Natural Resources, "A Study of State Petroleum Leasing Methods and Possible Alternatives" 1977; R. J. Kalter, W. E. Tyner, D. W. Hughes, "Alternative Energy Leasing Strategies and Schedules for the Outer Continental Shelf," National Science Foundation Grant No. SIA74-21846, 1975.
16. Pickett provides a discussion of the impact of alternative leasing regimes on the lessee's exploration decisions in the context of an option valuation model. See G. G. Pickett, Reference 9.
17. *Pure* royalty and *pure* profit-share leasing confer a *no-cost* option on the high bidder. For this reason, pure royalty and profit-share bidding must be supplemented by covenants such as work commitment and exploration commitment if they are to result in a *competitive equilibrium*. For a discussion of the necessary elements of equilibrium leasing contracts, see G. G. Pickett, Reference 9.
18. Mead et al., "Additional Studies," p. 155.

9. Nuclear Power Economics and Prospects, Bernard L. Cohen

1. Tennessee Valley Authority, *Summary of Capital Cost per kw for U.S. Nuclear Plants,* Knoxville, Tenn., 1982.
2. United Engineers and Constructors, *Final Report and Initial Update of the Energy Economic Data Base Program,* 1979, see also Arthur G. Bernstein, United Engineers, private communications, 1982.
3. A. Reynolds, *Cost of Coal vs. Nuclear in Electric Power Generation,* U.S. Energy Information Administration, Washington, D.C., 1982.
4. W. W. Brandfon, *The Economics of Nuclear Power* (Cincinnati: American Ceramic Society, 1982).
5. I. Spiewak and D. F. Cope, "Overview Paper on Nuclear Power," Oak Ridge National Laboratory Report ORNL/TM-7425.
6.W. B. Behnke, *Economic and Technical Experience of Nuclear Power Production in the U.S.,* International Conference on Nuclear Power Experience, Vienna, 1982.
7. *Regulatory Reform Case Studies* (Philadelphia: United Engineers and Constructors, August 1982).
8. R. M. Eshbach, *Is Criteria as Written Enough?* (Reading, Pa.: Gilbert Associates, 1981).
9. Consumer's Power Company Position Papers (Jackson, Mich.,

1981); R. Wheeler, *The Midland Chronicles* (Jackson, Mich.: Consumer's Power Company, 1981). Long Island Lighting Co., *The Shoreham Nuclear Power Plant: An Overview* (Long Island, N.Y.: 1982). M. R. Copulos, *Confrontation at Seabrook* (Washington, D.C.: Heritage Foundation, 1978).
10. R. Burtch (Niagara Mohawk), private communication.
11. M. L. Russel, C. W. Solbrig, and G. D. McPherson, "LOFT Contribution to Nuclear Power Reactor Safety and PWR Fuel Behavior", *Proceedings of the American Power Conference* 41 (1979), 196.
12. *Reactor Safety Study* (Rasmussen Report), Nuclear Regulatory Commission Document WASH-1400, NUREG 74/014 (1975).
13. Union of Concerned Scientists, "The Risks of Nuclear Power Reactors," (Cambridge, Mass.: 1977).
14. H. W. Lewis, Risk Assessment Review Group Report to the U.S. Nuclear Regulatory Commission, NUREG/CR-400 (1978).
15. U.S. Senate Committee on Public Works, "Air Quality and Stationary Source Emission Control," 1975. See also R. Wilson, S. D. Colome, J. D. Spengler, and D. G. Wilson, "Health Effects of Fossil Fuel Durning" (Cambridge, Mass.: Ballinger, 1980).
16. Harvard University Energy and Environment Policy Center, *Analysis of Health Effects Resulting from Population Exposures to Ambient Particulate Matter*, 1982.
17. National Academy of Sciences Committee on Biological Effects of Ionizing Radiation, "The Effects on Populations of Exposure to Low Levels of Ionizing Radiation" (Washington, D.C.: BEIR, 1980).
18. National Academy of Sciences Committee on Biological Effects of Ionizing Radiation, "The Effects on Populations of Exposure to Low Levels of Ionizing Radiation" (Washington, D.C.: BEIR, 1972).
19. United Nations Scientific Committee on Effects of Atomic Radiation (UNSCEAR), "Sources and Effects of Ionizing Radiation" (New York: United Nations, 1972).
20. United Nations Scientific Committee on Effects of Atomic Radiation (UNSCEAR), "Sources and Effects of Ionizing Radiation" (New York: United Nations, 1982).
21. International Commission on Radiological Protection, "Recommendations of the International Commission on Radiological Protection," ICRP Publication No. 26 (Oxford: Pergamon Press, 1977).
22. F. J. Rahn and M. Levenson, "Radioactivity Releases Following Class-9 Reactor Accidents," Health Phys. Soc., Las Vegas, June 1982; C. D. Wilkinson, NSAC Workshop on Reactor Accident Iodine Release, Palo Alto, July 1980; H. A. Morewitz, "Fission Product and Aerosol Behavior Following Degraded Core Accidents," *Nuclear Technology* (1981).
23. NRC (Draft), "Evaluation of Pressurized Thermal Shock," September 13, 1982.
24. "Steam Generator Tube Experience," U.S. Nuclear Regulatory Commission Document NUREG-0886, 1982.
25. Opinion Research Corporation, "Public Attitudes toward Nuclear

Power vs. Other Energy Sources," *ORC Public Opinion Index* 38: 17 (Princeton, N.J.: September 1980).

26. National Academy of Sciences Committee on Nuclear and Alternative Energy Systems, "Energy in Transition, 1985–2010" (San Francisco: W. H. Freeman, 1980); American Medical Association Council on Scientific Affairs, "Health Evaluation of Energy Generating Sources," *Journal of American Medical Association* 240 (1978), 2193; Nuclear Energy Policy Study Group, "Nuclear Power—Issues and Choices" (Cambridge, Mass.: Ballinger, 1977); Union of Concerned Scientists, "Risks"; United Kingdom Health and Safety Executive, "Comparative Risks of Electricity Production Systems" (1980); Norwegian Ministry of Oil and Energy, "Nuclear Power and Safety" (1978); Science Advisory Office, State of Maryland, "Coal and Nuclear Power" (1980); Legislative Office of Science Advisor, State of Michigan, "Coal and Nuclear Power" (1980). At least fifteen other references available on request.

27. T. F. Mancuso, A. Stuart, and G. Kneale, *Health Physics* 33: 369 (1977). I. D. J. Bross, M. Ball, and S. Falen, *Annual Journal of Public Health* 69: 130 (1979). J. W. Gofman, *Radiation and Human Health,* (San Francisco: Sierra Club Press, 1981).

28. A large number of these are cited in B. L. Cohen, "The Cancer Risk of Low Level Radiation," *Health Physics* 39: 659 (1980).

29. J. B. Kemeny, "Report of the President's Commission on the Accident at Three Mile Island" (Washington, D.C.: Oct. 1979); M. Rogovin, "Three Mile Island, A Report to the Commissioners and to the Public" (Washington, D.C.: Jan. 1980).

30. OECD Nuclear Energy Agency, Tenth Annual Activity Report (1982).

31. "The World List of Nuclear Power Plants," *Nuclear News,* February 1982, p. 83.

32. *Nuclear Engineering International,* August 1982 Supplement.

33. *Journal Official des debats de l'Assemblée National,* May 10, 1982.

34. Projected Costs of Electricity from Nuclear and Coal-Fired Power Plants, U.S. Energy Information Administration Document DOE.EIA-3056/1, August 1982.

35. OECD Nuclear Energy Agency, Tenth Annual Report (1982).

36. Atomic Industrial Forum, *Nuclear Power: Facts and Figures,* July 1982.

37. *Electricité de France/Production Thermique-Annuel Report* (1982).

38. International Union of Producers and Distributors of Electrical Energy Conference in Brussels, June 1982. Reported in *Nuclear News,* July 1982, p. 48.

39. R. H. Drake, M. L. Stein, C. A. Mangeng, and G. R. Thayer, "Quantitative Analysis of Nuclear Power Plant Licensing Reform," Los Alamos National Laboratory Report LA-9519-MS (1982).

40. B. L. Cohen, "Nuclear Journalism," *Policy Review* September 1983, p. 70.

41. United Nations Scientific Committee on the Effects of Atomic Radiation, "Sources and Effects of Ionizing Radiation," 1977.

42. A. M. Kellerer and H. H. Rossi, *Current Topics on Radiation Research* 8: 85 (1972).

43. D. C. Lloyd et al., "The Relationship between Chromosome Aberrations and Low LET Radiation Dose to Human Lymphocytes," *International Journal of Radiation Biology* 28: 75 (1975).

44. M. Terzaghi and J. B. Little, "Radiation Induced Transformation in a C3H Mouse Embryo-Derived Cell Line," *Cancer Research* 36: 1367 (1976).

45. R. L. Ullrich et al., "The Influence of Dose and Dose Rate on the Incidence of Neoplastic Disease in R. F. Mice after Neutron Irradiation," *Radiation Research* 68: 115 (1976).

46. R. E. Rowland, A. F. Stehney, and H. F. Lucas, *Radiation Research* 76: 368 (1978).

47. B. L. Cohen, "Failures and Critique of the BEIR-III Lung Cancer Risk Estimates," *Health Physics* 42: 267 (1982).

48. Most of these are listed in B. L. Cohen, "The Cancer Risk of Low Level Radiation," *Health Physics* 39: 659 (1980). See also U.S. General Accounting Office, *Problems in Assessing the Cancer Risks of Low Radiation Exposure*, Report EMD-81 (Washington, D.C.: U.S. GAO, 1981).

49. B. L. Cohen, "Analysis, Critique, and Reevaluation of High Level Waste Water Intrusion Scenario Studies," *Nuclear Technology* 48: 63 (1980).

50. B. L. Cohen, "Effects of ICRP-30 and BEIR-III on Hazard Estimates for High Level Radioactive Waste," *Health Physics* 42: 133 (1982).

51. B. L. Cohen, "Long Term Consequences of the Linear-No Threshold Dose Response Relationship for Chemical Carcinogens," *Risk Analysis* 1: 267 (1981)

52. B. L. Cohen, "The Role of Radon in Comparisons of Environmental Effects of Nuclear Energy, Coal Burning, and Phosphate Mining," *Health Physics* 40: 19 (1981).

10. International Nuclear Policy, Petr Beckmann

1. "World List of Nuclear Power Plants," *Nuclear News* 26: 2 (February 1982), 24, 71–90.

2. Bertrand Goldschmidt, *The Atomic Complex* (La Grange Park, Ill.: American Nuclear Society, 1982).

3. M. Willrich and T. B. Taylor, *Nuclear Theft: Risks and Safeguards,* (Cambridge, Mass.: Ballinger, 1974).

4. C. Starr and E. Zebroski, "Nuclear Power and Weapons Proliferation, American Power Conference, April 1977 (Electric Power Research Inst., Palo Alto). The table given here has been simplified.

5. B. Wolfe, "Nuclear Energy and Non-Proliferation," *Foreign Affairs* (Fall 1980), pp. 1–5.

6. B. L. Cohen, "Hazards from Plutonium Toxicity," *Health Physics,* January 1976, pp. 1–21.

7. W. Meyer, S. K. Loyalka, et al., *The Homemade Nuclear Bomb Syndrome,* (Hinsdale, Ill.: American Nuclear Society Public Information Committee, 1976).

8. Some of these methods could, in theory, also be used for separating plutonium-239 from the unwanted isotopes, but this would first involve chemical separation from the fission products and uranium in the spent fuel rods—a highly complicated and dangerous process. The uranium route is simpler, quicker, less hazardous, and more easily concealed.

9. *State Department Summary Sheet on Tarapur,* June 1980.

10. F. Hoyle and G. Hoyle, *Commonsense in Nuclear Energy,* (San Francisco: W. H. Freeman, 1980).

11. *The Windscale Inquiry. Report by the Hon. Mr. Justice Parker.* (London: Her Majesty's Stationery Office, 1978).

12. H. Gruemm, "Safeguards and Tamuz: Setting the Record Straight," *IAEA Bulletin,* December 1981, pp. 10–14.

13. D. O. Graham, *High Frontier—A New National Strategy,* (Washington, D.C.: Heritage Foundation, 1982).

14. France is cleverly taking advantage of public opposition to reprocessing in other countries, accepting their fuel for reprocessing in its new large plant at La Hague under long-term contract. By the time the contracts run out, German, Japanese, and other rate-payers will have paid for the French plant.

11. Electric Utility Issues, Rene H. Males

1. J. C. Bonbright, *Principles of Public Utility Rates* (New York: Columbia University Press, 1970); A. E. Kahn, *The Economics of Regulation: Principles and Institutions* (New York: Wiley, 1970).

2. E. Vennard, *Management of the Electric Energy Business* (New York: McGraw-Hill, 1979).

3. Electric Power Research Institute, *EPRI Journal,* Palo Alto (March 1979).

4. J. M. Gould, *Output and Productivity in the Electric and Gas Utilities: 1899–1942* (New York: National Bureau of Economic Research, 1946), p. 20.

5. Ibid., p.22.

6. Ibid., p.22.

7. Eminent domain allows a utility, after public service commission review, to condemn property for public use.

8. See Kahn.

9. Federal Power Act, 41 Stat. 1077 (1920).

10. Rural Electrification Act of 1936, 49 Stat. 1363 (1936).

11. Tennessee Valley Authority Act of 1933, 48 Stat. 58 (1933).

12. Bonneville Power Act, 50 Stat. 731 (1937).

13. Federal Power Commission, *The 1970 National Power Survey, Part 1* (Washington D.C.: U.S. Government Printing Office, December 1971).

14. Ibid.

15. R. H. Males and R. G. Uhler, "Load Management: Issues, Objectives, and Options," *Electric Utility Rate Design Study*, Electric Power Research Institute, February 1982, p. 63.

16. Applied Decision Analysis, Inc., "An Analysis of Power Plant Construction Lead Times, Vol. II: Analysis and Results," Electric Power Research Institute Report EA-2880, February 1983.

17. Strategic Decisions Group, "Generating Capacity in U.S. Electric Utilities," Electric Power Research Institute Report EA-2639-SR, October 1982.

18. Public Utility Regulatory Policies Act of 1978 (PURPA), 92 Stat. 3117 (1978).

19. Power Plant and Industrial Fuel Use Act of 1978, 92 Stat. 3289 (1978).

20. PURPA.

21. For example, most of the conservation programs in California were put in place under the strong direction of the California Public Utility Commission and the Energy Resources Conservation and Development Commission.

22. Allowance for Funds Used during Construction (AFUDC) is a noncash component that contributes to net income. AFUDC represents the carrying charges associated with financing a construction program.

23. C. Starr, "The Electric Power Research Institute," *Science* 219, (March 11, 1983), 1190–94.

24. Department of Energy Organization Act, 91 Stat. 565 (1977).

25. The noninvestor-owned segment of the industry benefits equally if not more than the investor-owned industry from this turn of events.

26. D. H. Geraghty and J. Lyneis, "A Dynamic Computer Simulation Model for Electric Utility Investment Decision-Making in a Changing External Environment," paper delivered before the Summer Computer Simulation Conference, Vancouver, British Columbia, July 11–13, 1982.

27. D. M. Geraghty and J. Lyneis, "Regulatory Policy and the Performance of Electric Utilities: A System Dynamics Analysis," paper delivered before the Department of Energy Workshop on Regulatory/Financial Models in the U.S. Electric Utility Industry, Los Alamos National Laboratory, Los Alamos, New Mexico, May 1982.

28. For example, the Coolwater demonstration facility converts coal to electricity by gasification with combined cycle power generation. This development facility is jointly funded by utilities, EPRI, and equipment suppliers.

29. "National Marketing—A New Strategy for the Second Century," Edison Electric Institute, (Washington, D.C.: 1983).

30. R. H. Males, "Tomorrow's Electrification," *Public Utilities Fortnightly* V 111: 5 (March 3, 1983), 21–25.

31. Ibid.
32. New York and Nebraska both have state agencies that own generating facilities. In addition, a number of municipals and cooperatives have created generation and transmission entities.
33. It took American Electric Power over ten years to complete its merger with Columbus & Southern Utilities Company.
34. National Association of Regulatory Utility Commissioners, "Official Report," (Washington, D.C.: NARUC Ad Hoc Committee on Utility Diversification, October 1982); "Electric Utility Diversification: A Guide to the Strategic Issues and Options," (Washington, D.C.: Edison Electric Institute, forthcoming).
35. Ibid.
36. "Deregulation of Electric Utilities: A Survey of Major Concepts and Issues," (Washington, D.C.: Edison Electric Institute, June 1981).
37. W. W. Berry, "The Case for Competition in the Electric Utility Industry," *Public Utilities Fortnightly* 110: 6 (September 16, 1982), 13–20.
38. A. J. Dowd and J. R. Burton, "Deregulation Is Not an Answer For Electric Utilities," *Public Utilities Fortnightly* 110: 6 (September 16, 1982), 21–27.
39. J. D. Hughes and M. B. Rosenzweig, "Sweden: A Model for a Segmented Electric Utility System," *Public Utilities Fortnightly* 109: 13 (June 24, 1982) 13–17.
40. Perhaps the best discussion is in P. L. Joskow and R. Schmalensee, *Deregulation of Electric Power: A Framework for Analysis* (Springfield, Va.: National Technical Information Service, September 1982). This work was sponsored by DOE's policy study on utility problems. Kahn also devoted several dozen pages to this issue when writing over a decade ago.
41. D. H. Geraghty and J. Lyneis, "Utility Investment Strategies: Reconciling the Objectives of Different Stakeholders," February 1983, unpublished.
42. Such conclusions also depend on other assumptions such as the allowed rate of return, the return earned, the rate of inflation, and economies of scale.

12. Advanced Energy Technology, Richard S. Greeley

1. Energy R & D Administration (ERDA), "Creating Energy Choices, ERDA-48," (Springfield, Va.: National Technical Information Service 1974). See also Project Independence Blueprint #4118-00629 (Springfield, Va.: U.S. GPO, 1974).
2. See Energy Information Administration (EIA) Monthly Review (February 1973) for ten-year *conventional* energy production and consumption statistics including coal, oil, natural gas, hydroelectric power, nuclear power, and "other" (which includes dry geothermal steam and electricity from wood and waste), U.S. Department of

Energy, National Energy Information Center, EI-20, Washington, D.C. 20585.

3. Energy Security Act of 1980, P.L. 96-294 and the Crude Oil Windfall Profits Tax Act of 1980, P.L. 96-223.

4. For a discussion of "reserves," "potential resources," and "speculative resources"; and for more detailed estimates see Resources for the Future, *Energy: The Next Twenty Years*, (Cambridge, Mass., Ballinger, 1979), pp. 225–47.

5. E.C. Mangold, and M.A. Muradaz, *State-of-the-Art of Selected Coal Liquefaction and Gasification Technologies*, (McLean, Va.: Mitre Corp., 1980), Vols. I and II. Also R.C. Corey et al., *Elements of a Federal Surface Coal Gasification Program*, (McLean, Va.: Mitre Corp., 1980).

6. *Energy Users Report* 10: 49 (December 9, 1982), 1247, Bureau of National Affairs, Washington, D.C.

7. *Energy Users Report* 11: 1 (January 6, 1983), 12.

8. One such project, the Great Plains Coal Gasification project, a $2 billion plant to convert North Dakota coal to pipeline-quality gas, is being supported by a loan guarantee from the Department of Energy (initiated before the SFC became operational in 1981). The GPCG project is scheduled to start producing gas in 1987, but already losses of up to $750 million are projected for the first ten years of plant operation due to falling oil prices to which the product price is tied, *New York Times*, April 11, 1983, p. D4.

9. See Chapter 6, this volume.

10. A successful atmospheric fluidized-bed boiler is currently in operation at Georgetown University in Washington, D.C., providing steam heat throughout the campus. Another unit has been installed by Iowa Beef Processors, the first in a line of retrofit units designed by Wormser Engineering Co. to replace fuel oil in existing boilers.

11. See, for example, Pittsburgh Energy Technology Center, Fourth International Symposium on Coal Slurry Combustion, May 10–12, 1982, Orlando, Florida. Proceedings, Vols. I–IV.

12. See Chapter 9 and Chapter 10, this volume.

13. *Energy Users Report* 11: 11 (March 17, 1983), 324.

14. U.S. Department of Energy, Los Alamos Laboratories, *A Thorium-Uranium Cycle Inertial Confinement Fusion Hybrid Concept*, conference proceedings of The Technology of Controlled Nuclear Fusion, May 9-11, 1978, Santa Fe, N.M.

15. R.S. Greeley et al., *Solar Heating and Cooling of Buildings*, (Ann Arbor, Mich.: Ann Arbor Science Publishers, 1981), p. 184.

16. See Chapter 6, this volume.

17. *Energy Users Report* 10: 27 (July 8, 1982), 719.

18. *Energy Users Report*, 10: 39 (September 30, 1982), 987.

19. R.E. Gerstein, *Geothermal Progress Monitor*, (GPM) Report #7. McLean Va.: Mitre Corp., 1983), p. 89.

20. For an up-to-date report on all current geothermal program activities see Gerstein.

21. J. Leigh, et al., *Site-Specific Analysis of Geothermal Development* (McLean, Va.: Mitre Corp., 1978).
22. Gerstein, p. 63.
23. S. Fred Singer, "Summary of ERDA/MITRE Workshops on Energy Conservation R & D in Industry" (McLean, Va.: Mitre Corp., 1975).
24. For a full discussion of the results of these acts, see Chapter 2, this volume.
25. Energy Reorganization Act of 1974 and the Non-Nuclear Energy Research and Development Act of 1974, P.L. 93-577; Solar Energy Research, Development and Demonstration Act of 1974, P.L. 93-473; Solar Heating and Cooling Demonstration Act of 1974, P.L. 93-409; and the Geothermal Energy Research, Development and Demonstration Act of 1974, P.L. 93-410.
26. Department of Energy Organization Act, P.L. 95-91.
27. P.L. 95-91 Title I, Sec. 102 (6).
28. *Energy Outlook, 1980-2000* (New York: Exxon Corporation, 1979).
29. A typical oil shale project from the third call, scheduled for funding in the fall of 1983, would require production of at least 10,000 barrels of product per day before 1990. The lowest bidder out of six projects submitted for the western shale resource in the Green River Formation in Colorado, Wyoming, and Utah would receive up to $1.6 billion in loan guarantees, price guarantees, or a combination. The average product bid price must be less than $67 per barrel in 1983 dollars, and any single product price must range between $30 and $95 per barrel (taken from a statement by Edward E. Noble, chairman of the Synthetic Fuels Corporation, March 16, 1983, regarding six ventures for western oil shale, *Energy Users Report* 11: 11 (March 17, 1983), 324.
30. Proceedings of the Energy Technology Conference and Exposition, reported in *Energy Users Report* 11: 11 (March 17, 1983).
31. See Chapter 10, this volume, for a full discussion of these points.
32. *Energy Users Report* 11: 11 (March 17, 1983), 324. See also *Report to the Congress on Alternative Financing for the CRBR Plant Project,* U.S. Department of Energy, Washington, D.C.
33. *Energy Users Report* 11: 5 (February 3, 1983), 115.
34. *Commercializing Inertial Confinement Fusion,* report on workshop supported by the Laser Fusion Branch, U.S. Department of Energy. (McLean, Va.: Mitre Corp., 1978).
35. See, for example, Greeley, Appendix C, pp. 325-349, for a list of 187 manufacturers of solar heating equipment.
36. B. Cone, et al., "Long Term Solar Parity Considerations Based on an Analysis of Incentives to Energy Production" (Richland, Wash.: Battelle Pacific Northwest Laboratory, 1978).
37. G. Bennington, et al., *Towards a National Plan for the Accelerated Commercialization of Solar Energy.* (McLean, Va.: Mitre Corp., 1980).

38. Alliance to Save Energy, *Third Party Financing: Increasing Investment in Energy Efficient Industrial Projects,* Washington, D.C.
39. Office of Management and Budget, *Budget of the United States, 1984* (Washington, D.C.: Superintendent of Documents, 1983).
40. *Chemical and Engineering News* 57: 35 (August 27, 1979), 22.
41. R. S. Greeley, *Western Hemisphere Energy Resources* (McLean, Va.: Mitre Corp., 1980).

FOR FURTHER READING

Adelman, M. A. *The World Petroleum Market.* Baltimore: Johns Hopkins University Press, 1972.

Arrow, Kenneth J., and Joseph P. Kalt. *Petroleum Price Regulation: Should We Decontrol?* Washington, D.C.: American Enterprise Institute, 1979.

Bohi, Douglas R., and Milton Russell. *Limiting Oil Imports.* Baltimore: John Hopkins University Press, 1978.

Brannon, G. M. *Studies in Energy Tax Policy,* Cambridge, Mass.: Ballinger, 1974.

Cabinet Task Force on Oil Import Control. *The Oil Import Question: A Report on the Relationship of Oil Imports to the National Security.* Washington, D.C. U.S. Government Printing Office, February 1970.

Committee on Nuclear and Alternative Energy Systems, National Research Council. *Energy in Transition: 1985–2010.* San Francisco: W. H. Freeman, 1980.

Dam, Kenneth W. *Oil Resources.* Chicago: University of Chicago Press, 1976.

Dasgupta, P. S., and G. M. Heal. *Economic Theory and Exhaustible Resources.* Cambridge: Cambridge University Press, 1979.

Eden, R. J., et al. *Energy Economics: Growth, Resources and Policies.* Cambridge: Cambridge University Press, 1981.

Feith, D. J. "The Oil Weapon De-mystified," *Policy Review* 15 (Winter 1981).

International Energy Agency. *World Energy Outlook.* Paris: Organization for Economic Cooperation and Development, 1982.

Johnson, William A. "The Impact of Price Controls on the Oil Industry: How to Worsen an Energy Crisis," in Gary D. Eppen, ed., *Energy: The Policy Issues.* Chicago: University of Chicago Press, 1975, pp. 99–121.

Kalt, Joseph P. *The Economics and Politics of Oil Price Regulation.* Cambridge, Mass.: MIT Press, 1981.

Kelly, J. B. *Arabia, the Gulf and the West.* New York: Basic Books, 1980.

Landsberg, H. H., et al. *Energy: The Next Twenty Years*. Cambridge, Mass.: Ballinger, 1979.

MacAvoy, Paul W., ed. *Federal Energy Administration Regulation: Report of the Presidential Task Force*. Washington, D.C.: American Enterprise Institute, 1977.

—— et al. "The Federal Energy Office as Regulator of the Energy Crisis," *Technology Review*, MIT 77, 39–45.

Mancke, Richard. *Performance of the Federal Energy Office*. Washington, D.C.: American Enterprise Institute, 1975.

Merklein, H. A., and William P. Murchison. *Those Gasoline Lines and How They Got There*. Dallas: Fisher Institute, 1980.

Mitchell, Edward J. *U.S. Energy Policy: A Primer*. Washington, D.C.: American Enterprise Institute, 1974.

Niskanen, William A., Jr. *Bureaucracy and Representative Government*. Chicago: Aldine-Atherton, 1971.

Odell, Peter. *Oil and World Power*. New York: Penguin, 1979.

Pejovich, Svetozar, ed. *Government Controls and the Free Market*. College Station: Texas A&M University Press, 1976.

Rustow, Dankwart A. *Oil and Turmoil*. New York: Norton, 1982.

Sampson, Anthony. *The Seven Sisters, the Great Oil Companies, and the World They Shaped*. New York: Viking, 1975.

Singer, S. Fred. "Limits to Arab Oil Power," *Foreign Policy* 30 (Spring 1978).

——. "Saudi Arabia's Oil Crisis." *Policy Review* 21 (Summer 1982), 87–100.

Stobaugh, R. B. "The Oil Companies in the Crisis," in Raymond Vernon, ed., *The Oil Crisis*. New York: Norton, 1976.

—— and Daniel Yergin, eds. *Energy Future*. Report of the Energy Project at the Harvard Business School. New York: Random House, 1979.

U.S. Department of Energy. *Energy Technologies and the Environment*. Washington, D.C.: U.S. Department of Energy, 1981.

U.S. Senate, Committee on Foreign Relations. *Multinational Oil Corporations and U.S. Foreign Policy*. Washington, D.C.: U.S. Government Printing Office, January 1975.

ABOUT UNITS

In order to discuss energy issues in a sensible way, it is necessary to use numbers to express energy quantities. But you will find a great many different units, such as British thermal units (Btu), kilowatt-hours, barrels of oil, etc.—all used to measure energy. Our purpose here will be to explain the profusion of units and how to convert one to the other.

An Aside on Exponential Notation

Exponential notation has the advantages of convenience and speed. Since $10 \times 10 = 100$, we can write 100 as 10^2. (The little 2 on top means that 10 is multiplied by itself 2 times.) Similarly, $10^3 = 1,000$, $10^6 = 1,000,000$, and so on.* Exponential notation can be used for anything, of course, not just for energy. For example, our budgetary deficits are measured in billions of dollars, or 10^9 dollars. Our GNP is measured in trillions, or 10^{12} dollars.

A thousand trillion equals 1 quadrillion. While we don't use quadrillions yet for measuring budgetary deficits, we do use them in the energy business. A quadrillion is 10^{15}, in other words, 1 followed by 15 zeros. A quadrillion Btu (10^{15} Btu) is referred to as 1 quad.

So far, so good. One advantage of exponential notation is speed, but it has other nice features. Consider, for example, that 1,000 million equal 1 billion, (or $10^3 \times 10^6 = 10^9$). Now you will note that $9 = 3 + 6$. In other words, to multiply two powers of 10, we simply add up the exponents. Another example is: $100 \times 1,000 = 100,000$, or $10^2 \times 10^3 = 10^5$.

The addition of exponents is obvious, since 10^5 simply means 10 multiplied 5 times by itself and that clearly is $(10 \times 10) \times (10 \times 10 \times 10)$, or 10^{2+3}.

Exponential notation also works the other way: $1,000 = 100,000/100$. This can be written as $10^3 = 10^5/10^2$ or $10^3 = 10^{5-2}$. From this we can figure out that 10^{-2} means $1/10^2$ or $1/100$. Obviously then, 10^{-6} means $1/10^6$ or 1 millionth, and so on.

---

* Read these as "ten to the third (power)" and "ten to the sixth," and note that they express a 1 with 3 zeros, and a 1 with 6 zeros, respectively.

Energy Versus Power

To the general public, terms like "energy" and "power" sort of mean the same thing. To the physicist and engineer their meanings are quite different. Energy and power are as different as distance and speed, which are measured in different units, like miles and miles per hour. Power is the *rate* at which energy is used. This means that the unit for power is an energy unit divided by time—but sometimes we use specially named units for power such as "watt." Of course we cannot have a unit like "watts per second" since watts already measure a rate ("rate" means "something per unit time").* I can give other examples: energy is sometimes measured in Btu's, but power would be in Btu per second (or per hour, or per day, or per year). On the other hand, to turn the power unit into an energy unit you *multiply* it by time; so if kilowatts (1 kw = 1,000 watt) measures power, a kilowatt-hour (kwh) measures energy—the energy expended if the power level is 1 kw over one whole hour (or 0.5 kw over 2 hours).

Types of Energy

There are many different types of energy, but for our purpose the most important are heat energy, mechanical energy, and chemical energy. We will learn at least some of the units for each of these types of energy. Keep in mind, however, that energy units can be used to measure any kind of energy; but it has become a matter of custom to use units like Btu's and calories for heat energy and other units (like joules) for mechanical energy.

One other matter that complicates life is the use of British units and metric units—often in the same equation. This means one has to remember twice as many units if one lives in the United States.

Let's start with heat energy. The British unit is the Btu (British thermal unit), while the metric unit is the calorie. (We will use the kilocalorie, written kcal, which is 1,000 gram-calories). Heat can be measured then in two different ways, depending on whether we use British or metric units. Obviously the two units must be related and convertible into each other. Let's find the conversion factor from the fundamental definition.

---

* By analogy, a "knot" is a unit of speed (1 nautical mile per hour), not of distance. The unit "knots per hour" makes no sense for measuring speed.

Definitions

One Btu is the energy required to heat one pound of water one degree Fahrenheit (°F).

One kcal is the energy required to heat one kilogram of water one degree centigrade (°C).

To convert British units into metric units, and vice versa, we need first some of the simple conversion factors (with all numbers approximate):

1 meter = 3.3 ft
1 kg (kilogram) = 2.2 lb. (pound)
1 degree C = 1.8 degree F (for measuring temperature differences)

As you can see, the kilocalorie must contain more energy than the Btu. After all, it has to heat a quantity of water that is 2.2 times greater over a temperature interval that is 1.8 times greater. Now 2.2 × 1.8 = 4. Therefore, 1 kcal = 4 Btu.

Mechanical energy is measured in quite a different way. The standard metric unit is one joule. A "joule per second" is, of course, the familiar "watt." The watt is the power unit not only for mechanical energy, but also for electrical energy where it corresponds to voltage times ampere. A light bulb on a 110-volt circuit, which draws a current of 0.5 amp, uses 110 × 0.5 = 55 watts of electric power (converting it into light and heat).

Since 1 watt = 1 joule/sec, 1 joule = 1 watt-sec. Therefore 1 kilowatt-hour is 1 kw × 3600 sec, or 3,600,000 joules (written as 3.6×10^6 joules).

We need one more conversion factor, between heat energy and mechanical energy. It is useful to remember that 1 Btu = 1,055 joules (or approximately 1 Btu = 1 kilojoule). From this relation a useful conversion can be derived for electrical problems, namely that 1 kilowatt hour = 3,412 Btu:

$$10^3 \text{ w-hr} \times (\frac{3600 \text{ j}}{1 \text{ w-hr}}) \times \frac{(1 \text{ Btu})}{(1055 \text{ j})} = 3412 \text{ Btu}$$

Just a reminder: 1 kw = 1,000 watt = 10^3 watt.

1 mw (megawatt) = 10^6 watt; 1 Gw (gigawatt) = 10^9 watt.

We don't bother here with the British units, except the popular "horsepower": 1 hp = 746 watt.

Let's take a quick look at *chemical* energy and particularly at the energy contained in the three most important fuels: gas, oil, and coal.

Gas

Natural gas is metered in units of cubic feet (CF) at standard temperature and pressure. The heat content of a cubic foot of gas is 1,000 Btu. The heat content of 1,000 cubic feet (often written 1 MCF) is one million Btu, or 10^6 Btu. That is a convenient number to remember. Another convenient number to remember is that a trillion cubic feet, 10^{12} CF, (often written 1 TCF) contains 10^{15} Btu, or 1 quad. The United States uses about 18 TCF of gas per year; in other words, 18 quad out of a total of 70 quad is in the form of gas.

Oil

The heat content of oil varies slightly depending on its composition. The standard barrel of oil (containing 42 U.S. gallons) has a heat content of 5.8×10^6 Btu (or 5.8 mbtu).

Now barrels and gallons are both volume units: the weight of the oil depends on its specific gravity. But on the average, 7.3 barrels weigh 1 metric ton* (or 10^3 kg). We can calculate the heat content of a tonne of oil as $(5.8 \times 10^6 \times 7.3) = (40 \times 10^6)$, or 4×10^7 Btu. But this is exactly 10^7 kcal. Therefore 1 metric ton of oil has a heat content of 10 million kcal, a convenient number to remember.

Another convenient number to remember, derived by simple calculation, is that oil used at the rate of 1 million barrels per day (mbd) is used at a rate of 50 million tonnes per year (mty). The calculation is quite direct:

$$10^6 \text{ b/d} \times \left(\frac{1 \text{ tonne}}{7.3 \text{ b}}\right) \times \left(\frac{365 \text{ d}}{1 \text{ yr}}\right) = 50 \text{ mt/yr}$$

Another useful conversion is that 1 mbd corresponds to an energy consumption of 2.1 quad per year. Again, the calculation is very simple:

$$10^6 \text{ b/d} \times (5.8 \times 10^6 \text{ Btu/b}) \times (365 \text{ d/yr}) = 2.1 \times 10^{15} \text{ Btu/yr}$$

The present U.S. oil consumption of 15 mbd corresponds, therefore, to ~ 30 quad per year, about 44 percent of total energy consumption.

---

* *Metric* ton is often written as "tonne"; i.e., 1 tonne = 1,000 kg or 2,200 lbs, while 1 short ton = 2,000 lbs.

Coal

Coal has a heat content that depends very much on its quality. Lignite coal, for example, has only half the heat content of high-rank bituminous coal. A good average value for the latter is 25 million Btu (mbtu) per short ton, or 12,500 Btu per pound. The annual U.S. consumption of coal of 600 million tons therefore corresponds to an energy use of 15 quad per year, roughly 20 percent of the total energy consumption of the United States.

ENERGY STATISTICS

Table ES-1

Approximate Conversion Factors

For Crude Oil*

FROM \ INTO	Tonnes	Long Tons	Short Tons	Barrels	Kilolitres (cub.metres)	1 000 Gallons (Imp.)	1 000 Gallons (U.S.)
				MULTIPLY BY			
Tonnes (metric tons)	1	0.984	1.102	7.33	1.16	0.256	0.308
Long Tons	1.016	1	1.120	7.45	1.18	0.261	0.313
Short Tons	0.907	0.893	1	6.65	1.05	0.233	0.279
Barrels	0.136	0.134	0.150	1	0.159	0.035	0.042
Kilolitres (cub. metres)	0.863	0.849	0.951	6.29	1	0.220	0.264
1 000 Gallons (Imp.)	3.91	3.83	4.29	28.6	4.55	1	1.201
1 000 Gallons (U.S.)	3.25	3.19	3.58	23.8	3.79	0.833	1

For Crude Oil and Products

TO CONVERT	Barrels to Tonnes	Tonnes to Barrels	Barrels/Day to Tonnes/Year	Tonnes/Year to Barrels/Day
	MULTIPLY BY			
Crude Oil*	0.136	7.33	49.8	0.0201
Motor Spirit	0.118	8.45	43.2	0.0232
Kerosine	0.128	7.80	46.8	0.0214
Gas/Diesel	0.133	7.50	48.7	0.0205
Fuel Oil	0.149	6.70	54.5	0.0184

*based on world average gravity (excluding natural gas liquids).

Approximate Calorific Equivalents (Oil = 10 000 kcal/kg)

One million tonnes of oil equals approximately:	Heat units and other fuels expressed in terms of million tonnes of oil.	million tonnes of oil
Heat Units		
40 million million Btu	10 million million Btu approximates to	0.25
397 million therms	100 million therms approximates to	0.25
10 000 teracalories	10 000 teracalories approximates to	1.00
Solid Fuels†		
1.5 million tonnes of coal	1 million tonnes of coal approximates to	0.67
3.0 million tonnes of lignite	1 million tonnes of lignite approximates to	0.33
Natural Gas (1 cub. ft. = 1 000 Btu) (1 cub. metre = 9 000 kcal)		
1.111 thousand million cub. metres (BCM)	1 thousand million cub. metres (BCM) approximates to	0.90
39.2 thousand million cub. ft.	10 thousand million cub. ft. approximates to	0.26
0.0392 million million cub. ft. (TCF)	10 million million cub. ft. (TCF) approximates to	260
107 million cub. ft./day for a year	100 million cub. ft./day for a year approximates to	0.93
Town Gas (1 cub. ft. = 470 Btu) (1 cub. metre = 4 200 kcal)		
2.4 thousand million cub. metres	1 thousand million cut. metres (BCM) approximates to	0.42
84.1 thousand million cub. ft.	10 thousand million cub. ft. approximates to	0.12
230 million cub. ft./day for a year	100 million cub. ft./day for a year approximates to	0.43
Electricity (1 kWh = 3 412 Btu) (1 kWh = 860 kcal)		
12 thousand million kWh	10 thousand million kWh approximates to	0.86

One million tonnes of oil produces about 4 000 million units (kWh) of electricity in a modern power station.

†Solid fuels: the equivalents now stated represent the relative calorific values of coal and lignite as produced.

Source: British Petroleum Co. *BP Statistical Review of World Energy 1982.* London: BP, 1983, p. 33.

Table ES-2

World Primary Energy Consumption

Country/Area	1972	1973	1974	1975	1976	1977	1978	1979	1980	1981	1982	Yearly Change 1982 over 1972	1982 over 1977
	Million Tonnes Oil Equivalent												
U.S.A.	1 767.8	1 822.7	1 769.1	1 722.1	1 804.3	1 846.8	1 896.6	1 918.6	1 851.4	1 806.3	1 728.4	-0.2%	-1.3%
Canada	179.7	190.9	197.6	195.0	206.7	213.0	217.9	221.8	226.1	219.8	207.4	+1.5%	-0.5%
Total North America	1 947.5	2 013.6	1 966.7	1 917.1	2 011.0	2 059.8	2 114.5	2 140.4	2 077.5	2 026.1	1 935.8	-0.1%	-1.2%
Latin America	226.1	244.6	257.8	263.7	273.9	291.5	306.2	325.7	341.1	347.8	362.2	+4.8%	+4.4%
Total Western Hemisphere	2 173.6	2 258.2	2 224.5	2 180.8	2 284.9	2 351.3	2 420.7	2 466.1	2 418.6	2 373.9	2 298.0	+0.6%	-0.5%
Western Europe													
Austria	22.1	23.8	23.5	23.4	23.7	24.0	25.0	26.4	26.5	25.2	24.8	+1.2%	+0.6%
Belgium & Luxembourg	51.3	50.8	49.9	46.7	49.8	50.2	51.4	51.1	50.4	48.0	44.8	-1.3%	-2.2%
Denmark	20.7	19.7	18.1	18.2	19.2	19.9	19.5	20.1	19.5	18.2	18.1	-1.3%	-1.9%
Finland	16.3	17.9	17.2	17.4	18.4	19.4	20.0	21.6	21.6	21.6	21.4	+2.8%	+1.9%
France	171.4	186.1	184.1	171.1	183.2	186.6	191.8	192.9	189.3	186.5	180.6	+0.5%	-0.7%
Greece	13.4	15.3	14.9	17.0	18.7	19.4	19.8	17.4	17.4	16.7	16.8	+2.3%	-2.9%
Iceland	1.8	2.1	1.8	2.0	2.1	1.8	1.9	1.9	1.9	1.8	1.3	-3.0%	-5.9%
Republic of Ireland	7.1	7.4	7.4	7.1	7.2	7.7	8.0	8.7	8.5	8.3	8.0	+1.3%	+1.0%
Italy	132.5	137.9	136.9	132.9	142.7	142.7	145.4	149.7	145.8	144.1	141.4	+0.7%	-0.2%
Netherlands	72.3	76.9	71.2	71.3	76.3	75.2	74.1	78.5	73.7	70.7	63.4	-1.3%	-3.4%
Norway	26.9	28.1	28.3	28.8	30.7	28.1	29.2	31.3	29.1	28.6	31.8	+1.7%	+2.6%
Portugal	8.3	9.0	9.0	8.8	8.8	10.1	10.5	11.3	11.4	11.5	12.0	+3.9%	+3.5%
Spain	52.6	57.8	61.3	61.9	67.4	69.7	71.1	75.4	76.5	76.9	75.0	+3.6%	+1.5%

Sweden	44.0	46.2	43.0	45.1	48.7	47.6	42.2	44.1	41.1	40.0	37.6	−1.5%	−4.6%
Switzerland	21.5	23.9	22.5	23.5	22.6	25.3	24.9	24.5	26.5	26.4	25.7	+1.7%	+0.3%
Turkey	15.4	18.0	17.6	19.7	22.7	24.4	23.1	23.4	24.5	25.8	27.9	+6.1%	+2.7%
United Kingdom	215.6	224.8	214.5	204.3	207.4	211.9	211.4	221.9	203.8	196.1	193.4	−1.1%	−1.8%
West Germany	247.7	264.9	256.0	242.6	259.6	259.7	270.0	287.0	270.5	258.5	250.4	+0.1%	−0.7%
Yugoslavia	26.7	28.9	30.8	32.0	32.6	34.4	35.8	38.9	39.9	40.0	41.5	+4.5%	+3.7%
Cyprus/Gibraltar/Malta	1.3	1.4	1.3	1.1	1.3	1.4	1.5	1.5	1.5	1.4	1.4	+1.0%	+0.1%
Total Western Europe	1 168.9	1 240.9	1 209.3	1 174.9	1 243.1	1 259.5	1 276.6	1 327.6	1 279.4	1 246.3	1 217.3	+0.4%	−0.7%
Middle East	78.7	87.1	96.3	94.6	102.5	108.7	112.6	107.4	117.1	121.0	125.7	+4.8%	+3.0%
Africa	96.0	98.2	101.7	106.2	117.8	124.7	139.2	153.0	164.7	176.7	181.3	+6.6%	+7.8%
South Asia	99.2	103.3	104.0	105.7	114.5	118.9	125.3	130.3	142.9	152.4	163.0	+5.1%	+6.5%
South East Asia	114.3	127.4	133.3	136.8	145.3	155.3	169.4	190.3	195.3	199.9	200.2	+5.8%	+5.2%
Japan	310.8	347.7	346.5	330.5	344.7	348.1	354.5	369.9	359.6	353.5	340.2	+0.9%	−0.5%
Australasia	63.9	66.6	69.9	71.5	76.0	80.3	82.1	86.1	85.9	88.5	89.5	+3.4%	+2.2%
U.S.S.R.	836.5	874.1	924.1	969.8	1 016.7	1 060.6	1 049	1 134.0	1 169.0	1 200.6	1 242.0	+4.0%	+3.2%
Eastern Europe	353.9	357.2	366.8	387.7	408.5	423.6	434.5	443.2	453.2	443.2	455.4	+2.6%	+1.5%
China	334.9	362.5	388.1	410.4	437.4	459.6	487.9	536.4	518.0	506.3	522.1	+4.5%	+2.6%
Total Eastern Hemisphere	3 457.1	3 665.0	3 740.0	3 788.1	4 006.5	4 139.3	4 287.0	4 478.2	4 485.1	4 488.4	4 536.7	+2.8%	+1.9%
World (Excl. U.S.S.R., E. Europe & China)	4 105.4	4 329.4	4 285.5	4 201.0	4 428.8	4 546.8	4 680.4	4 830.1	4 763.5	4 712.2	4 615.2	+1.2%	+0.3%
World	5 630.7	5 923.2	5 964.5	5 968.9	6 291.4	6 490.6	6 707.7	6 944.3	6 903.7	6 862.3	6 834.7	+2.0%	+1.0%

Table ES-3
World Crude Oil and Natural Gas Production and Proven Reserves

Country/area	Crude oil production (million barrels per day)[a]																Natural gas production 1981 (tcf)[b]	Proven reserves[c] Crude oil (bb)	Proven reserves Natural gas (tcf)
	1967	1968	1969	1970	1971	1972	1973	1974	1975	1976	1977	1978	1979	1980	1981	1982			
North America																			
Canada	1.1	1.2	1.3	1.5	1.6	1.8	2.1	2.0	1.7	1.6	1.6	1.6	1.8	1.7	1.6	1.5	2.6	7	90
United States	10.2	10.6	10.8	11.3	11.2	11.2	11.0	10.5	10.0	9.7	9.8	10.3	10.1	10.2	10.2	10.2	20.1	29	202
Mexico	0.41	0.44	0.46	0.49	0.49	0.51	0.55	0.64	0.81	0.93	1.1	1.3	1.6	2.2	2.6	3.0	1.5	57	75
Total	11.7	12.2	12.6	13.3	13.3	13.5	13.6	13.1	12.6	12.3	12.5	13.2	13.5	14.1	14.4	14.7	24.2	94	367
South America																			
Ecuador	—	—	—	—	—	0.10	0.21	0.20	0.20	0.20	0.20	0.21	0.26	0.21	0.22	0.21	0.1	1	4
Venezuela	3.6	3.6	3.6	3.8	3.6	3.3	3.5	3.1	2.4	2.4	2.3	2.2	2.4	2.2	2.2	1.9	1.0	20	47
Total	15.9	16.5	16.8	17.6	17.4	17.5	17.8	16.8	15.4	15.0	15.1	15.5	15.8	15.6	15.5	15.2	3.1	25	101
Western Europe																			
Norway	—	—	—	—	—	—	—	—	0.20	0.30	0.30	0.40	0.40	0.53	0.51	0.49	1.0	8	49
United Kingdom	—	—	—	—	—	—	—	—	—	0.24	0.80	1.1	1.6	1.7	1.8	2.1	1.5	15	26
Total	0.45	0.50	0.50	0.50	0.43	0.43	0.44	0.45	0.62	0.91	1.4	1.8	2.4	2.6	2.7	3.0	7.0	25	151
Middle East																			
Iran	2.6	2.8	3.4	3.8	4.6	5.1	5.9	6.1	5.4	5.9	5.7	5.3	3.2	1.5	1.3	2.0	0.1	57	484
Iraq	1.2	1.5	1.5	1.6	1.7	1.5	2.0	2.0	2.3	2.3	2.3	2.6	3.5	2.6	0.90	0.98	0.1	30	27
Kuwait	2.3	2.4	2.6	2.7	2.9	3.0	2.8	2.3	1.8	2.0	1.8	1.9	2.3	1.4	0.97	0.70	0.34	68	35
Neutral Zone	0.42	0.41	0.42	0.51	0.55	0.57	0.54	0.54	0.5	0.47	0.36	0.47	0.57	0.54	0.37	0.31	na	7	8
Oman	0.06	0.24	0.33	0.33	0.29	0.28	0.30	0.29	0.34	0.37	0.34	0.32	0.30	0.29	0.32	0.34	0.01	3	3
Qatar	0.33	0.34	0.36	0.37	0.43	0.49	0.57	0.52	0.44	0.49	0.44	0.49	0.51	0.47	0.41	0.34	0.21	3	60

Saudi Arabia	2.6	2.8	3.0	3.6	4.5	5.8	7.3	8.4	6.9	8.5	9.2	8.1	9.6	9.9	9.9	6.7	0.44	168	118
UAE	0.38	0.50	0.61	0.78	1.06	1.21	1.53	1.65	1.66	1.91	1.91	1.91	1.82	1.70	1.50	1.2	0.23	32	23
Total	10.0	11.2	12.4	13.8	16.2	18.0	21.0	21.9	19.7	22.2	22.4	21.1	21.9	18.8	16.0	12.8	1.52	368	762
Africa																			
Algeria	0.86	0.92	0.96	1.0	0.78	1.1	1.1	1.0	0.98	1.1	1.2	1.2	1.2	1.1	1.0	0.84	1.1	8	131
Egypt	0.13	0.22	0.34	0.47	0.42	0.35	0.26	0.23	0.30	0.33	0.42	0.48	0.53	0.59	0.69	0.71	0.03	3	3
Gabon	0.10	0.10	0.10	0.11	0.12	0.13	0.16	0.20	0.21	0.22	0.23	0.21	0.21	0.18	0.15	0.14	0.01	0.5	0.5
Libya	1.7	2.6	3.1	3.3	2.8	2.2	2.2	1.5	1.5	1.9	2.1	2.0	2.1	1.8	1.1	1.2	0.67	23	23
Nigeria	0.32	0.15	0.54	1.1	1.5	1.8	2.1	2.3	1.8	2.1	2.1	1.9	2.3	2.1	1.4	1.3	0.66	17	41
Total	3.2	4.1	5.2	6.2	5.8	5.8	6.0	5.6	5.0	5.9	6.3	6.1	6.7	6.1	4.9	4.6	2.5	56	212
Asian area																			
Indonesia	0.51	0.60	0.75	0.86	0.89	1.1	1.3	1.4	1.3	1.5	1.7	1.6	1.6	1.6	1.6	1.3	1.1	10	27
Total	0.62	0.73	0.89	1.0	1.1	1.3	1.7	1.7	1.6	1.9	2.1	2.1	2.1	2.1	2.0	1.9	2.0	39	152
World total (excluding USSR, E. Europe, and China)	30.68	33.55	36.4	39.9	41.9	44.1	48.2	47.6	43.8	47.3	49.1	43.9	51.4	47.9	44.3	41.3	38.0	605	1745
USSR, E. Europe, and China	6.32	6.75	7.20	7.90	8.40	8.90	9.60	10.6	11.40	12.3	13.10	14.2	14.4	14.8	14.8	14.9	19.0	66d	1170
World	37.0	40.3	43.6	47.8	50.3	53.0	57.8	58.2	55.2	59.6	62.2	63.1	65.8	62.7	59.1	56.2	57.0	671	2915

a Published in British Petroleum Co., BP Statistical Review of World Oil Industry 1977. London: BP, 1982.
b Published in World Energy Industry 2:4 (1982).
c Published in U.S. Dept. of Energy, Energy Information Administration. International Energy Annual, Washington, D.C.: DOE, 1982; bb = billion barrels, tcf = trillion cubic feet.
d Includes Albania, Bulgaria, Cuba, Czechoslovakia, East Germany, Hungary, Mongolia, North Korea, Poland, Romania, and Yugoslavia.

Table ES-4a

World Oil Consumption

Country/Area	1972	1973	1974	1975	1976	1977	1978	1979	1980	1981	1982	Yearly Change 1982 over 1972	1982 over 1977
						Thousand Barrels Daily							
U.S.A.*	15 990	16 870	16 150	15 875	16 980	17 925	18 255	17 910	16 460	15 675	14 905	− 1.0%	− 4.1%
Canada	1 655	1 755	1 785	1 735	1 790	1 810	1 835	1 915	1 855	1 760	1 580	− 0.8%	− 3.1%
Total North America	**17 645**	**18 625**	**17 935**	**17 610**	**18 770**	**19 735**	**20 090**	**19 825**	**18 315**	**17 435**	**16 485**	**− 1.0%**	**− 4.0%**
Latin America	3 205	3 515	3 660	3 695	3 755	4 010	4 175	4 440	4 615	4 725	4 900	+ 4.3%	+ 4.0%
Total Western Hemisphere	**20 850**	**22 140**	**21 595**	**21 305**	**22 525**	**23 745**	**24 265**	**24 265**	**22 930**	**22 160**	**21 385**	**∅**	**− 2.4%**
Western Europe													
Austria	215	235	210	215	230	225	240	250	245	225	215	− 0.4%	− 1.1%
Belgium & Luxembourg	620	635	560	535	560	565	585	560	540	500	465	− 3.0%	− 4.0%
Denmark	385	360	320	315	335	330	325	320	275	260	240	− 5.0%	− 6.8%
Finland	235	260	230	235	255	250	250	265	255	245	230	− 0.5%	− 2.0%
France	2 315	2 585	2 460	2 255	2 430	2 350	2 445	2 440	2 265	2 060	1 935	− 2.1%	− 4.2%
Greece	170	200	185	195	210	215	230	245	245	240	245	+ 3.3%	+ 2.0%
Iceland	10	15	15	10	10	10	10	10	10	10	10	− 0.6%	− 2.5%
Republic of Ireland	100	110	110	105	105	115	120	130	120	110	95	− 0.8%	− 4.1%
Italy	1 965	2 070	2 015	1 895	2 065	1 920	2 015	2 080	1 975	1 940	1 845	− 0.8%	− 1.1%
Netherlands	805	835	725	710	795	770	780	860	795	735	640	− 2.7%	− 4.2%
Norway	170	175	160	165	180	180	175	185	160	155	160	− 0.8%	− 2.4%
Portugal	120	130	135	140	145	150	150	165	170	180	185	+ 4.5%	+ 5.0%
Spain	655	790	820	865	970	930	955	1 015	1 070	1 040	970	+ 3.6%	+ 0.4%

Sweden	565	585	540	535	590	560	535	565	505	445	400	− 3.6%	− 6.9%
Switzerland	280	300	270	260	270	270	280	270	265	250	235	− 1.9%	− 3.0%
Turkey	200	250	255	275	310	340	315	300	300	315	335	+ 5.1%	− 0.2%
United Kingdom	2 230	2 300	2 135	1 875	1 860	1 885	1 930	1 950	1 670	1 555	1 580	− 3.7%	− 3.8%
West Germany	2 885	3 070	2 760	2 655	2 855	2 855	2 960	3 050	2 725	2 465	2 360	− 2.2%	− 3.9%
Yugoslavia	205	225	235	245	265	280	295	315	300	285	280	+ 3.3%	+ 0.3%
Cyprus/Gibraltar/Malta	25	25	25	20	25	25	25	25	30	30	30	+ 1.0%	+ 0.1%
Total Western Europe	14 155	15 155	14 165	13 505	14 465	14 225	14 620	15 000	13 920	13 045	12 455	− 1.5%	− 2.9%
Middle East	1 115	1 210	1 320	1 320	1 475	1 565	1 620	1 495	1 625	1 685	1 720	+ 4.3%	+ 1.9%
Africa	920	1 010	1 035	1 050	1 155	1 205	1 290	1 360	1 485	1 565	1 635	+ 5.9%	+ 5.9%
South Asia	585	635	600	610	665	705	760	775	825	890	925	+ 4.5%	+ 5.5%
South East Asia	1 415	1 540	1 585	1 630	1 765	1 915	2 130	2 400	2 440	2 480	2 435	+ 5.5%	+ 4.8%
Japan	4 735	5 460	5 270	5 020	5 190	5 350	5 420	5 485	4 935	4 690	4 380	− 1.2%	− 4.5%
Australasia	660	725	750	735	770	800	795	805	770	760	755	+ 1.1%	− 1.3%
U.S.S.R.	6 115	6 595	7 280	7 520	7 780	8 125	8 480	8 640	8 795	8 985	9 075	+ 4.0%	+ 2.3%
Eastern Europe	1 360	1 510	1 575	1 675	1 840	1 985	2 040	2 085	2 110	2 110	2 085	+ 4.2%	+ 1.0%
China	855	1 065	1 225	1 350	1 530	1 630	1 705	1 835	1 765	1 705	1 660	+ 6.7%	+ 0.1%
Total Eastern Hemisphere	31 915	34 905	34 805	34 415	36 635	37 505	38 860	39 880	38 670	37 915	37 125	+ 1.3%	− 0.4%
World (excl. U.S.S.R., E. Europe & China)	44 435	47 875	46 320	45 175	48 010	49 510	50 900	51 585	48 930	47 275	45 690	Ø	− 1.9%
World	52 765	57 045	56 400	55 720	59 160	61 250	63 125	64 145	61 600	60 075	58 510	+ 0.8%	− 1.1%

* U.S. processing gain has been deducted from local domestic product demand.

Ø less than + or − 0.05%.

Source: British Petroleum Co. BP Statistical Review of World Energy 1982. London: BP, 1983, p. 21.

Table ES-4b

World Oil Consumption

Country/Area	1972	1973	1974	1975	1976	1977	1978	1979	1980	1981	1982	Yearly Change 1982 over 1972	Yearly Change 1982 over 1977
					Million Tonnes Oil								
U.S.A.	775.8	818.0	782.6	765.9	822.4	865.9	888.8	868.0	794.1	746.0	703.0	– 1.0%	– 4.1%
Canada	79.3	83.7	84.8	83.1	85.9	85.6	86.9	90.1	87.6	81.7	73.0	– 0.8%	– 3.1%
Total North America	**855.1**	**901.7**	**867.4**	**849.0**	**908.3**	**951.5**	**975.7**	**958.1**	**881.7**	**827.7**	**776.0**	**– 1.0%**	**– 4.0%**
Latin America	154.5	168.3	174.7	176.0	181.6	193.4	201.4	214.2	222.8	227.2	235.4	+ 4.3%	+ 4.0%
Total Western Hemisphere	**1 009.6**	**1 070.0**	**1 042.1**	**1 025.0**	**1 089.9**	**1 144.9**	**1 177.1**	**1 172.3**	**1 104.5**	**1 054.9**	**1 011.4**	**Ø**	**– 2.4%**
Western Europe Austria	10.9	11.8	10.6	10.7	11.6	11.1	12.0	12.5	12.2	11.0	10.5	– 0.4%	– 1.1%
Belgium & Luxembourg	31.1	31.5	28.1	26.5	28.0	28.0	29.0	27.6	26.6	24.5	22.8	– 3.0%	– 4.0%
Denmark	19.4	17.9	16.0	15.7	16.7	16.6	16.1	15.9	13.6	12.8	11.7	– 5.0%	– 6.8%
Finland	11.9	13.3	11.6	11.9	12.8	12.5	12.5	13.3	12.8	12.3	11.3	– 0.5%	– 2.0%
France	114.1	127.3	121.0	110.4	119.5	114.6	119.0	118.3	109.9	99.0	92.4	– 2.1%	– 4.2%
Greece	8.6	10.0	9.4	9.9	10.6	10.8	11.7	12.4	12.4	11.9	11.9	+ 3.3%	+ 2.0%
Iceland	0.6	0.7	0.6	0.6	0.6	0.6	0.6	0.6	0.6	0.5	0.5	– 0.6%	– 2.5%
Republic of Ireland	5.0	5.4	5.4	5.2	5.3	5.7	6.0	6.3	5.9	5.2	4.6	– 0.8%	– 4.1%
Italy	98.2	103.6	100.8	94.5	98.8	96.1	99.8	103.2	97.9	95.7	90.7	– 0.8%	– 1.1%
Netherlands	40.1	41.3	35.4	34.8	39.2	37.6	37.5	41.3	38.5	35.7	30.4	– 2.7%	– 4.2%
Norway	8.5	8.6	7.7	8.0	9.0	8.9	8.7	9.0	7.9	7.5	7.8	– 0.8%	– 2.4%
Portugal	5.9	6.3	6.6	6.8	7.1	7.1	7.4	8.1	8.5	8.8	9.1	+ 4.5%	+ 5.0%
Spain	32.5	39.1	41.1	42.7	48.3	45.5	46.4	49.1	52.2	50.4	46.5	+ 3.6%	+ 0.4%

Sweden	28.6	29.4	27.1	26.6	29.6	28.1	26.7	28.4	25.0	21.8	19.7	− 3.6%	− 6.9%
Switzerland	13.6	14.7	13.0	12.5	13.0	13.1	13.4	12.9	12.8	11.9	11.2	− 1.9%	− 3.0%
Turkey	10.0	12.4	12.5	13.4	15.4	16.6	15.3	14.7	14.8	15.4	16.5	+ 5.1%	− 0.2%
United Kingdom	110.5	113.2	105.3	92.0	91.4	92.0	94.0	94.5	80.8	74.7	75.6	− 3.7%	− 3.8%
West Germany	140.9	149.7	134.3	128.9	138.9	137.1	142.7	147.0	131.1	117.6	112.4	− 2.2%	− 3.9%
Yugoslavia	10.1	11.3	11.5	12.2	13.2	13.9	14.8	15.8	15.1	14.4	14.1	+ 3.3%	+ 0.3%
Cyprus/Gibraltar/Malta	1.3	1.4	1.3	1.1	1.3	1.4	1.5	1.5	1.5	1.4	1.4	+ 1.0%	+ 0.1%
Total Western Europe	701.8	748.9	699.3	664.4	710.3	697.3	715.1	732.4	680.1	632.5	601.1	− 1.5%	− 2.9%
Middle East	55.9	62.2	67.1	66.8	74.7	78.9	81.5	75.4	82.0	84.7	86.7	+ 4.3%	+ 1.9%
Africa	44.7	49.5	50.4	51.5	56.9	59.3	62.8	65.9	71.9	75.9	79.0	+ 5.9%	+ 5.9%
South Asia	28.9	31.3	29.6	30.1	32.6	34.5	37.1	37.7	40.4	43.4	45.1	+ 4.5%	+ 5.5%
South East Asia	71.2	77.6	79.2	81.2	88.7	95.8	105.9	119.2	121.9	123.3	121.0	+ 5.5%	+ 4.8%
Japan	234.4	269.1	258.9	244.0	253.5	260.4	262.7	265.1	237.7	223.9	207.0	− 1.2%	− 4.5%
Australasia	31.7	34.8	35.8	35.1	36.5	38.0	37.6	38.1	36.4	35.8	35.5	+ 1.1%	− 1.3%
U.S.S.R.	302.9	325.7	358.5	375.1	384.9	399.6	419.2	427.0	436.0	444.1	448.5	+ 4.0%	+ 2.3%
Eastern Europe	67.2	75.1	77.5	83.3	89.7	96.2	98.9	101.1	102.6	102.4	101.1	+ 4.2%	+ 1.0%
China	43.1	53.3	61.9	68.3	76.9	82.0	84.7	91.1	88.0	84.8	82.4	+ 6.7%	+ 0.1%
Total Eastern Hemisphere	1 582.8	1 728.0	1 718.2	1 699.8	1 804.7	1 842.0	1 905.5	1 953.0	1 897.0	1 850.8	1 807.4	+ 1.3%	− 0.4%
World (excl. U.S.S.R., E. Europe & China)	2 179.2	2 343.4	2 262.4	2 198.1	2 343.1	2 409.1	2 479.8	2 506.1	2 374.9	2 274.4	2 186.8	Ø	− 1.9%
World	2 592.4	2 798.0	2 760.3	2 724.8	2 894.6	2 986.9	3 082.6	3 125.3	3 001.5	2 905.7	2 818.8	+ 0.8%	− 1.1%

Ø less than + or − 0.05%.

Source: British Petroleum Co. *BP Statistical Review of World Energy 1982*. London: BP, 1983, p. 20.

Table ES-5

World Natural Gas Consumption

Country/Area	1972	1973	1974	1975	1976	1977	1978	1979	1980	1981	1982	Yearly Change 1982 over 1972	Yearly Change 1982 over 1977
					Million Tonnes Oil Equivalent								
U.S.A.	587.4	572.3	555.1	508.7	510.0	502.3	504.2	520.8	516.5	501.8	463.0	− 2.4%	− 1.6%
Canada	39.3	41.8	42.2	43.1	46.1	45.9	47.3	50.1	49.3	47.8	43.0	+ 0.9%	− 1.3%
Total North America	**626.7**	**614.1**	**597.3**	**551.8**	**556.1**	**548.2**	**551.5**	**570.9**	**565.8**	**549.6**	**506.0**	**− 2.1%**	**− 1.6%**
Latin America	36.4	36.5	37.8	39.2	40.5	39.6	42.3	49.1	53.0	54.4	60.0	+ 5.1%	+ 8.6%
Total Western Hemisphere	**663.1**	**650.6**	**635.1**	**591.0**	**596.6**	**587.8**	**593.8**	**620.0**	**618.8**	**604.0**	**566.0**	**− 1.6%**	**− 0.8%**
Western Europe													
Austria	3.1	3.4	3.7	3.6	4.1	4.2	4.4	4.3	4.2	3.9	3.8	+ 2.1%	− 2.1%
Belgium & Luxembourg	6.7	8.2	9.8	9.6	10.3	10.1	9.9	10.3	10.3	9.5	7.2	+ 0.8%	− 6.5%
Denmark	—	—	—	—	—	—	—	—	—	—	0.5	*	*
Finland	—	—	0.4	0.7	0.8	0.7	0.8	0.8	0.8	0.6	0.6	*	− 3.0%
France	13.2	15.7	17.2	17.0	19.0	20.4	20.9	23.3	23.6	24.5	23.0	+ 5.7%	+ 2.4%
Greece	—	—	—	—	—	—	—	—	—	—	—	—	—
Iceland	—	—	—	—	—	—	—	—	—	—	—	—	—
Republic of Ireland	—	—	—	—	—	—	—	0.3	0.5	0.9	1.0	*	*
Italy	12.3	14.4	15.8	18.0	22.0	21.6	22.5	22.9	22.9	22.8	22.9	+ 6.4%	+ 1.2%
Netherlands	29.0	32.2	32.1	33.2	33.0	33.4	32.6	33.1	30.3	30.0	27.3	− 0.6%	− 4.0%
Norway	—	—	—	—	—	—	—	—	—	—	—	—	—
Portugal	—	—	—	—	—	—	—	—	—	—	—	—	—
Spain	1.0	1.0	1.3	1.3	1.5	1.4	1.5	1.4	1.8	2.1	3.2	+11.9%	+17.4%

Sweden	—	—	—	—	—	—	—	—	—	—	—	—	—
Switzerland	0.1	0.2	0.3	0.5	0.5	0.6	0.7	0.8	0.8	0.8	0.8	+20.0%	+ 4.6%
Turkey	—	—	—	—	—	—	—	—	—	—	—	—	—
United Kingdom	25.2	26.1	31.8	32.9	34.6	36.9	37.9	41.9	41.4*	42.4	41.6	+ 5.1%	+ 2.4%
West Germany	21.5	27.0	32.5	34.4	36.3	38.9	41.7	46.2	44.4	41.4	38.6	+ 6.0%	− 0.2%
Yugoslavia	1.4	1.7	2.3	2.2	1.5	1.7	1.7	2.3	3.4	3.7	3.8	+10.6%	+17.5%
Cyprus/Gibraltar/Malta	—	—	—	—	—	—	—	—	—	—	—	—	—
Total Western Europe	**113.5**	**129.9**	**147.2**	**153.4**	**163.6**	**169.9**	**174.6**	**187.6**	**184.4**	**182.6**	**174.3**	**+ 4.4%**	**+ 0.5%**
Middle East	21.0	24.1	27.6	26.2	26.9	28.3	30.1	31.0	34.1	35.3	38.0	+ 6.1%	+ 5.7%
Africa	2.4	3.2	3.6	4.3	5.1	6.9	11.1	16.5	17.5	19.9	18.0	+22.2%	+21.1%
South Asia	6.8	8.0	7.9	8.1	8.8	9.4	9.7	6.4	7.3	8.9	11.8	+ 5.6%	+ 4.7%
South East Asia	2.4	3.6	4.3	4.1	4.2	4.6	5.9	6.4	7.1	7.3	7.7	+12.3%	+10.8%
Japan	3.7	5.3	7.0	7.7	9.3	10.9	15.8	20.3	23.4	24.2	24.7	+20.9%	+17.8%
Australasia	3.2	3.9	4.6	4.9	6.1	7.4	7.8	8.7	9.8	11.6	12.0	+14.2%	+10.3%
U.S.S.R.	182.8	198.8	210.8	230.0	253.1	271.2	289.2	307.0	328.0	353.7	380.0	+ 7.6%	+ 7.0%
Eastern Europe	41.2	42.1	46.0	51.3	57.5	58.4	60.0	61.5	64.0	69.4	70.1	+ 5.5%	+ 3.7%
China	4.9	6.4	7.7	8.7	9.8	10.9	11.7	12.4	11.7	10.4	9.5	+ 6.8%	− 2.7%
Total Eastern Hemisphere	**381.9**	**425.3**	**466.7**	**498.7**	**544.4**	**578.4**	**615.9**	**657.8**	**687.3**	**723.3**	**746.1**	**+ 6.9%**	**+ 5.2%**
World (excl. U.S.S.R., E. Europe & China)	**816.1**	**828.6**	**837.3**	**799.7**	**820.6**	**825.7**	**848.8**	**896.9**	**902.4**	**893.8**	**852.5**	**+ 0.4%**	**+ 0.6%**
World	**1 045.0**	**1 075.9**	**1 101.8**	**1 089.7**	**1 141.0**	**1 166.2**	**1 209.7**	**1 277.8**	**1 306.1**	**1 327.3**	**1 312.1**	**+ 2.3%**	**+ 2.4%**

* greater than 300%.

Source: British Petroleum Co. *BP Statistical Review of World Energy 1982.* London: BP, 1983, p. 27.

Table ES-6
World Hard Coal Production (1981)

	Production
	million metric tons
North, South America	
Canada	34
U.S.	692
Others	20
Total	**746**
Europe	
Belgium	6
Czechoslovakia	27
France	19
Poland	163
Spain	14
United Kingdom	125
USSR	544
West Germany	96
Others	12
Total	**1,006**
Africa	
South Africa	125
Others	6
Total	**131**
Asia	
China	600
India	123
Japan	18
North, South Korea	67
Others	14
Total	**822**
Oceania	
Australia	95
Others	2
Total	**97**
World Total	**2,803**

Source: Institute of Gas Technology. *Energy Statistics,* 1983, p. 132.

Table ES-7
Nuclear Electricity Generation: U.S. and Selected Countries[a]

	1973	1974	1975	1976	1977	1978	1979	1980	1981	1982
					10^9 kwh					
Argentina	0	1.0	2.5	2.6	1.6	2.9	2.7	2.3	2.8	1.9
Belgium	0	0.1	6.8	10.0	11.9	12.5	11.4	12.5	12.8	15.6
Canada	18.3	15.4	13.2	18.0	26.8	32.9	38.4	40.4	43.3	42.6
Finland	0	0	0	0	2.7	3.3	6.7	7.0	14.5	16.5
France	11.6	14.7	18.3	15.8	17.9	30.5	39.9	61.2	105.2	108.9
India	1.9	2.5	2.5	3.2	2.8	2.3	3.2	2.9	3.1	2.2
Italy	3.1	3.4	3.8	3.8	3.4	4.4	2.6	2.2	2.7	6.8
Japan	9.4	18.1	22.2	36.8	28.1	53.2	62.0	82.8	86.0	104.5
Netherlands	1.1	3.3	3.3	3.9	3.7	4.1	3.5	4.2	3.7	3.9
Pakistan	0.5	0.6	0.5	0.5	0.3	0.2	0[b]	0.1	0.2	0.1
South Korea	0	0	0	0	0.1	2.3	3.2	3.5	2.9	3.8
Spain	6.5	7.2	7.5	7.6	6.5	7.6	6.7	5.2	9.4	8.8
Sweden	2.1	1.6	12.0	16.0	19.9	23.8	21.0	26.7	37.7	38.8
Switzerland	6.2	7.0	7.7	7.9	8.1	8.3	11.8	14.3	15.2	15.0
Taiwan	0	0	0	0	0.1	2.7	6.3	8.2	10.7	13.1
United Kingdom	28.0	34.0	30.5	36.8	38.1	36.7	38.5	37.2	38.9	44.1
U.S.	88.0	104.5	181.8	201.7	263.3	292.7	270.6	265.5	288.5	298.6
West Germany	11.9	12.0	21.7	24.5	35.8	35.9	42.2	43.7	53.4	63.4

r = revised
a. U.S. Department of Energy, Energy Information Administration, *Monthly Energy Review*, April 1983. Reported in billion gross kilowatt-hours.
b. Less than 50,000,000 gross kilowatt-hours.

Table ES-8

World Coal Consumption

Country/Area	1972	1973	1974	1975	1976	1977	1978	1979	1980	1981	1982	Yearly Change 1982 over 1972	1982 over 1977
					Million Tonnes Oil Equivalent								
U.S.A.	316.6	335.0	331.9	322.9	346.5	351.9	348.9	380.7	393.2	406.3	394.6	+ 2.2%	+ 2.3%
Canada	15.2	15.6	15.9	15.5	18.3	23.4	19.2	18.2	22.6	22.9	22.7	+ 4.1%	− 0.6%
Total North America	**331.8**	**350.6**	**347.8**	**338.4**	**364.8**	**375.3**	**368.1**	**398.9**	**415.8**	**429.2**	**417.3**	**+ 2.3%**	**+ 2.1%**
Latin America	11.3	11.7	13.1	14.2	15.2	14.1	15.2	16.0	16.6	16.9	17.2	+ 4.3%	+ 4.1%
Total Western Hemisphere	**343.1**	**362.3**	**360.9**	**352.6**	**380.0**	**389.4**	**383.3**	**414.9**	**432.4**	**446.1**	**434.5**	**+ 2.4%**	**+ 2.2%**
Western Europe													
Austria	3.6	3.6	3.4	3.2	3.1	2.9	2.8	3.1	3.3	3.1	3.2	− 1.2%	+ 1.8%
Belgium & Luxembourg	13.1	11.0	11.9	8.9	9.3	9.4	9.7	10.6	10.7	11.1	11.4	− 1.4%	+ 3.9%
Denmark	1.3	1.8	2.1	2.5	2.5	3.3	3.4	4.2	5.9	5.4	5.9	+16.2%	+12.1%
Finland	1.8	2.0	2.1	1.8	2.5	2.6	3.5	3.2	3.8	1.8	1.9	+ 0.4%	− 6.6%
France	29.5	29.5	30.3	26.5	30.0	29.8	30.5	28.5	27.7	25.1	26.8	− 1.0%	− 2.1%
Greece	4.0	4.7	4.9	6.5	7.6	8.1	7.2	4.0	4.0	3.9	4.0	Ø	−13.33%
Iceland	†	†	†	†	—	—	—	—	—	—	—	—	—
Republic of Ireland	2.0	1.8	1.8	1.7	1.7	1.8	1.8	1.9	1.9	2.0	2.2	+ 1.3%	+ 4.4%
Italy	8.9	9.0	9.8	9.8	9.7	9.6	9.8	10.7	12.6	12.9	14.4	+ 5.0%	+ 8.5%
Netherlands	3.1	3.1	2.9	2.5	3.2	3.2	3.1	3.3	3.9	4.1	4.7	+ 4.4%	+ 8.0%
Norway	0.8	0.5	0.6	0.6	0.5	0.5	0.5	0.5	0.5	0.5	0.3	− 8.3%	− 8.5%
Portugal	0.4	0.6	0.4	0.4	0.4	0.4	0.4	0.4	0.4	0.4	0.4	+ 0.5%	+ 2.0%
Spain	8.7	8.5	8.8	9.2	9.9	10.6	10.5	10.9	14.1	17.0	17.8	+ 7.5%	+10.9%
Sweden	0.8	0.8	0.6	0.9	1.2	1.0	1.5	1.7	1.6	1.4	1.7	+ 8.3%	+11.6%

Switzerland	0.4	0.2	0.1	0.1	0.2	0.3	0.3	0.5	0.5	0.7	0.4	− 0.9%	+ 9.8%
Turkey	4.5	4.6	4.8	4.9	5.3	5.5	5.4	6.5	7.5	7.4	8.2	+ 6.2%	+ 8.4%
United Kingdom	72.2	78.4	69.1	71.9	72.7	73.4	70.4	76.1	72.6	69.6	65.4	− 1.0%	− 2.3%
West Germany	80.2	82.1	82.5	70.7	75.9	71.7	72.9	79.8	80.2	81.4	79.7	− 0.1%	+ 2.1%
Yugoslavia	10.4	11.4	11.7	12.6	12.6	12.5	12.8	13.9	14.3	14.6	15.9	+ 4.3%	+ 4.8%
Cyprus/Gibraltar/Malta	†	†	†	†	†	†	†	†	†	†	†	—	—
Total Western Europe	**245.7**	**253.6**	**247.8**	**234.7**	**248.3**	**246.6**	**246.5**	**259.8**	**265.5**	**262.4**	**264.3**	**+ 0.7%**	**+ 1.4%**
Middle East	†	†	†	†	†	†	†	†	†	†	†	+ 2.3%	+10.8%
Africa	42.5	38.6	40.4	43.1	46.7	47.3	53.3	58.1	62.2	67.1	70.1	+ 5.1%	+ 8.2%
South Asia	54.6	55.0	56.8	59.1	62.5	64.0	67.2	74.9	82.2	86.0	91.5	+ 5.3%	+ 7.4%
South East Asia	34.5	39.8	42.4	44.2	44.7	47.2	47.2	55.1	55.9	57.2	58.3	+ 5.4%	+ 4.3%
Japan	50.5	53.7	57.3	54.4	52.5	52.5	46.5	50.4	57.6	63.6	62.0	+ 2.1%	+ 3.4%
Australasia	22.1	22.9	24.4	26.1	24.9	26.8	28.2	30.1	30.9	31.5	33.3	+ 4.2%	+ 4.4%
U.S.S.R.	317.0	315.0	316.2	326.2	335.7	341.2	344.4	342.5	342.5	336.8	345.0	+ 0.9%	+ 0.2%
Eastern Europe	241.3	235.0	237.6	246.9	253.9	260.7	266.7	270.0	274.0	258.0	270.0	+ 1.1%	+ 0.7%
China	278.1	292.5	307.4	321.8	337.5	354.4	380.0	420.0	403.3	394.2	412.2	+ 4.0%	+ 3.1%
Total Eastern Hemisphere	**1 286.3**	**1 306.1**	**1 330.3**	**1 356.5**	**1 406.7**	**1 440.7**	**1 480.0**	**1 560.9**	**1 574.1**	**1 556.8**	**1 606.7**	**+ 2.3%**	**+ 2.2%**
World (excl. U.S.S.R., E. Europe & China)	793.0	825.9	830.0	814.2	859.6	873.8	872.2	943.3	986.7	1 013.9	1 014.0	+ 2.5%	+ 3.0%
World	**1 629.4**	**1 668.4**	**1 691.2**	**1 709.1**	**1 786.7**	**1 830.1**	**1 863.3**	**1 975.8**	**2 006.5**	**2 002.9**	**2 041.2**	**+ 2.3%**	**+ 2.2%**

Ø less than + or − 0.05%.

† less than 0.05 million tonnes oil equivalent.

Source: British Petroleum Co. *BP Statistical Review of World Energy 1982.* London: BP, 1983, p. 28.

Table ES-9
U.S. Energy Production by Primary Energy Type[a]

	Total	Coal	Crude Oil	Natural Gas Plant Liquids
	---------------------------------- 10^{15} Btu ----------------------------------			
1972	62.812	14.485	20.041	2.597
1973	62.433	14.366	19.493	2.569
1974	61.229	14.468	18.575	2.471
1975	60.059	15.189	17.729	2.374
1976	60.091	15.853	17.262	2.327
1977	60.293	15.829	17.454	2.327
1978	61.231	15.037	18.434	2.245
1979	63.851	17.651	18.104	2.286
1980	65.499	19.209	18.249	2.254
1981	65.183^r	19.071	18.146	2.298
1982	64.174	19.076	18.357	2.215

	Natural Gas Dry	Hydro	Electricity[b] Nuclear	Other[c]
	---------------------------------- 10^{15} Btu ----------------------------------			
1972	22.208	2.861	0.584	0.035
1973	22.187	2.861	0.910	0.046
1974	21.210	3.177	1.272	0.056
1975	19.640	3.155	1.900	0.072
1976	19.480	2.976	2.111	0.081
1977	19.565	2.333	2.702	0.082
1978	19.485	2.937	3.024	0.068
1979	20.076	2.931	2.715	0.089
1980	20.112	2.890	2.672	0.114
1981	19.907^r	2.732	2.901	0.127
1982	18.178	3.223	3.013	0.111

r = revised
a. U.S. Department of Energy, Energy Information Administration, *Monthly Energy Review*, March 1979, for 1972 data; ibid., February 1983, for 1973–82 data.
b. Based on equivalent heat input.
c. Includes electricity produced from wood and waste and geothermal power.

Table ES-10
U.S. Energy Consumption by Primary Energy Type[a]

	Total	Coal	Petroleum	Natural Gas (Dry)[b]
	----------------------------- 10^{15} Btu -----------------------------			
1972	71.625	12.446	32.947	22.699
1973	74.609	13.300	34.840	22.512
1974	72.759	12.876	33.455	21.732
1975	70.707	12.823	32.731	19.948
1976	74.510	13.733	35.175	20.345
1977	76.332	13.964	37.122	19.931
1978	78.175	13.846	37.965	20.000
1979	78.910	15.109	37.123	20.666
1980	75.910	15.461	34.202	20.391
1981	74.123	16.118	32.113	19.911
1982	70.923	15.655	30.400	18.305

	Electricity[c]			
	Hydro	Nuclear	Other[d]	Net Imports of Coal Coke
	----------------------------- 10^{15} Btu -----------------------------			
1972	2.941	0.584	0.035	(0.027)
1973	3.010	0.910	0.046	(0.008)
1974	3.309	1.272	0.056	0.059
1975	3.219	1.900	0.072	0.014
1976	3.066	2.111	0.081	0.000
1977	2.515	2.702	0.082	0.015
1978	3.141	3.024	0.068	0.131
1979	3.141	2.715	0.089	0.066
1980	3.107	2.672	0.114	(0.037)
1981	2.970	2.901	0.127	(0.017)
1982	3.462	3.013	0.111	(0.023)

a. U.S. Department of Energy, Energy Information Administration, *Monthly Energy Review,* March 1979, for 1972 data; ibid., February 1983, for 1973–82 data.
b. Beginning in 1972, natural gas transmission losses and unaccounted for gas are no longer included.
c. Based on equivalent heat input.
d. Includes electricity produced from wood and waste and geothermal power.

Table ES-11
U.S. Oil Consumption[a]

	Residential/ Commercial	Industrial	Transportation	Electric Utilities	Total
			--------- 10¹⁵ Btu ---------		
1973	4.321	9.103	17.745	3.671	34.840
1974	3.932	8.707	17.317	3.499	33.455
1975	3.760	8.192	17.547	3.231	32.731
1976	4.160	9.092	18.469	3.454	35.175
1977	4.148	9.789	19.157	4.028	37.122
1978	4.062	10.046	20.044	3.813	37.965
1979	3.687	10.294	19.786	3.357	37.123
1980	3.280	9.272	18.996	2.654	34.202
1981	3.122	8.203	18.561r	2.226r	32.112r
1982	2.966	7.621	17.996	1.817	30.400

r = revised
a. U.S. Department of Energy, Energy Information Administration, *Monthly Energy Review*, February 1983.

GLOSSARY

Active solar heating: Solar heat collected through a circulating system of water or air.

Backstop: A substitute technology that becomes economically viable when the price of an existing technology (e.g., oil) rises beyond a certain limit.

Biomass: Organic material such as corn, wood, or sugarcane that can be turned into an energy fuel.

Break-even: The point at which more power is produced than is required to keep a nuclear fusion power plant operating.

Breeder reactor: A nuclear reactor that produces more fissile material than it consumes. In fast breeder reactors, high-energy (fast) neutrons produce most of the fissions, while in thermal breeder reactors, fissions are principally caused by low-energy (thermal) neutrons.

British thermal unit (Btu): The amount of heat required to raise the temperature of 1 pound of water 1° F.

Ceiling price: Maximum allowable price permitted under regulation.

Coal slurries: Mixtures of coal with water, oil, or other liquid.

Cogeneration: The generation of both steam and electric energy in the same facility.

Combined cycle: A power plant that typically combines a high-temperature gas turbine with a steam boiler.

Crude oil: A mixture of hydrocarbons that exists in the liquid phase in natural underground reservoirs and remains liquid at atmospheric pressure after passing through surface separating facilities. Statistically, crude oil reported at refineries, in pipelines, at pipeline terminals, and on leases may include lease condensates.

Depletion allowance: A tax credit based on the permanent reduction in value of a depletable resource that results from removing or using some part of it.

Distillate fuel oil: A light fuel oil distilled off during the refining process. Included are products known as No. 1 and No. 2 heating oils, diesel fuels, and No. 4 fuel oil. These products are used primarily for space heating, on- and off-highway diesel engine fuel (including railroad engine fuel), and electric power generation.

District heating: A process whereby excess steam generated at a facility is distributed to heat buildings in the surrounding area.

Elasticity: The fractional change in the quantity demanded of a good, divided by the fractional change in an economic variable, such as price or income.

Enrichment: A process whereby the percentage of the naturally present fissile uranium isotope (U-235) is artificially increased.

Expensing: Writing off of certain expenses immediately for tax purposes, rather than over the life of an energy facility or piece of capital equipment.

Exploratory well: A well drilled to find oil or gas in an unproved area, to find a new reservoir in a field known to contain productive oil or gas reservoirs, or to extend the limit of a known oil or gas reservoir.

Fossil fuel: Any naturally occurring fuel such as coal, oil, or natural gas, derived from the remains of ancient plants and animals. These sometimes are called conventional fuels or conventional energy sources (as compared with nuclear, solar, and wind energy) because they provide most of today's energy for the world's industrial economy.

Fuel cell: A "reverse battery" that typically uses hydrogen (usually from natural gas) and oxygen (from air) in an electrolytic cell to produce electricity (and water).

Fusion: The combining of atomic nuclei of very light elements by high-speed collision to form new and heavier elements, the result being the release of energy.

Gas-centrifuge process: A method of isotopic separation in which heavy gaseous atoms or molecules are separated from light atoms or molecules by centrifugal force.

Geopressured brines: Salt water found within the earth at high pressure; it often contains appreciable quantities of methane (natural gas).

Geothermal energy: Energy from the internal heat of the earth, which may be residual heat, friction heat, or a result of radioactive decay. The heat is found in rocks and fluids at various depths and can be extracted by drilling and/or pumping.

Gigawatt: One million kilowatts.

Gross national product (GNP): The total market value of the goods and services produced in a national economy during a given year for final consumption, capital investment, and governmental use.

Heat pump: A mechanically driven device that uses a refrigeration cycle to raise a low-grade heat source to a higher temperature.

High-temperature gas reactor (HTGR): A nuclear power plant operated at high temperature, typically using helium to transfer the heat from the reactor core to the steam boiler. Experiments have been conducted using thorium and uranium-233 in an HTGR.

Hybrid fusion-fission: A proposed power plant using the nuclear fusion process to generate neutrons to convert fertile uranium-238 into fissionable fuel.

Hydroelectric power: Electricity generation using water flow to drive a turbine.

Incremental pricing: A provision of the NGPA under which selected customers are charged a price for natural gas that is calculated by including the total cost of all new high-priced gas purchased by the distribution company.

Inertial confinement: An experimental nuclear fusion device that compresses and heats the reacting gases (heavy isotopes of hydrogen—deuterium and tritium) using focused high-powered laser beams.

Interstate gas: Natural gas that enters interstate commerce and therefore is subject to federal controls.

Intrastate gas: Natural gas that is produced and consumed within the same state. Before the NGPA, intrastate gas was not regulated by the federal government.

Laser: A source of coherent (in phase), essentially monochromatic light.

Liquefied natural gas (LNG): Natural gas that has been cooled to about − 160°C for storage or shipment as a liquid in high-pressure cryogenic containers.

Liquid metal fast breeder reactor (LMFBR): A nuclear power plant using liquid metal (typically sodium or a mixture of sodium and potassium) to transfer heat from the reactor core to the steam boiler. The reactor is designed to use "fast" (high-energy) neutrons to maintain the nuclear reaction, in contrast to "slow" or thermal (low-energy) neutrons in a conventional nuclear power plant, in order to make efficient use of the neutrons in breeding (converting) fertile uranium-238 into fissionable plutonium-239).

Long-term elasticity: An elasticity, usually a price elasticity, reflecting the changes in demand for a good occurring over a time span long enough to allow for adjustments in capital stocks. Usually long-term elasticities are larger in absolute value than short-term elasticities.

Low-priority user: FERC establishes priorities for different classes of users of natural gas. When natural gas shortages occur, service to low-priority users is curtailed. Gas curtailment to low-priority users can be expected during high-demand periods, such as severe winters. The NGPA requires that FERC pass on some of the costs of expensive natural gas to low-priority users.

Magnetic confinement: An experimental nuclear fusion plant that uses strong magnetic fields to contain the very high-temperature plasma of reacting deuterium and tritium.

Magnetohydrodynamics: An experimental process for converting a high-temperature ionized gas (typically the combustion gases from coal "seeded" with cesium to provide the ionization) directly into electricity by passing the gas through a magnetic field.

Market-clearing level: The estimated price of a commodity at which its demand equals its supply.

Molten salt reactor (MSR): A nuclear power plant using a molten salt such as lithium fluoride to dissolve and circulate the fissionable uranium chloride.

National Energy Act of 1978 (NEA): A package of five bills affecting U.S. energy markets. The five acts are:
- The National Energy Conservation Policy Act
- The Power Plant and Industrial Fuel Use Act (FUA)
- The Public Utilities Regulatory Policy Act (PURPA)
- The Natural Gas Policy Act (NGPA)
- The Energy Tax Act

Nominal prices: Those prices actually observed in the marketplace at any point in time. Nominal prices are sometimes referred to as market prices.

Nuclear fuel reprocessing: The chemical separation of spent (used) nuclear fuel into salvageable fuel material and radioactive waste.

Nuclear waste: The radioactive products formed by fission and other nuclear processes in a reactor. Most nuclear waste is initially in the form of spent fuel. If this material is reprocessed, new categories of waste result: high-level, transuranic, and low-level wastes (as well as others).

Oil shale: A range of shale materials containing organic matter (kerogen) that can be converted into shale oil, gas, and carbonaceous residue by destructive distillation.

Organization for Economic Cooperation and Development (OECD): A twenty-four-member body composed of representatives of the United States, Canada, Japan, the Western European countries, Australia, and New Zealand. The organization's purpose is to promote mutual economic development and to contribute to the development of the world economy.

Organization of Petroleum Exporting Countries (OPEC): A cartel of oil-exporting nations consisting of Venezuela, Ecuador, Indonesia, Algeria, Libya, Nigeria, Gabon, Iran, Kuwait, Saudi Arabia, Iraq, the United Arab Emirates, and Qatar.

Passive solar heating: Systems that use location, absorption, insulation, or other natural processes to collect and transfer heat. South-facing windows and greenhouses are two examples.

Petrochemical: Any chemical derived from petroleum or natural gas, such as polyethylene.

Photovoltaics: Devices that directly generate electrical current when exposed to sunlight. They are constructed of semiconductor materials.

Plasma: An ionized gas; in other words, a gas containing elements or compounds from which the electrons have been stripped so that the elements or compounds will move in response to an electric or magnetic field.

Plutonium: A radioactive man-made metallic element with atomic number 94, created by absorption of neutrons in uranium-238. Its most important isotope is plutonium-239, which is fissionable and can be used as a reactor fuel.

Price tier: Classes of crude oil production established for purposes of governmental price controls.

Radioactive waste disposal: The placement of waste radioactive material in a safe place. (Current regulations require isolation of radioactive materials from possible contamination of human society for at least a thousand years.)

Reactor core: The central portion of a nuclear reactor, containing the fuel elements and the control rods.

Real prices: Nominal prices adjusted for the effects of inflation on the purchasing power of the dollar. These prices are always used in reference to a particular year, such as real 1979 dollars, and sometimes are referred to as "constant" prices.

Renewable resources: Sources of energy not subject to exhaustion, such as wood and other biomass, solar radiation, hydroelectric power, and wind.

Reprocessing: A generic term for the chemical and mechanical processes applied to fuel elements discharged from a nuclear reactor; the purpose is to recover for reuse fissile materials such as plutonium-239, uranium-235, and uranium-233, and to isolate the undesirable fission products.

Reserves: Resources that are known in location, quantity, and quality, and that are economically recoverable using currently available technologies.

Residual fuel oil: Topped crude oil obtained in refinery operations, including ASTM grades No. 5 and No. 6, heavy diesel, Navy Special, and Bunker C oils used for generation of heat and/or power.

Scrubbers: Equipment used to remove sulfur dioxide from flue gases.

Secondary recovery: Methods of obtaining oil and gas by the augmentation of reservoir energy, often by the injection of air, gas, or water into a production formation. (See Tertiary recovery.)

Shale oil: A liquid similar to conventional crude oil that is obtained by processing an organic mineral (kerogen) in oil shale.

Short-term elasticity: An elasticity, usually a price elasticity, reflecting the change in demand for a good that occurs over a time span too short to allow changes in capital stock.

Spot market: Sales available for immediate delivery, not generally recurring under fixed-term contracts.

Spot prices: The price of a commodity (such as oil) applying to immediate delivery, as distinguished from future delivery under a long-term contract.

Synthetic fuels: Generally oil or gas from coal or shale. The definition is sometimes expanded to include heavy oils and tars that cannot be pumped in the conventional manner from the earth, and fuels from biomass, such as gasohol.

Tar sands: Consolidated or unconsolidated rocks with interstices containing bitumen that ranges from very viscous to solid. In a natural state, tar sands cannot be recovered through primary methods of petroleum production.

Tertiary recovery: Use of heat and chemical methods other than air, gas, or water injection to augment oil recovery (presumably occurring after secondary recovery).

Tight formations: Layers within the earth containing natural gas that does not flow readily to the well; they must be artificially fractured or otherwise stimulated to release the gas.

Uranium: A radioactive element of atomic number 92. Naturally occurring uranium is a mixture of 99.28 percent uranium-238, 0.71 percent uranium-235, and 0.0058 percent uranium-234. Uranium-235 is a fissile material and is the primary fuel of light water reactors. When bombarded with slow or fast neutrons, it will undergo fission. Uranium-238 is a fertile material that is transmuted to plutonium-239 upon the absorption of a neutron.

Wellhead: The point at which oil or natural gas is transferred from the well to a pipeline or another non-well facility. This term is used in referring to "wellhead price," which is the price producers of oil and natural gas receive.

Windfall profits tax: An excise tax paid by producers and royalty owners on domestically produced oil.

Yellowcake: A uranium concentrate that results from the milling (concentrating) of uranium ore. It typically contains 80 to 90 percent uranium oxide (U_3O_8).

ABOUT THE AUTHORS

CONNIE C. BARLOW is president of ARTA Alaska Inc. in Juneau.

She holds a B.S. in geology from Michigan State University and has held senior staff positions at the Federal-State Land-Use Planning Commission for Alaska, the Alaska State Senate, and the Alaska Department of Natural Resources. She is the author or co-author (with Arlon R. Tussing) of a number of articles and reports on petroleum, natural gas, and other energy and environmental policy issues.

PETR BECKMANN is Professor (Emeritus) of Electrical Engineering at the University of Colorado in Boulder, and publisher and editor of "Access to Energy."

He received his Ph.D from Prague Technical University and his D.Sc. from the Czechoslovak Academy of Sciences. A proponent of advanced technology and free enterprise, he founded Golem Press in 1967 and began the widely read newsletter, "Access to Energy," in 1973. A nationally known supporter of nuclear power, he was appointed by Ronald Reagan to the Energy Task Force in August 1980. Dr. Beckmann is the author of many technical, scientific, and popular papers and books, including *The Health Hazards of Not Going Nuclear* (1976).

BERNARD L. COHEN is professor of physics at the University of Pittsburgh where he also served as director of the Scaife Nuclear Laboratory.

He has received degrees from Case–Western Reserve University and the University of Pittsburgh, and earned the D.Sc. from Carnegie–Mellon University. His experience includes positions with the Oak Ridge National Laboratory and short-term appointments with the Electric Power Research Institute, the Argonne National Laboratory, as well as a number of consulting positions for various energy firms and government agencies. Professor Cohen was the recipient of the Landauer Award (Health Physics) in 1980 and the American Physical Society's Bonner Prize in 1981. He is the author of many publications, including *Nuclear Science and Society* (1974) and

ABOUT THE AUTHORS • 421

Before It's Too Late: A Scientist's Case for Nuclear Power (1983).

STEPHEN D. EULE is a researcher in Natural Resources Policy for The Heritage Foundation.
He holds a degree in biology from Southern Connecticut State College, has done graduate work in botany at the University of Maryland, and received his M.A. in geography from the George Washington University. His research interests are in the economic and spatial aspects of energy development, resource utilization and the environment, and urban, transportation, and electrical transmission system evolution and regional development.

RICHARD S. GREELEY is president of the Greeley Technical Group, Inc., a consulting firm providing technical and economic analyses in the fields of energy, environment, and resource planning located in St. Davids, Pennsylvania.
He is a physical chemist with degrees from Harvard University and Northwestern University, and a Ph.D. from the University of Tennessee. His experience includes positions as director of research and technology at the Mitre Corporation, and vice-president for economics and environmental science at R. F. Weston, Inc. He has managed projects involving fossil, nuclear, and solar energy research and development. He has also directed projects involving the control of air and water pollution from toxic substances.

RICHARD N. HOLWILL is Deputy Assistant Secretary of State for Inter-American Affairs and previously was vice-president for government information of The Heritage Foundation in Washington, D.C.
He is a graduate of Louisiana State University with a degree in history. He has served as vice-president of Energy Decisions, Inc., a Washington-based consulting firm specializing in analysis of energy economics and regulation. He was also White House Correspondent for National Public Radio and has worked in the oil and gas industry. Mr. Holwill has written extensively on energy issues and edited *The First Year* and *Agenda '83* for The Heritage Foundation.

LESTER B. LAVE is professor of economics and public policy at Carnegie-Mellon University.
A graduate of Reed College, he received his Ph.D. in eco-

nomics from Harvard University. He has taught at Harvard
University, Northwestern University, and the University of
Pittsburgh, and was a senior fellow in the Economics Studies
Program at the Brookings Institution. He has been consultant
to almost every cabinet-level department and many firms, and
has served on committees of the National Academy of Sciences
and the American Association for the Advancement of Science.
Professor Lave is the author of many publications, including
Air Pollution and Human Health (with E. Seskin, 1977), and
Clearing the Air: Reforming the Clean Air Act (with G. S.
Omenn, 1981).

RENE MALES is director of the Energy Analysis and Environment Division of the Electric Power Research Institute in
Palo Alto, California.

A graduate of Ripon College, he studied mathematics at the
University of Chicago before obtaining his M.B.A. from Northwestern University. At Commonwealth Edison from 1956 to
1976, he worked primarily in the financial, economic, and
operation areas of the company as director of economic research, and later as manager of general services. He has served
as a member of advisory committees to the U.S. Departments
of Energy and Commerce, the Oak Ridge and Brookhaven
National Laboratories, the National Petroleum Council, the
Federal Power Commission, and the National Academy of
Engineering. Mr. Males is the author of several articles and
publications on electric energy.

WALTER J. MEAD is professor of economics at the University
of California at Santa Barbara.

He was educated at the University of Oregon and Columbia
University, and earned his Ph.D. from the University of Oregon. He has served as senior economist for the Ford Founda-
tion Energy Policy Project (1972–73), and worked with the
Committee for Economic Development. He has also consulted
for the Stanford Research Institute, the U.S. Departments of
Agriculture and Interior, the Naval Petroleum and Oil Shale
Reserve, and the Alaska Legislature Resource Committee,
among others. He is the author of many publications including
U.S. Energy Policy: Errors of the Past, Proposals for the Future
(co-edited with A. E. Utton, 1978) and *Price Controls and
International Petroleum Product Prices* (with R. Deacon and V.
B. Agarwal, 1980).

GREGORY G. PICKETT is assistant professor of economics at the University of Connecticut at Storrs.

He has earned his B.A., M.A., C.Phil., and Ph.D. degrees in economics from the University of California at Santa Barbara. He was awarded an Earhart Foundation Fellowship (1978–79) and was a research fellow for the Energy Institute Project (1981–82). His fields of specialization include financial economics, energy and natural resource economics, and options and auction markets. His current research interests are in competitive bidding for offshore oil and gas leases and joint ventures, and energy exploration finance.

S. FRED SINGER was senior fellow for natural resources policy at The Heritage Foundation while on leave from the University of Virginia, where he is professor of environmental sciences.

He is a geophysicist, with degrees from Ohio State University and a Ph.D. in physics from Princeton University. He has been in the federal service at various times, including a term as director of the U.S. Weather Satellite Center, and later as deputy assistant secretary of the interior. Currently he serves as vice-chairman of the U.S. National Advisory Committee for Oceans and Atmosphere. Dr. Singer has been a consultant on energy policy to the U.S. Treasury, the state of Alaska, and private corporations. He is the author of some 200 scholarly publications, as well as numerous articles in *The Wall Street Journal* and other newspapers and magazines.

ARLON R. TUSSING is president of Arlon R. Tussing and Associates, Inc., in Seattle, Washington, and professor of economics in the Institute of Social and Economic Research of the University of Alaska.

He holds B.A. degrees from University of Chicago and Oregon State College, and a Ph.D. in economics from the University of Washington. His experience includes positions with the U.S. Senate Committee on Interior and Insular Affairs, the U.S. Senate National Fuels and Energy Study, and the Federal Field Committee for Development Planning in Alaska. He has served on the Economic Advisory Board of the U.S. Department of Commerce and the Council of the International Association of Energy Economists. Dr. Tussing is the author of more than 200 books, monographs, and articles on oil and gas, electric power and other energy issues, environmental management and policy, public finance, and economics.

JAMES W. WETZLER is deputy chief of staff for the Joint Committee on Taxation, U.S. Congress.

He was educated at the Wharton School, University of Pennsylvania, and earned his Ph.D. in economics at Harvard University. As a supervisor of a nonpartisan staff of experts, he has developed tax legislation for members of the House Ways and Means Committee and the Senate Finance Committee, has worked with the Treasury and other administration officials in formulating administration tax proposals, and has analyzed tax proposals for members of Congress. Dr. Wetzler has published numerous articles, reviews, and comments in a variety of tax journals.

BENJAMIN ZYCHER was until recently a senior staff economist at the Council of Economic Advisers. He is now a senior economist at the Jet Propulsion Laboratory, California Institute of Technology.

He has earned degrees from the University of California at Los Angeles and the University of California at Berkeley, and received his Ph.D. from UCLA. He is also the recipient of two fellowship awards. He has been associated with The Economics Group, National Economic Research Associates, and the Rand Corporation. Dr. Zycher was a contributor to *The Annual Report of the Council of Economic Advisers* (1977, 1982, 1983) and has written extensively on energy issues.

INDEX

accelerated cost recovery system (ACRS), 159–65
acid rain, 186
ad valorem tariff, 110
advanced energy technology, 308–42. *See also* energy conservation, geothermal energy, nuclear fission breeder reactors, nuclear fusion reactors, solar energy, synthetic fuels
Advisory Committee of Reactor Safeguards, 223
Afghanistan, Soviet invasion of, 25, 36
air pollution: acid rain from, 186; from coal use, 180–86, 250; fatalities from, 248; government regulation of, 180; inefficient decrease of, 186–88; local and regional problems, 183–85; reduction of sulfur emission, 183–84. *See also* Clean Air Act
Alaskan natural gas: benefits of removal of export ban, 118–19, 142–43; changing natural gas market, 137–39; discovery of, 119–20; export limitations on, 118, 135, 142–43; export markets, 140–41; transportation options, 135–37
Alaskan Natural Gas Transportation Act of 1976, 135
Alaska Natural Gas Transportation System (ANGTS), 136; alternatives to, 136, 140; reshaping of, 139
Alaskan oil: discovery of, 119–20; effects of export ban removal, 118–19, 133–35, 354–55; export ban, 97, 118, 121–24; reasons for export restrictions, 122; transportation alternatives, 135–37; West Coast glut of, 124
Alaskan oil pipeline, 22
All-Alaska Pipeline System (AAPS), 140
American Bulk Ship Owners Committee, 122

American Gas Association, 119
American Maritime Association, 122
Atlantic Richfield Company (ARCO), 122
Atomic Safety and Licensing Board, 223

biomass, as a fuel substitution, 39
black lung disease, 76–77, 174–75. *See also* coal mining; health hazards
Bonneville Power Administration (BPA), 276
British Windscale Enquiry, 263

cash bonus bidding, 189–90, 195, 200, 201, 207, 213, 214
Clean Air Act: 1970 amendments to, 172; 1977 amendments to, 172; alternatives to, 186–87
Clinch River Breeder Reactor (CRBR), 316–17, 329, 337, 338. *See also* nuclear reactors
coal: government regulation of, 172, 176; mining of, 76–82, 176–88; problems of, 173–76; U.S. reserves of, 173
coal mining: occupational hazards of, 174–75 (*See also* black lung disease); pollution from, 180–88; public health concerns of, 174–77; regulation of, 176–78; strip mining, 179
coal workers pneumocariosis (CWP), 174. *See also* black lung disease
common carrier, 68, 73
Connally "Hot Oil" Act, 105
consumption taxes, types of, 110–11
contract carrier, 68, 73
Cost of Living Council (CLC), 82–83, 86
Council on Environmental Quality, 84
coupon rationing, 104

425